Building Regulations Pocket Book

The new edition of the *Building Regulations Pocket Book* has been fully updated with recent changes to the UK Building Regulations and Planning Law. This handy guide provides you with all the information you need to comply with the UK Building Regulations and Approved Documents. On site, in the van, in the office – wherever you are – this is the book you'll refer to time and time again to check the regulations on your current job.

- Part 1 provides an overview of the Building Act.
- Part 2 offers a handy guide to the dos and don'ts of gaining the Local Council's approval for Planning Permission and Building Regulations Approval.
- Part 3 presents an overview of the requirements of the Approved Documents associated with the Building Regulations.
- Part 4 is an easy-to-read explanation of the essential requirements of the Building Regulations that any architect, builder or DIYer needs to know to keep their work safe and compliant on both domestic and non-domestic jobs.

Key new updates to this second edition include, but are not limited to: changes to the fire regulations as a result of the Hackitt Review, updates to Approved Document F and L, new Approved Documents covering *Overheating* (AD-O) and *Infrastructure for the charging of electric vehicles* (AD-S), amendments to Use Classes and the reinstatement of the *Manual to the Building Regulations*. This book is essential reading for all building contractors and sub-contractors, site engineers, building engineers, building control officers, building surveyors, architects, construction site managers as well as DIYers and those who are supervising work in their own home.

Ray Tricker is a Senior Consultant with over 50 years of continuous service in Quality, Safety and Environmental Management, Project Management, Communication Electronics, Railway Command, Control and Signalling Systems, Information Technology and the development of Molecular Nanotechnology.

He served with the Royal Corps of Signals (for a total of 37 years) during which time he held various managerial posts culminating in being appointed as the Chief Engineer of NATO's Communication Security Agency (ACE COMSEC). Most of Ray's work since leaving the Services has centered on the European Railways.

Samantha Alford (MSc, MCIPS) is an established compliance and business management consultant and technical author. She served in the Royal Air Force for 18 years and has over 35 years of experience in a range of fields including Governance, Oversight and Strategic Planning, and Compliance. Sam has worked in the public, private, charity and voluntary sectors and is currently Director of a consultancy providing data protection guidance to small businesses and technical authorship.

Building Regulations Pocket Book

Second Edition

Ray Tricker and Samantha Alford

Routledge
Taylor & Francis Group

LONDON AND NEW YORK

Second edition published 2023
by Routledge
4 Park Square, Milton Park, Abingdon, Oxon, OX14 4RN

and by Routledge
605 Third Avenue, New York, NY 10158

Routledge is an imprint of the Taylor & Francis Group, an informa business

First edition published by Routledge 2018

British Library Cataloguing-in-Publication Data
A catalogue record for this book is available from the British Library

Library of Congress Cataloging-in-Publication Data
Names: Tricker, Ray, author. | Alford, Samantha, author.
Title: Building regulations pocket book / Ray Tricker and Samantha Alford.
Description: Second edition. | Abingdon, Oxon ; New York, NY : Routledge, 2022. |
Series: Routledge pocket books | Includes bibliographical references and index. |
Summary: "The new edition of the Building Regulations Pocket Book
has been fully updated with new changes to the UK Building Regulations
and Planning Law"– Provided by publisher.
Identifiers: LCCN 2021058875 (print) | LCCN 2021058876 (ebook) |
ISBN 9781032003566 (hardback) | ISBN 9780367774172 (paperback) |
ISBN 9781003173786 (ebook)
Subjects: LCSH: Great Britain. Building Act 1984. |
Building laws–England. | Building law–Wales. |
LCGFT: Administrative regulations.
Classification: LCC KD1140 .T754 2022 (print) |
LCC KD1140 (ebook) | DDC 343.4207/869–dc23/eng/20220325
LC record available at https://lccn.loc.gov/2021058875
LC ebook record available at https://lccn.loc.gov/2021058876

ISBN: 978-1-032-00356-6 (hbk)
ISBN: 978-0-367-77417-2 (pbk)
ISBN: 978-1-003-17378-6 (ebk)

DOI: 10.1201/9781003173786

Typeset in Baskerville
by Newgen Publishing UK

Contents

Part 2: Requirements for Planning Permission and Building Regulations approval   **23**

Part 3: Requirements of the Approved Documents   87

Preface

Following the success of the first edition of the *Building Regulations Pocket Book* (which we produced in response to requests from readers asking for an abbreviated version of our increasingly popular *Building Regulations in Brief*, and the changes made to the Fire Regulations following the Grenfell Tower disaster), we have been asked to produce an updated version of our previous publication and this second edition includes, but is not limited to:

- Changes in light of the Hackitt Review.
- Updates to Approved Document B.
- Changes to fees, Use Classes and permitted development.
- The long-awaited updates to Approved Documents F and L (which came into effect on 15 June 2022).
- The introduction of two new Approved Documents O (*Overheating*) and S (*Electric vehicle charging points*).

The joy of a pocket edition is that it can easily be transported from office to site, put in a briefcase or just left in a van for reference purposes. It can even be downloaded onto your mobile device! The aim of this pocket book, therefore, is to provide the reader with a user-friendly, easy-to-read, résumé of the current requirements of the Building Regulations.

This handy guide provides you with all the information you need to comply with the UK Building Regulations and Approved Documents – no matter where you are, this is the book you'll refer to time and time again to double check the regulations on your current job. It is the most reliable and portable guide for compliance with the Building Regulations. The book has therefore become essential reading for all building contractors, sub-contractors, site engineers, building engineers, building control officers, building surveyors, architects, construction site managers, DIYers and homeowners having work done on their property.

In effect, the *Building Regulations Pocket Book* is a précis of the most important points contained in our more complete *Building Regulations in Brief* (tenth edition) and it is envisaged that the two books should complement each other. Where required, and by virtue of its extensive coverage, *Building Regulations in Brief* can be used to provide more detail on all aspects building projects – large or small.

Building inspectors, acting on behalf of Local Authorities, are primarily concerned with whether a building complies with the requirements of the Building Regulations, and this pocket book will provide them with a quick checklist that they can use during inspections. Designers, architects and builders, through experience, are normally aware of the overall requirements for building projects; however, they will still need a reminder when they come across a different situation for the first time – and this is where the pocket book will become invaluable. But it is not just the professional trade that this book is aimed at: it will also prove extremely useful to the average student, DIY enthusiast and homeowner!

The structure of this book

The *Building Regulations Pocket Book* (second edition) includes all the key features of *Building Regulations in Brief* (tenth edition – hereinafter referred to as BRIB10) and basically follows a similar format. It is divided into four parts:

Part 1 provides an overview of the Building Act (based on Chapter 1 of BRIB10).

Part 2 offers a handy guide to the dos and don'ts of gaining Local Council's approval for Planning Permission and Building Regulations Approval (based on Chapter 5 of BRIB10).

Part 3 presents an overview of the requirements of the Approved Documents associated with the Building Regulations (based on Chapter 3 of BRIB10).

And finally, probably what will become your primary guide for all building projects:

Part 4 is an easy-to-read explanation of the essential requirements of the Building Regulations that an architect, builder, student or DIYer needs to be reminded of concerning all aspects of a building, whether for domestic or non-domestic use (based on Chapters 6 and 7 of BRIB10).

Shaded boxes are used in Part 4 of the book to show either the full text of the Building Regulation's **legal requirements** or a paraphrased version of these requirements. Cross references to the **actual requirements** and explanatory drawings contained in the Government's Approved Documents are provided throughout Part 4.

For example, under 'Wall Type 1 – Junctions with ceiling and roof':

> • The junction between the separating wall and the roof   E 2.56
> should be filled with a flexible closer which is also suitable
> as a fire-stop. (See AD-E Diagram 2.13.)

And the actual Approved Documents can be downloaded free of charge from the Government's website (in case you want to know a little more about a particular regulation or requirement).

The following symbols (shown in the margins) will help you to get the most out of this book:

 An important requirement or point.

 A good idea, suggestion or something worth remembering.

 Note: Used to provide further amplification or information.

About the authors

Ray Tricker (during COVID lockdown!)

Ray is a Senior Consultant with over 50 years of continuous service in Quality, Safety and Environmental Management, Project Management, Communication Electronics, Railway Command, Control and Signalling Systems, Information Technology and the development of Molecular Nanotechnology.

He served with the Royal Corps of Signals (for a total of 37 years) during which time he held various managerial posts culminating in being appointed as the Chief Engineer of NATO's Communication Security Agency (ACE COMSEC).

Most of Ray's work since leaving the Services has centered on the European Railways. He has held a number of posts with the Union International des Chemins de fer (UIC) [e.g. Quality Manager of the European Train Control System (ETCS)] and with the European Union (EU) Commission [e.g. European Rail Traffic Management System (ERTMS) Users Group Project Coordinator, HEROE Project Coordinator]. He was also a UKAS Assessor (for the assessment of certification bodies for the harmonisation of the Trans-European, High-Speed, Railway Networks) and recently he was appointed as the Quality, Safety and Environmental Manager for the Project Management Consultancy responsible for overseeing the multi-billion-dollar Trinidad Rapid Rail System.

Currently, as well as writing well over 60 books on diverse subjects such as Quality, Safety and Environmental Management, Building, Wiring and Water Regulations for Taylor & Francis (under their Routledge imprint) and Elsevier (under their Butterworth-Heinemann and Newnes imprints), he is busy assisting small businesses from around the world (usually on a no-cost basis) to produce their own auditable Quality and/ or Integrated Management Systems to meet the requirements of ISO 9001, ISO 14001 and OHSAS 18001, etc.

Samantha Alford

Samantha is an established technical author, instructor, business management specialist and data protection officer. She has over 35 years of continuous experience in Compliance, Governance and Oversight, and has worked in the public, private, voluntary and charity sectors. Sam served in the Royal Air Force for 18 years and has a wide and varied skillset.

Her MSc included a specialisation in performance measurement and management and she is also a certified GDPR practitioner, an IOSH health and safety manager, a quality auditor (internal, external and third party) and an experienced instructor. Her book *GDPR: A Game of Snakes and Ladders* was published by Routledge in February 2020.

Sam has extensive experience in business management at both strategic and operational levels and her skillset includes planning, policy and business documentation, procedure development, contract management and performance measurement. She is a Director of Professional Procurement and Project Management Ltd, a project services company specialising in procurement and contract strategy, programme and project delivery, project support services and technical authorship. Outside of work she is a Chairman of the local Chamber of Commerce board.

Part 1
Background information

1.1 What is the Building Act 1984?

The Building Act 1984 imposes a set of requirements for England and Wales aimed at:

- Buildings under construction and during use.
- Conservation of fuel.
- Preventing and detecting crime.
- Preventing waste, undue consumption or contamination of water.
- Protection and enhancement of the environment.
- Providing a method for controlling the Design, Build, Operation and Maintenance (DBOM) of buildings.
- Providing a mechanism for the inspection and maintenance of services, fittings and equipment used.
- Providing guidance on how services, fittings and equipment may be used.
- Securing the health, safety, welfare and convenience of persons regarding buildings.

1.1.1 What about the rest of the UK?

The Building Act 1984 only applies to England and Wales. Separate Acts and Regulations apply in Scotland or Northern Ireland and these are shown in Table 1.1.

 Note: In Scotland, the building requirements are very similar to England and Wales, except that a building warrant is still required before work can start.

DOI: 10.1201/9781003173786-1

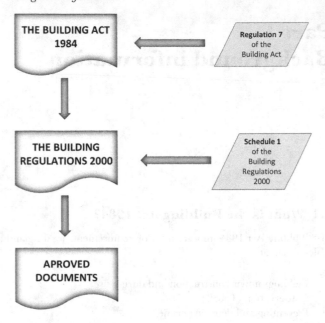

Figure 1.1 Implementing the Building Act

Table 1.1 Building legislation within the United Kingdom

	Act	*Regulations*	*Implementation*
England and Wales	Building Act 1984	Building Regulations 2010	Approved Documents
Scotland	Building (Scotland) Act 2003	Building (Scotland) Regulations 2004 (as amended)	Technical Handbooks
Northern Ireland	Building Act (Northern Ireland) Order 1979 (as amended)	Building Regulations (Northern Ireland) 2012	Technical Booklets

1.2 What does the Building Act 1984 contain?

The Building Act 1984 is made up of five parts, as shown in Table 1.2. These parts are then broken down into a number of sections and subsections as shown in Appendix 1A at the end of Chapter 1 of our associated publication, *Building Regulations in Brief* (10th edition – hereinafter referred to as BRIB10).

Table 1.2 The structure of the Building Act 1984

Part 1	The Building Regulations
Part 2	Supervision of building work, etc., other than by a Local Authority
Part 3	Other provisions about buildings
Part 4	General
Part 5	Supplementary

1.3 What are the Supplementary Regulations?

The Supplementary Regulations that make up Part 5 of the Building Act comprise seven schedules whose function is to list the principal areas that require regulation and to show how the Building Regulations are to be controlled by Local Authorities. These schedules are shown in Table 1.3.

Table 1.3 The contents of the Supplementary Regulations

Schedule	*Title*	*Description*
Schedule 1	Building Regulations	Schedule 1 describes the mandatory requirements for completing **all** building work.
Schedule 2	Relaxation of Building Regulations for existing work	Schedule 2 provides guidance in connection with work that has been carried out prior to a Local Authority (under the Building Act 1984 Section 36) dispensing with or relaxing some of the requirements contained in the Building Regulations. Schedule 2 is quite difficult to understand and if it affects you, then I would strongly advise that you discuss it with the Local Authority before proceeding any further.

(continued)

Table 1.3 (Continued)

Schedule	Title	Description
Schedule 3	Inner London	Schedule 3 applies to how Building Regulations are to be used in Inner London and as well as ruling which sections of the Act may be omitted. It also details how by-laws concerning the relation to the demolition of buildings (in Inner London) may be made.
Schedule 4	Provisions consequential upon Public Body's notice	Schedule 4 concerns the authority and ruling of Public Bodies' notices and certificates.
Schedule 5	Transitional provisions	Schedule 5 lists the transitional effect of the Building Act 1984 concerning existing Acts of Parliament.
Schedule 6	Consequential amendments	Schedule 6 lists the consequential amendments that will have to be made to existing Acts of Parliament owing to the acceptance of the Building Act 1984.
Schedule 7	Repeals	Schedule 7 lists the cancellation (repeal) of some sections of existing Acts of Parliament, owing to acceptance of the Building Act 1984.

1.4 What are 'Approved Documents'?

Approved Documents describe how the requirements of Schedule 1 and Regulation 7 of the Building Act 1984 can be met.

Each Approved Document reproduces the actual requirements contained in the Building Regulations relevant to the subject area (e.g. Approved Document P deals with electrical safety). This is then followed by practical and technical guidance (together with examples) showing how the requirements can be met in some of the more common building situations.

Note: All statutory instruments can be accessed on the UK government legislation website (www.legislation.gov.uk) and the Planning Portal, or may be purchased from The Stationery Office or RIBA.

The current sets of Approved Documents are in 17 Parts, A to S (less 'I' and 'N'), as shown in Table 1.4.

Table 1.4 Building Regulations for the United Kingdom

England and Wales	Scotland	Northern Ireland
Approved Document A Structure: 2013	Technical Handbook 2017 – Structure	Technical Booklet D: 2012 – Structure Technical Booklet B: 2013 – Materials and workmanship
Approved Document B (2 volumes) Fire safety: 2020	Technical Handbook 2017 – Fire	Technical Booklet E: 2012 – Fire safety
Approved Document C Site preparation and resistance to contaminants and water: 2013	Technical Handbook 2016 – Environment	Technical Booklet C: 2012 – Site preparation and resistance to contaminants and moisture
Approved Document D Toxic substances: 2013	Technical Handbook 2017 – Environment	Technical Booklet J: 2012 – Solid waste in buildings
Approved Document E Resistance to sound: 2015	Technical Handbook 2017 – Noise	Technical Booklet G: 2012 – Resistance to the passage of sound
Approved Document F (2 volumes) Ventilation: 2021	Technical Handbook 2017 – Environment	Technical Booklet K: 2012 – Ventilation
Approved Document G Sanitation, hot water safety and water efficiency: 2016	Technical Handbook 2017 Section 4 – Environment	Technical Booklet P: 2012 – Sanitary appliances and unvented hot water storage systems and reducing the risk of scalding

(continued)

Table 1.4 (Continued)

England and Wales		Scotland	Northern Ireland
Approved Document H	Drainage and waste disposal: 2015	Technical Handbook 2017 – Environment	Technical Booklet N: 2012 – Drainage
Approved Document J	Combustion appliances and fuel storage systems: 2013	Technical Handbook 2017 – Safety	Technical Booklet L: 2012 – Combustion appliances and fuel storage systems
Approved Document K	Protection from falling, collision and impact: 2013	Technical Handbook 2017 – Safety	Technical Booklet H: 2012 – Stairs, ramps, guarding and protection from impact
		Technical Handbook 2017 – Energy	Technical Booklet V: 2012 – Glazing
Approved Document L (2 volumes)	Conservation of fuel and power: 2021	Technical Handbook 2017 – Energy	Technical Booklet F1 and F2: 2014 – Conservation of fuel and power
Approved Document M (2 volumes)	Access to and use of buildings: 2021	Technical Handbook 2017 – Safety	Technical Booklet R: 2012 – Access to and use of buildings
Approved Document O	Overheating: 2021		
Approved Document P	Electrical safety: 2013	Technical Handbook 2017 – Safety	

Approved Document Q	Security in dwellings: 2015
Approved Document R	High-speed electronic communications networks: 2016
Approved Document S	Infrastructure for charging electric vehicles: 2021
Approved Document 7	Material and workmanship: 2018

Technical Booklet M: 2017 – Physical infrastructure for high-speed communications networks

Changes which came into effect in 2018 and 2020 included a major revision of Approved Document B and Regulation 7 (as a result of the Grenfell Tower inquiry). New editions of Approved Documents F and L (which were released in December 2021) and two new documents (O and S) have been introduced. **All** of these documents come into effect on **15 June 2022**.

Notes:

1. Approved Documents F and L have now been completely rewritten and the numbering and order of guidance has changed (F is currently in 2 volumes and L has been reduced from 4 volumes to 2 volumes).
2. Approved Document O only applies in respect of new residential buildings (including dwellings and institutional and student accommodation where people sleep on the premises).

1.4.1 Approved Document F – Ventilation: 2021

There are four main requirements to this Approved Document:

1. People in the building to be provided with an adequate means of ventilation.
2. Mechanical ventilation systems to be suitably commissioned.
3. The person carrying out work on the ventilation system must give the building owner sufficient information about the system and its maintenance so that it may be operated effectively.
4. The person carrying out the work should ensure that a test of the air-flow rate is carried out and that the Local Authority is advised of the results of this testing.

The document has two distinct parts, one for dwelling houses and a second for other buildings, and has been substantially revised in 2021.

1.4.2 Approved Document L – Conservation of fuel and power: 2021

This Approved Document has two distinct parts, one for dwelling houses and a second for other buildings, and has been substantially revised in 2021. Previously, it was split into four separate parts which covered the conservation of fuel and power in new dwellings (L1A), existing dwellings

(L1B), new buildings other than dwellings (L2A), and existing buildings other than dwellings (L2B).

There are three main aspects to this Approved Document, covering conservation of fuel and power in new dwellings, target CO_2 emissions and energy performance certificates and the on-site generation of electricity. This Approved Document requires that:

- Fuel and power is conserved by limiting heat gains and losses through thermal elements, other parts of the building fabric, and from pipes, ducts and vessels used for space heating, space cooling and hot water services. Building services are energy efficient, have effective controls and are commissioned by testing to ensure that they use no more power than is necessary.
- Where an on-site electricity generation system is installed, the system and its electrical output should be appropriately sized for the site and available infrastructure, have effective controls and must be commissioned by testing and adjusting as necessary to ensure that it produces the maximum electricity that is reasonable in the circumstances.

Note: Approved Document L also includes the following Regulations:

- Regulation 6: guidance on the requirements relating to material change of use.
- Regulation 22: requirements relating to a change in energy status.
- Regulation 23: the renovation or replacement of thermal elements.
- Regulation 24: the methodology for the calculation of energy performance.
- Regulation 25: minimum energy performance requirements for new buildings (target CO_2 emission, fabric efficiency and primary energy rates).
- Regulation 25A: high-efficiency alternative systems for new buildings.
- Regulation 25B: nearly zero-energy requirements for new buildings.
- Regulation 26: methodology for calculating CO_2 emission rates (new buildings).
- Regulation 26A: methodology for calculating target fabric efficiency rates (new dwellings).
- Regulation 26C: methodology for calculating target primary energy rates for new buildings.

- Regulation 27, 27A, 27C: details of when the Local Authority should be informed of the CO_2 emission rate, fabric efficiency rate for a new building and the primary energy rate for a new building.
- Regulation 28: consequential improvements to energy performance.
- Regulation 40: the provision of information about use of fuel and power in the building.
- Regulation 40A: the provision of information about any systems for on-site generation of electricity.
- Regulation 43: pressure testing of a building and the notification of the Local Authority of the results of these tests.
- Regulation 44: commissioning of a fixed ventilation system.
- Regulation 44AZ: commissioning of a system for on-site electricity generation.

1.4.3 Approved Document O – Overheating: 2021

This Approved Document comes into effect on 15 June 2022. There are two main aspects to the document:

1   That reasonable provision is made in dwellings, institutions or any other building containing one or more rooms for residential purposes (excluding hotels) to limit unwanted solar gains in summer and provide an adequate means to remove heat from the indoor environment.

2   That account is taken of the safety of any occupant, and their reasonable enjoyment of the residence; and that mechanical cooling is only used where insufficient heat is capable of being removed from the indoor environment without it.

1.4.4 Approved Document S – Infrastructure for charging electric vehicles: 2021

This Approved Document comes into effect in June 2022 and relates to electric vehicle charging provisions. It has six parts:

1   New residential buildings with associated parking must have access to electric vehicle charging points. (Regulation 44D contains options for fulfilling S1.)

2 How the provision of electric vehicle charging points in dwellings resulting from a material change of use is achieved. (Further guidance is contained in Regulation 44E.)

3 The provision of electric vehicle charging points in residential buildings undergoing major renovation which will have more than 10 associated parking spaces after the renovation is completed. (Further guidance in Regulation 44F.)

4 The provision of electric vehicle charging points in new buildings which are not residential buildings or mixed-use buildings that have more than 10 associated parking spaces after the renovation is completed. (More guidance in Regulation 44G.)

5 The provision of electric vehicle charging points for buildings undergoing major renovation which are not residential buildings or mixed-use buildings which will have more than 10 associated parking spaces after the renovation is completed. (Further guidance in Regulation 44H.)

6 The provision of electric vehicle charging points in new mixed-use buildings and mixed-use buildings undergoing major renovation where S1, S3, S4 or S5 may apply.

Any changes to the Approved Documents which come into effect in England have to be subsequently ratified by the Welsh Assembly. On occasions, therefore, earlier versions of the Approved Documents will apply in Wales until the assembly legislative process catches up with the English amendments.

 Free pdf copies of these Approved Documents are available from: www. gov.uk/government/collections/approved-documents.

1.5 How are buildings classified?

For the purpose of the Building Act there are six normal classifications for buildings:

1 by size (e.g. small/large);
2 by description (e.g. house/flat);
3 by design (e.g. prefabricated);
4 by purpose (e.g. place of assembly);
5 by location (e.g. in a Conservation Area);
6 'any other characteristic whatsoever'!

1.6 Who polices the Building Act?

Local Authorities are responsible for ensuring that any building work conforms to the requirements of the Building Regulations and the Building Act. They have the authority to:

*   Make you take down and remove or rebuild anything that contravenes a regulation.
*   Make you complete alterations so that your work complies with the Building Regulations.
*   Employ a third party to take down and rebuild any non-conforming buildings or parts of buildings **and then send you the bill**!

 Local Authorities can, in certain circumstances, even take you to court and have you fined – especially if you fail to complete the removal or rebuilding of any non-conforming work. It doesn't matter whether you are the owner or merely the occupier of the building; you can be prosecuted or ordered to carry out remedial work on a property.

1.7 How to comply with the Building Regulations?

The Local Authority is responsible for making regular checks that all building work being completed in their area is in compliance with the approved plan and the Building Regulations. Checks will be completed at specific stages of work (e.g. the excavation of foundations) and will include:

*   Tests of the soil or sub-soil of the site of the building.
*   Tests of any material, component or combination of components that has been, is being or is proposed to be used in the construction of a building.
*   Tests of any service, fitting or equipment that has been, is being or is proposed to be provided in or in connection with a building.

 The Local Authority has the power to ask the person responsible for the building work to complete some of these tests on their behalf and the cost of carrying out these tests will normally be charged to the owner or occupier of the building!

1.8 What are the duties of the Local Authority?

The Local Authority is responsible for ensuring that the requirements of the Building Act 1984 are carried out and that the appropriate associated Building Regulations are enforced, subject to:

- The Public Health Act 1961 (relating to United Districts and Joint Boards).
- The Local Government, Planning and Land Act 1980 (relating to urban development areas).
- The Public Health (Control of Disease) Act 1984 (as amended on 30 April 2021 owing to the COVID pandemic).

1.9 What are the powers of the Local Authority?

Under the Building Act 1984, the Local Authority has:

- The power to enter **any** premises at all 'reasonable hours' to check if there is (or has been) a contravention of the Building Act or Building Regulations.
- Overall responsibility for the construction and maintenance of sewers and drains, as well as the laying and maintenance of water mains and pipes.
- The power to sell any materials that have been removed, by them, from any premises when executing works under this Act (paying all proceeds, less expenses, from the sale of these materials, to the owner or occupier).
- The authority to make the owner or occupier of any premises complete essential and remedial work in connection with the Building Act 1984 (particularly with respect to the construction, laying, alteration or repair of a sewer or drain).
- The authority to complete remedial and essential work themselves (on repayment of expenses) if the owner or occupier refuses to do this work himself.

1.10 What is the 'Building Regulations Advisory Committee'?

The Building Act allows the Secretary of State to appoint a committee (known as the Building Regulations Advisory Committee) to review, amend, improve and produce new Building Regulations and the associated documentation.

1.11 Who are Approved Inspectors?

An Approved Inspector is a person who is approved and appointed by the Secretary of State to inspect, supervise and authorise building work. The Register of Approved Inspectors can be obtained on-line from: www.cic air.org.uk/approved-inspectors-register//.

1.12 Does the Fire Authority have any say in Building Regulations?

When a requirement 'encroaches' on something that is normally handled by the Fire Authority (such as the provision of a means of escape or any structural fire precautions) the Local Authority **must** consult the Fire Authority before making any decision.

1.13 What Notices and Certificates are required?

1.13.1 What are Initial Notices?

An Approved Inspector **and** the person intending to carry out the work is required to present an Initial Notice and a plan of work to the Local Authority. Acceptance of an Initial Notice by a Local Authority is treated as 'depositing plans of work'.

If, for any reason, the Local Authority rejects the Initial Notice, then the Approved Inspector can appeal to a Magistrates' court for a ruling. If still dissatisfied, they can appeal to the Crown Court.

If the Local Authority believes that work has not commenced within three years of the date on which it accepted an Initial Notice, it may cancel the notice.

1.13.2 What are Plans Certificates?

A Plans Certificate is issued to the Local Authority by an Approved Inspector when they have inspected a site and are satisfied that the plans of work specified in the Initial Notice do not contravene the Building Regulations in any way. This Plans Certificate:

- Can relate to the whole or part of the work specified in the Initial Notice.
- Does not have any effect unless the Local Authority accepts it.
- May only be rejected by the Local Authority 'on prescribed grounds'.

Once the development has commenced, the permission remains in place unless the Local Authority serve a completion notice. If such a notice is served, the development must then be completed within 12 months. In the absence of such a notice, the permission remains in place.

1.13.3 What are Final Certificates?

Once the Approved Inspector is satisfied that all work has been completed in accordance with the work specified in the Initial Notice, he will provide a Final Certificate to the Local Authority and the person who carried out the work.

1.13.4 What is a Public Bodies Notice?

If a Public Body (e.g. Local Authority or County Council) thinks that they (or local agents) have the expertise to supervise the completion of building work on one of its **own** buildings, then they can provide the Local Authority with a notice (referred to as a 'Public Body's Notice') together with their plan of work.

1.13.5 What is 'Type Approval'?

Type Approval is where the Secretary of State is empowered to approve a particular type of building matter as complying, either generally or specifically, with a particular requirement of the Building Regulations.

There are changes planned to this section of the Building Act, so be sure to check the latest legislation.

1.13.6 What causes some plans for building work to be rejected?

The Local Authority will reject all plans for building work that are defective or contravene any of the Building Regulations. If a plan for proposed building work is accompanied by a certificate from a person(s) approved by the Secretary of State, then only in **extreme** circumstances can the Local Authority reject the plans.

1.13.7 Who retains these records?

Local Authorities are required to keep a register of all Initial Notices, Amendment Notices, Public Body's Notices (including Plans Certificates, Final Certificates, and the Public Body's Final Certificates) which have been given. The Local Authority is required to make this register available for public inspection.

Note: Copies of these records are also available on most Councils' websites.

1.14 Are there any exemptions to the Building Regulations?

The following are exempt from the Building Regulations:

• A 'Public Body' (e.g. Local Authorities, County Councils and any other body 'that acts under an enactment for public purposes and not for its own profit').
• Buildings (or classes of buildings) that have been declared exempt by the Secretary of State.
• Buildings belonging to 'statutory undertakers' (e.g. the Civil Aviation Authority).
• Classes of buildings, services, fittings or equipment which are 'prescribed' by the Secretary of State.
• Educational buildings erected according to plans approved by the Secretary of State.

1.14.1 What about Crown Buildings?

Although the majority of the requirements of the Building Regulations are applicable to Crown Buildings (i.e. a building in which there is a Crown interest) or Government buildings (held in trust for Her Majesty), there are occasional deviations which may need further investigation.

1.14.2 What about buildings in Inner London?

You will find that the majority of the requirements found in the Building Regulations are also applicable to buildings in Inner London boroughs (e.g. Inner Temple).

 Note: There are changes planned to this section of the Building Act, so be sure to check the latest legislation.

1.14.3 What about the United Kingdom Atomic Energy Authority?

The Building Regulations do **not** apply to buildings belonging to, or occupied by, the United Kingdom Atomic Energy Authority (UKAEA), unless they are dwellings or offices.

1.15 Can appeals be made against a Local Authority's ruling?

An appeal against a Local Authority's ruling can be made if:

*   The Local Authority fails to notify you of their decision to reject an application to dispense with or relax a requirement in Building Regulations within two months of the application.
*   The Local Authority rejects an application to dispense with or relax a requirement in Building Regulations.
*   You disagree with a Local Authority notice requiring works or demolition.
*   You disagree with a notice to remove or alter offending work which has been issued by the Local Authority **after** the work has been completed.
*   You disagree with a Local Authority's rejection of an Initial Notice, Amendment Notice, Public Body's Notice, Plans Certificate, Final Certificate, Public Body's Certificate or Public Body's Final Certificate.

The appeal against the Local Authority's decision is made to the Magistrates' Court. If your appeal is unsuccessful at the Magistrates' Court, you may appeal to the Crown Court.

Where the Secretary of State has given a ruling, this ruling may, in certain circumstances, be appealed to the High Court on a point of law.

1.15.1 Can I apply for a relaxation of the Building Regulations in certain circumstances?

The Building Act allows the Local Authority to dispense with, or relax, a Building Regulation if they believe that that requirement is unreasonable in relation to a particular type of work being carried out. In the majority of cases, applications to dispense with or relax Building Regulations can be settled locally. In more complicated cases, the Local Authority can seek guidance from the Secretary of State who will decide whether the requirement may be relaxed or dispensed with (either unconditionally or subject to certain conditions).

The Local Authority may charge a fee for reviewing and deciding on these matters.

1.15.2 Can I change a plan of work once it has been approved?

If a person intending to carry out building work has already had their plan (or plans) passed by the Local Authority and wants to change them, they may submit a set of revised plans to the Local Authority, showing precisely how they intend to deviate from the approved plan and asking for approval.

1.15.3 Must I complete 'approved work' in a certain time?

Once a building plan has been passed by the Local Authority, then **'work must commence within three years'** from the date that it was approved. Failure to start work within this time could result in the Local Authority cancelling the approved plans and you will have to resubmit them if you want to carry on with your project. If the Local Authority advise you that the deposit of plans is of 'no effect', it is as if the plans had **never** been deposited.

Note: The meaning of the term 'work must commence' can vary, but normally means physically laying foundations of the building.

1.15.4 What happens if I contravene any of these requirements?

If you contravene the Building Regulations or wilfully obstruct a person acting 'in the execution of the Building Act 1984 or of its associated Building Regulations', then on conviction you could be liable to a fine of up to Level 5 on the Standard Scale (at the time of writing, this is £5,000), plus £50 for every day afterwards until the matter is rectified plus costs. In exceptional circumstances, contravention of the regulations can lead to a short holiday in one of HM Prisons!

 Note: Following the Grenfell Tower fire, Approved Document B (*Fire safety*) has been substantially revised with a focus on build quality, preventing fire spread, protecting exits and providing access for the emergency services.

1.15.5 What about civil liability?

All building work should be completed safely and without risk to people employed on the site or visiting the site. Any contravention of the Building Regulations that causes injury (or death) to any person is liable to prosecution in the normal way.

1.15.6 What about compensation?

If an owner or occupier considers he has sustained damage because of something that the Local Authority has done, they can appeal to the local Magistrates' Court.

1.16 What about dangerous buildings?

If a building, or part of a building or structure, is in a dangerous condition (or is used to carry loads which would make it dangerous) then the Local Authority may apply to a Magistrates' court to make an order requiring the owner to:

- carry out work to avert the danger;
- demolish the building, structure or any dangerous part of it, and remove any rubbish resulting from the demolition.

If the actual ownership is not clear then the Authority may carry out the work.

1.16.1 Emergency measures

In emergencies, the Local Authority can make the owner take immediate action to remove the danger or they can complete the necessary action themselves (recovering from the owner such expenses reasonably incurred).

1.16.2 Can I demolish a dangerous building?

You must have good reasons for knocking down a building. Making way for rebuilding or improvement will in most cases be incorporated in a planning application. You are **not** allowed to begin any demolition work (even on a dangerous building) unless you have given the Local Authority notice of your intention **and** this has either been acknowledged by the Local Authority or the relevant notification period has expired. In this notice you will have to:

*   Show how you intend to demolish the building.
*   Specify the building to be demolished.
*   State the reason(s) for wanting to demolish the building.

Copies of this notice will then have to be sent to:

*   Any public electricity supplier in whose area the building is located.
*   Any public gas supplier in whose area the building is located.
*   The Local Authority.
*   The occupier of any building adjacent to the building in question.

This regulation does not apply to the demolition of an internal part of an occupied building, or a greenhouse, conservatory, shed or prefabricated garage (that forms part of that building) or an agricultural building.

1.16.3 Can I be made to demolish a dangerous building?

If the Local Authority considers that a building is so dangerous that it should be demolished, it is entitled to issue a notice to the owner requiring them (in accordance with the Housing Act 2004) to arrange with the relevant statutory undertakers for the disconnection of gas, electricity and water supplies to the building, and to:

- Disconnect, seal and remove any sewer or drain in or under the building.
- Leave the site in a satisfactory condition following completion of all demolition work.
- Make good the surface of the ground that has been disturbed in connection with the removal of drains.
- Remove material or rubbish resulting from the demolition and clear the site.
- Repair and make good any damage to an adjacent building caused by the demolition or by the negligent act or omission of any person engaged in it.
- Shore up any building adjacent to the building to which the notice relates.
- Weatherproof any surfaces of an adjacent building that are exposed by the demolition.

Before complying with this notice, the owner must give the Local Authority 48 hours' notice of commencement of the work. Otherwise, you can be fined up to £2,500.

In certain circumstances, the owner of an adjacent building may be liable to assist in the cost of shoring up their part of the building and waterproofing the surfaces. It could be worthwhile checking this point with the Local Authority.

1.17 What about defective buildings?

If a building or structure is, because of its ruinous or dilapidated condition, liable to cause damage to (or be a nuisance to) the amenities of the neighbourhood, then the Local Authority can require the owner to:

- carry out necessary repairs and/or restoration;
- demolish the building or structure (or any part of it) and to remove all the rubbish or other material resulting from this demolition.

1.18 What are the rights of the owner or occupier of the premises?

When a person has been given a notice by a Local Authority to complete work, he has the right to appeal to a Magistrates' court on any of the following grounds that:

- The notice or requirement is not justified by the terms of the provision under which it claimed to have been given.
- There has been some informality, defect or error in (or in connection with) the notice.
- The authority has refused (unreasonably) to approve completion of alternative works, or the works required by the notice to be executed are unreasonable or unnecessary.
- The time limit set to complete the work is insufficient.
- The notice should lawfully have been served on the occupier of the premises in question instead of on the owner (or vice versa).
- Some other person (who is likely to benefit from completion of the work) should share in the expense of the works.

Part 2
Requirements for Planning Permission and Building Regulations approval

2.1 Introduction

There are some changes that you can legally make to your own home without having to get permission as long as the work does not affect the external appearance of the building (this is called 'permitted development'), but outside these rights, you will find that the majority of building work requires you to have Planning Permission and/or Buildings Regulations approval **prior** to work actually commencing.

2.1.1 Community Right to Build and assets of community value

The 2020 Localism Act allows certain community organisations to put forward small-scale developments on a specific site, without the need for Planning Permission. This gives the community the opportunity to put forward facilities that they want to see and which will benefit them as a group.

Suitable projects include maintaining affordable housing stock and providing or maintaining local facilities such as a village hall or playground. Community Right to Build Orders are subject to a limited number of exclusions, such as proposals needing to fall below certain thresholds so that an Environmental Impact Assessment is not required. Proposals are subject to testing by an independent person and a community referendum.

 More information is available on the Community Right to Build website: https://mycommunity.org.uk/.

DOI: 10.1201/9781003173786-2

2.1.2 Planning requirements

The intention of this part of the book is to show whether Planning Permission and/or Building Regulations apply to particular projects (listed in alphabetical order). There may be variations in the planning requirements, and to some extent the Building Regulations, from one area of the county to another, so you should consult your Local Authority for specific guidance which relates to your particular area.

In Part 2 of the book:

- All measurements are taken externally.
- Designated areas include:
  - Areas of Outstanding Natural Beauty;
  - Conservation Areas;
  - National Parks;
  - The Norfolk or Suffolk Broads;
  - World Heritage Sites.
- The regulations depend on previous works on the site: where the word 'original' is used in planning Regulations, it means the house as it was first built or as it was on 1 July 1948. Any extensions added since that date count towards the allowance (it is, however, always best to check with Planning Control staff if you are unsure).

Following UK devolution, this guidance currently applies to England, whilst advice specific to Northern Ireland, Scotland or Wales is available via applicable links on their Planning Portals:

- https://epicpublic.planningni.gov.uk/publicaccess/
- https://beta.gov.scot/building-planning-and-design/
- www.planningportal.co.uk/wales_en/

 Permitted development rights do not apply to flats, maisonettes, Listed Buildings or in areas known as a 'designated areas'.

 Note: BS 7913 provides guidance on the conservation of historic buildings.

2.2 Additional storeys/extending upwards

Adding an additional storey will be considered permitted development as long you meet the following conditions:

Table 2.1 Authorisation requirements for additional storeys

Planning Permission	Building Regulations approval
No Provided it has Prior Approval and meets the limits and conditions listed below.	**Yes**

- You have obtained your Local Authority's Prior Approval.
- There is no more than one flat in the current house.
- The current house was not converted from a previous non-residential use under permitted development rights.
- The house was constructed between 1 July 1948 and 5 March 2018.
- The house is not on designated land or a Site of Special Scientific Interest.
- There have been no additional storeys added to it in the past.

The Building Regulations approval is required to confirm that:

- Stairs providing access to the new floor are safely designed.
- The structure is stable (including the existing roof) and allows safe escape from a fire.
- The structural strength of the new floor is sufficient.
- There is reasonable sound insulation between the conversion and the rooms below.

In addition to the conditions above, proposed developments involving additional storeys are subject to the following limitations:

- Additional storeys:
    - cannot add more than 3.5m to the total height;
    - must be constructed on the principal part of the house;
    - must not be greater than 3m in height or the height of any existing storey in the principal part of the house.
- Engineering operations (such as strengthening existing walls and foundations) can only include works within the existing curtilage of the house.
- In a terrace or semi-detached property, the total height cannot be more than 3.5m higher than the next highest building attached to, adjoining, or in the same row as the property.

- Materials of a similar appearance to those used in the construction of the exterior of the current house should be used.
- Only one storey can be added to a single-storey house.
- Two storeys can be added if the house has more than one storey.
- The house cannot exceed 18m in total height.
- Windows cannot be placed in any wall or roof slope on the side elevation of the house.

After the development is complete there should be no visible support structures on or attached to the exterior of the house and the roof pitch of the principal part of the house must be the same as it was prior to the development. The house must remain in use as a domestic residential property.

 Requirements for the addition of extra storeys to houses over three storeys, or flats, are likely to be similar but more extensive than those outlined above.

2.3 Advertising and signs

Advertisements on buildings and on land often need planning consent. Professional and business types of display and permanent signs may come under the category of 'advertising control' for which planning consent is required.

You are permitted to display certain small signs at the front of residential premises, such as election posters, notices of meetings, jumble sales, car for sale, etc., and information such as the house name or number, as well as information signs such as 'Beware of the dog'. Temporary notices relating to local events, such as fêtes and concerts, may be displayed for a short period, as long as they are under 0.6m^2. Estate agents' boards are covered by different rules but, in general, these should not be bigger than 0.5m^2 on each side.

Table 2.2 Authorisation requirements for advertising and signs

Planning Permission		*Building Regulations approval*	
No	If the advertisement is less than 0.3m^2, as long as it is not illuminated.	**Possibly**	Domestic adverts and signs are not normally subject to building control, although, if lit, any electrical work would be.

It is illegal to post notices on empty shop windows, doors and buildings, and also on trees. This is commonly known as 'fly posting' and can carry heavy fines under the Town and Country Planning Act.

Note: Professional and business displays and permanent signs come under the category of 'advertising control' for which planning consent is required.

2.4 Aerials, satellite dishes and flagpoles

Before you go to any expense by purchasing or renting an antenna or satellite dish, check whether you need Planning Permission, Listed Building Consent, or permission from the landlord or owner. You are responsible for placing antennas in the appropriate position.

Table 2.3 Authorisation requirements for aerials, satellite dishes and flagpoles

Planning Permission		*Building Regulations approval*	
Satellite dishes, aerials and antennae on buildings up to 15m high			
No	It would be exempt if under 4m high, and possibly if attached to the house and below the roof height. Otherwise, you probably need Planning Permission.	No	But make sure that the fixing point is stable and the installation is safe.
Possibly	If the building is listed, in a designated area or it is intended to install the antenna on a chimney, wall or roof slope visible from a road or Broads waterway.		
Satellite dishes, aerials and antennae on buildings over 15m high			
No	Unless: • There are more than four antennas on the property overall.	No	But make sure that the fixing point is stable and the installation is safe.

(continued)

Table 2.3 (Continued)

Planning Permission		Building Regulations approval
	• The size of any antenna is more than 130cm in any linear dimension.	
	• The cubic capacity of each antenna is more than 35 litres.	
	• The antenna is fitted to a chimney stack and is greater than 60cm in any linear dimension.	
	• The antenna sticks out more than 300cm above the highest part of the roof line.	
	• The building is in a designated area and it is intended to install the antenna on a chimney, wall or roof slope visible from a road or Broads waterway.	
Flagpole		
No	If you intend to fly the Union flag or national flag of any country, then you do not need Planning Permission to install a flagpole. However, generally speaking, Planning Permission will be required if you intend to fly flags used for advertising. Portable flagpoles are an alternative solution as they do not require Planning Permission.	**No** But make sure that the fixing point is stable and the installation is safe.

2.5 Basements

 Check if underpinning or additional foundation work is required for your basement project.

The creation of living space in basements and its planning regime remains under constant review. In all circumstances before you carry out any work you are advised to contact your Local Planning Authority to discuss any local policy changes.

 For further advice on basement conversions, see: www.planningportal. co.uk/info/200244/project_advice/142/basement_conversions.

Table 2.4 Authorisation requirements for basements

Planning Permission		Building Regulations approval	
Existing basements			
No	Provided it is not a separate unit or the usage is changed significantly (or a light well added).	Yes	Ensuring adequate fire escape routes, ventilation, ceiling height, damp proof, electrical wiring and water supplies are in place.
New basements			
Yes	If you will be excavating to create a new basement and it involves major works, creates a new separate unit of accommodation and/or alters the external appearance of the house (e.g. adding a light well).	Yes	Ensuring adequate fire escape routes, ventilation, ceiling height, damp proof, electrical wiring and water supplies are in place.

2.6 Biomass-fuelled appliances

Table 2.5 Authorisation requirements for biomass-fuelled appliances

Planning Permission		Building Regulations approval	
No	Planning Permission is not normally required if: • The biomass installation is in a house. • The work is all internal to the building. • Any outside flue on the rear or side elevation is no more than 1m above the highest part of the roof.	**Yes**	Particularly in relation to electrical and plumbing work, ventilation, noise and general safety.
Yes	If the building is listed or in a designated area, Conservation Area or World Heritage Site, flues may **not** be visible from the highway.	**Possibly**	Possibly, as the Clean Air Act 1993 requires that in designated Smoke Control Areas only 'authorised fuels' can be used (including gas, electricity, anthracite and specified manufactured smokeless fuels).

Note: If the project also requires an outside building to store fuel or related equipment, then the same rules apply to that building as for other extensions and garden outbuildings.

2.7 Ceilings and floors

Table 2.6 Authorisation requirements for ceilings and floors

Planning Permission	Building Regulations approval

No Planning Permission is not generally required to replace a floor or ceiling unless you live in a Listed Building.	**Yes** • If you plan to replace more than 25% of a ceiling below a cold loft space or flat roof, Building Regulations approval will be required and the thermal insulation in the ceiling would need to be improved.
	
	• The condensation risk of the roof space should be assessed and provisions made in accordance with Approved Document C.
	
	• In a dwelling, replacing a pitched roof would require new or additional loft insulation:
	  • at least 250mm mineral fibre or cellulose fibre, quilt laid between and across ceiling joists **or** loose fill or equivalent
	  • in a dwelling's flat roof, the insulation is placed between and over joists as required to achieve the target U-value set out in the Approved Document.

If a dwelling has a pitched roof you will need to provide new or additional loft insulation unless the loft is already boarded out and the boarding is not going to be removed as part of the work.

Note: You will find guidance on ceilings and floors in Sections 4.6 and 4.8 of this book.

2.8 Central heating and boilers

Table 2.7 Authorisation requirements for central heating

Planning Permission	Building Regulations approval
No • Unless it is a Listed Building or in a Conservation Area. If an external flue is required, provided that: • any flues on the rear or side elevation are no more than 1m above the highest part of the roof • no flue is fitted to the principal or side elevation if the building fronts onto a highway in a Conservation Area, World Heritage Site, National Park, Area of Outstanding Natural Beauty or the Broads • you seek consent for internal and external work if the building is listed or in a designated area.	**No** • If electric and it is installed by an approved person and complies with Approved Document P. • If gas, solid fuel or oil fired and it is installed by an approved person and complies with Approved Document J.

If the project requires an outside building to store fuel or related equipment, the same rules apply to that building as for other extensions and garden outbuildings. Planning Permission is not normally required for the installation or replacement of a boiler or heating system if all the work is internal. Where permission is required, it is not always necessary to apply for this in advance if an emergency occurs as long as the repairs comply with the regulations and you apply for retrospective approval (and a completion certificate).

 Building Regulations approval, however, is required for new boilers or a cooker supplying heating such as an Aga/ Raeburn, etc., because of the safety issues and the need for energy efficiency. This is generally achieved by employing an installer who is registered under an approved scheme.

2.9 Change of use

The use of buildings or land for a different purpose may need consent even if no building or engineering works are proposed.

A material change of use is where there is a change in the purposes for which, or the circumstances in which, a building is used, so that after that change:

- The building is used as a dwelling, where previously it was not.
- The building contains a flat, where previously it did not.
- The building is used as a hotel or a boarding house, where previously it was not.
- The building is used as an institution, where previously it was not.
- The building is used as a public building, where previously it was not.
- The building is not a building described in Classes 1 to 6 in Schedule 2, where previously it was.
- The building, which contains at least one dwelling, contains a greater or lesser number of dwellings than it did previously.
- The building contains a room for residential purposes, where previously it did not.
- The building, which contains at least one room for residential purposes, contains a greater or lesser number of such rooms than it did previously.
- The building is used as a shop, where previously it was not.

Whenever such changes occur and there is a substantial change in the use of the whole of a building, the building must be brought up to the standards within the Approved Documents, in particular:

- access and facilities for the fire service, preventing fire spread, means of warning and escape;
- combustion appliances;
- conservation of fuel and power – buildings other than dwellings and dwellings;
- electrical safety;
- foul water drainage;
- resistance to moisture;
- sanitary conveniences and washing facilities;
- solid waste storage;
- ventilation.

The table below summarises the changes to use classes and types of use from 1 September 2020.

Table 2.8 Use classes and types of use

Class	Sub Class	Description	Comments
Class A			Class A revoked on 1 September 2020.
Class B	B2	General industrial	B1 Business revoked on 1 September 2020.
	B8	Storage	
Class C	C1	Hotels	
	C2	Residential institutions	
	C2A	Secure residential institutions	
	C3	Dwelling houses. This class is in three parts: • C3(a) – use by a single person or a family, an employer and domestic employee (e.g. *au pair*), or a carer/foster parent and the person receiving their care. • C3(b) – up to six people living together as a single household and receiving care (e.g. supported housing schemes). • C3(c) – up to six people living together as a single household such as a small religious community or a homeowner and their lodgers.	
	C4	Houses in multiple occupation – small shared houses occupied by between three and six unrelated individuals sharing a kitchen or bathroom. Houses in multiple occupation with more than six residents become a *sui generis*.	
Class D			Class D revoked on 1 September 2020.

Table 2.8 (Continued)

Class	Sub Class	Description	Comments
Class E		Commercial, business and service: • E(a) – display or retail sale of goods, other than hot food. • E(b) – sale of food and drink for consumption (mostly) on the premises. • E(c); • c(i) – financial services; • c(ii) – professional services; • c(iii) – other appropriate services in a commercial, business or service locality. • E(d) – indoor sport and recreation/fitness. • E(e) – medical or health services. • E(f) – creche, day nursery or day centre (not including a residential use). • E(g) – uses which can be carried out in a residential area without detriment to the environment: • g(i) – offices to carry out any operational/administrative functions; • g(ii) – research and development of products or processes; • g(iii) – industrial processes.	Introduced on 1 September 2020. Covering uses previously defined in the revoked Classes A1/2/3, B1, D1(a–b) and D2(e).

(continued)

 Note: There were significant changes made to use classes in September 2020. Classes A1/2/3 and B1 are now treated as Class E. Planning applications submitted before 1 September 2020 may still use the old categories.

Table 2.8 (Continued)

Class	Sub Class	Description	Comments
Class F		Local community and learning	Class F introduced on 1 September 2020.
	F1	Learning and non-residential institutions: • F1(a) – provision of education. • F1(b) – display of works of art (not for sale or hire). • F1(c) – museums. • F1(d) – public libraries or public reading rooms. • F1(e) – public halls or exhibition halls. • F1(f) – public worship or religious instruction. • F1(g) – law courts.	
	F2	Local community: • F2(a) – shops selling essential goods, including food, where the shop's premises do not exceed 280m^2 and there is no other such facility within 1,000m. • F2(b) – halls or meeting places for the principal use of the local community. • F2(c) – areas or places for outdoor sport or recreation (not involving motorised vehicles or firearms). • F2(d) – indoor or outdoor swimming pools or skating rinks.	

Table 2.8 (Continued)

Class	Sub Class	Description	Comments
	Sui generis is a Latin expression that covers all forms of "industrial pollution"	'Alkali work' involving the discharge of industrial pollutants in, amusement arcades/centres or funfairs, betting offices/shops, bingo halls, casinos, cinemas, concert halls, dance halls, drinking and food establishments, fuel stations, hiring/ selling and/ or displaying motor vehicles, hostels, hot food takeaways, launderettes, night clubs, pay day loan shops, public houses, wine bars or drinking establishments, retail warehouse clubs, scrap yards or yards for the storage/distribution of minerals and/or the breaking of motor vehicles, taxi businesses, theatres, venues for live music performance, and waste disposal installations for the incineration, chemical treatment or landfill of hazardous waste.	Other uses become *sui generis* when they fall outside the defined limits of any other use class.

Note: The fee for Prior Approval for change of use is £96 (at the time of writing).

2.9.1 Material alterations

Where work results in a building, or controlled service or fitting, being no longer in compliance with a relevant requirement where previously it was, or if a previous compliance is made unsatisfactory, it is deemed to be a 'material alteration'. Any material alteration needs to comply with the requirements for conservation of heat and energy.

2.9.1.1 Material alterations in dwellings

A building is subject to a material alteration by:

- making an existing element part of the thermal envelope of the building (where previously it was not);
- providing (or extending) a controlled service;
- providing a controlled fitting;
- renovating a thermal element;
- substantially replacing a thermal element.

2.9.1.2 Material alterations in buildings other than dwellings

When an existing element becomes part of the thermal element of a building (where previously it was not) and it has a U-value worse than $3.3W/(m^2 \cdot K)$, it should be replaced (unless it is a display window or high-usage door).

2.9.2 Temporary change of use

As part of the Coronavirus legislation, certain buildings or land (other than A4 drinking establishments) were permitted to be used as a state-funded school for up to two academic years, provided that it returns to its previous use when no longer required or at the end of term (GDPO Schedule 2, Part 4, Class C). Previously vacant commercial land, subject to certain limitations and conditions (highways, noise, contamination and siting) was permitted to be used as a state-funded school for up to three academic years (GDPO Schedule 2, Part 4, Class CA).

2.9.3 Temporary flexible use

As part of the Coronavirus legislation, buildings in class A1, A2, A3, A5, B1, D1, D2 or betting offices or pay day loan shops (*sui generis*) uses were permitted to change their use for a single continuous period of up three years to A1, A2, A3, B1 and D1 (a,d,e,f or g) uses (GDPO Schedule 2, Part 4, Class D).

2.9.4 Temporary use for commercial film making

Buildings/land may be used for commercial film making for a up to 9 months in any 27-month period. This permits the erection of temporary structures, works, plants or machinery, but it is subject to limitations and conditions and requires Prior Approval (GDPO Schedule 2, Part 4, Class E).

2.9.5 Temporary use for the provision of takeaway food

Between 24 March 2020 and 23 March 2022, restaurants and drinking establishments, including those with expanded food provision, can change to a use for the provision of takeaway food and/or any use for the provision of hot or cold food that has\been prepared for consumers for collection or delivery to be consumed, reheated or cooked by consumers off the premises.

The Local Authority must be notified of the change of use, and the use must revert back to its original one at the end of the time period above, or earlier if the provision of takeaway food ceases (GPDO Schedule 2, Part 4, Class DA).

2.9.6 Emergency development by a Local Authority or Health Service Body

Until 31 December 2021, a change of use can occur on land owned, leased, occupied or maintained by a Local Authority or Health Service Body to prevent an emergency, or to reduce, control or mitigate the effects of an emergency, or take other action in connection with an emergency.

2.10 Conservatories

A conservatory has to be separated from the rest of the house to be exempt (e.g. with patio doors between it and the main house). Conservatories and sun lounges attached to a house are classed as permitted development, subject to the conditions in the table below and provided that the glazing complies with Approved Document K.

Table 2.9 Authorisation requirements for conservatories

Planning Permission	Building Regulations approval
No **Note:** Conservatories are now subject to the same limitations as other extensions. The most up to date guide for this can be found in 'mini guides' on the Planning Portal.	**No** Provided that: • Glazing and any fixed electrical installations comply with the applicable Building Regulations requirements. • The conservatory is built at ground level and has a floor area less than 30m². • The conservatory is separated from the house by external quality walls, doors or windows. • There is an independent heating system with separate temperature and ON/OFF controls.
	Any new structural opening between the conservatory and the existing house will require Building Regulations approval, even if the conservatory itself is exempt.

2.11 Conversions

Work on a loft conversion or a roof can affect bats and you **will** need to consider protected species when planning work of this type. Licence may be needed. You will also probably need Planning Permission whether or not actual building work is proposed!

 Note: A survey will probably be required if bats are using the building.

Table 2.10 Authorisation requirements for conversions

Planning Permission		Building Regulations approval	

Flats

Yes	Even where construction works may not be intended.	**Yes**	Unless you are not proposing any building work to make the change.

Loft conversions

No	Provided you do not alter or extend the roof space and they are subject to: • A volume allowance of 40m³ additional roof space for terraced houses (or 50m³ for detached and semi-detached houses). • Materials used are similar in appearance to the existing house. • The extension is not beyond the plane of the existing roof slope that fronts the highway. • No extension is higher than the highest part of the roof. • No verandas, balconies or raised platforms or chimneys, vents or flues are affected. • Roof extensions, apart from hip to gable ones, are set back as far as practicable, and at least 20cm from the original eaves. • Side-facing windows are obscure-glazed.	**Yes**	To ensure: • Reasonable sound insulation between the conversion and the rooms below. • Safe escape from fire. • Safely designed stairs to the new floor. • The stability of the structure (including the existing roof) is not endangered. • The structural strength of the new floor is sufficient. For further advice you should contact Building Control to discuss your proposal. You will also need to find out whether the work you intend carrying out falls within The Party Wall Act 1996.

Conversion to shops and offices

Yes	**Yes**

The areas that you should consider include:

- Ensuring that the stability of the structure (including the existing roof) is not endangered.
- Ensuring there is reasonable sound insulation between the conversion and adjacent rooms.
- Providing safely designed stairs to the new floor.
- Making sure the structural strength of the new floor is sufficient.
- Providing a safe escape from fire.

2.11.1 Converting an old building

Throughout the UK there are many underused or redundant buildings, particularly farm buildings which may no longer be required, or are unsuitable, for agricultural use. Such buildings, which are usually made of weathered stone and slate, contribute substantially to the character and appearance of the landscape and the built environment of an area.

Planning Permission will not normally be granted for reconstruction if substantial collapse occurs during work on the conversion of a building. Applicants are strongly advised to employ qualified architects or designers in preparing conversion proposals. Informal discussions with a planning officer at an early stage in considering design solutions are also encouraged.

Table 2.11 Authorisation requirements for the conversion of old buildings

Planning Permission	Building Regulations approval
Yes	**Yes**

2.12 Decoration and repairs inside and outside a building

Table 2.12 Authorisation requirements for decoration and repairs inside and outside a building

Planning Permission		Building Regulations approval	

Repairs to a house, shop or office

No	Provided the repairs are of a minor nature (e.g. replacing the felt to a flat roof, repointing brickwork, or replacing floorboards).	No	For minor repairs.
Yes	For major works such as removing a substantial part of a wall and rebuilding it, or underpinning a building.	Yes	For major works.

Internal decoration, repair and maintenance

No	If you are only decorating, repairing or completing maintenance work.	No	If you are only decorating, repairing or completing maintenance work.

External decoration, repair and maintenance

No	Provided the building is not listed and as long as it does not make the building any larger or affect the thermal element of the building.	No	Provided it does not make the building any larger or affect the thermal element of the building.
Yes	For work on external walls.	Yes	For some types of repair work on external walls.

To alter the position of a WC/bath/kitchen within a house, shop or flat

No	If you are refitting a kitchen or bathroom with new units and fittings.	No	If you are refitting.
Yes	If you are fitting a bathroom or kitchen where there was not one before, or drainage and electrical work relating to a refit.	Yes	All electrical installations must comply with the requirements of the Wiring Regulations (i.e. BS 7671:2018)

(continued)

Table 2.12 (Continued)

Planning Permission	Building Regulations approval

To alter the construction of fireplaces, hearths or flues within a house, shop or flat

No   To fit or replace an external flue, chimney or soil and vent pipe.   **Yes**   If you are installing a flue.

If the property is a Listed Building or in a Conservation Area

Yes   • For **any** external work, especially if it will alter the visual appearance, or use alternative materials.   **Yes**
   • To alter, repair or maintain a gate, fence, wall or other means of enclosure.

To insert cavity wall insulation

No   Provided there is no change in external appearance or the building is not listed or in a Conservation Area.   **Yes**   • All insulation has to comply to the relevant Building Regulations both when installed during construction and when fitted retrospectively. If such an upgrade is not technically or functionally feasible, the element should be upgraded to the best standard which can be achieved within a simple payback of no greater than 15 years.
   • If you are installing loft insulation as part of a roof renovation project, where more than 25% of the roof is being renewed, then the level of insulation should meet the standards required by Building

Table 2.12 (Continued)

Planning Permission		Building Regulations approval
		Regulations Approved Documents. Care should be taken not to block any ventilation at the edges.
To apply cladding		
No Provided it is not a Listed Building or in a Conservation Area.	**Yes**	To re-render or replace timber cladding to external walls, depending on the extent of the work. Where 25% or more of the wall is re-clad, re-rendered or re-plastered internally, or 25% or more of the external wall is rebuilt, the thermal insulation will need to be improved and Building Regulations will apply.

2.13 Demolition

You must have good reasons; you will require a Prior Approval Application before knocking a building down; and you should get formal confirmation of the Council's agreement on how you propose carrying out the demolition.

 The penalties are severe for demolishing something illegally!

Although you do not need to make a planning application to demolish a Listed Building or to demolish a building in a Conservation Area, you may need Listed Building or Conservation Area Consent. You will not automatically get Planning Permission to build any replacement structure or to change the use of the site if you demolish a building (even if it has been damaged).

Table 2.13 Authorisation requirements for demolition

Planning Permission		Building Regulations approval	
No	Unless the Local Council has made an Article 4 direction restricting permitted development rights in respect of demolition.	**Yes**	• Six weeks' prior notice must be given to the Local Authority building control. • Demolition must comply with the Construction (Design and Management) Regulations 2015 and have a health and safety plan. • For a partial demolition, to ensure that the remaining part of the house (or adjoining buildings/extensions) remain structurally sound.
Perhaps	The Local Council may wish to agree with you how you propose to carry out the demolition. This is called a Prior Approval Application.		
Listed Buildings and Conservation Areas			
Yes	An application for planning permission is required or any demolition work in a conservation area. Listing Building consent may also be required.	**Yes**	

2.14 Doors and windows

External windows and doors are known as 'controlled fittings' and require certain standards to be met when they are replaced. As FENSA does not apply to commercial premises or new-build properties, the replacement of windows in offices and other commercial premises (including the replacement of shop fronts) will, therefore, **all** require Local Authority Building Control approval.

Table 2.14 Authorisation requirements for doors and windows

Planning Permission		Building Regulations approval	
Repairs, maintenance and minor improvements			
No		**Yes**	Unless you use an installer registered with a competent person scheme
Conservation areas, designated areas or those with an Article 4 direction			
Yes		**No**	As long as they are installed by an installer registered with a competent person scheme
New windows and doors			
No	If their appearance is similar to those used in the house	**No**	As long as they are installed by an installer registered with a competent person scheme
Yes	If it is a new bay window	**Maybe**	
Internal secondary glazing			
No		**Maybe**	
Replacing shop windows			
Yes	Further information is available from local Building Control or from the Glass and Glazing Federation (www.ggf.org.uk).	**Yes**	

If you are a leaseholder, you may first need to get permission from your landlord or management company to carry out work on or replace a door or window.

2.15 Drains and sewers

Table 2.15 Authorisation requirements for drains and sewers

Planning Permission		Building Regulations approval	
No	For repairs, maintenance and very minor works.	Yes	Work on drains and sewers should be carried out in compliance with Approved Document H.
Perhaps	If permitted development rights in the area extend beyond repair, maintenance and very minor works.		
Yes	Internal or external works to Listed Buildings are likely to require Listed Building Consent.		

Sewers can be publicly or privately owned. Public sewers are maintained by the sewerage undertaker (you will find their address on your water bill) whilst private sewers are owned by the properties they serve. Building work on and around a sewer needs permission of the sewer owner. If you fail to confirm who owns the drain or sewer, or fail to comply with relevant legislation, you could be faced with the bill for legal and remedial action.

2.16 Electrical work in the home and garden

Whilst the Building Regulations only 'set' standards for electrical installation work in dwellings, AD-P (i.e. written in compliance with BS 7671) applies to **all** electrical work carried out in domestic buildings and is a mandatory requirement.

Although repairs, replacements and maintenance work or extra power points, etc., are classed as non-notifiable work and **can** be completed by a DIY enthusiast (family member or friend), the installation will **still** need to be in accordance with the manufacturer's instructions, the BS 7671 requirements **and** be done in such a way that it does not present a

Table 2.16 Authorisation requirements for electrical work in the home and garden

Planning Permission		Building Regulations approval	
No	Unless it is in a Listed Building or in a Conservation Area.	**No**	Provided that: • You comply with AD-P (for dwellings) and other relevant Building Regulations Approved Documents. • **All** electrical work complies with the requirements of BS 7671:2018 (i.e. The Wiring Regulations).

safety hazard. It is, therefore, advisable always to seek the advice of (or employ) an electrician certified under the Competent Person Scheme to certify the work carried out is safe, without you having to notify Building Control.

If you need to make a Building Regulations application to Building Control, they will arrange to have the electrical installation inspected at first fix stage and tested upon completion. Alternatively, you could use an Approved Inspector.

Note: You will find more information on Electrical matters in Part 4.17.

2.17 Extensions

You will need to apply for Planning Permission if you live in a Conservation Area, a National Park, an Area of Outstanding Natural Beauty or the Norfolk Broads. You **will** also require Planning Permission if you want to make additions or extensions to a flat or maisonette.

You will require approval to build an extension if it would 'materially alter the appearance of the building' and although major alterations and extensions nearly always need approval, some extensions (such as porches, garages and conservatories) may be classified as permitted development and, therefore, do not need planning consent.

There are a number of classes of new buildings or extensions of existing buildings that are exempt (some partially and other completely) from the regulations (see the table below). They may, however, require Planning Permission.

Table 2.17 Authorisation requirements for extensions

Planning Permission	Building Regulations approval
No Provided that:	**Yes** Most extensions of properties require approval under the Building Regulations.

No Provided that:
- Extensions of more than one storey do not extend beyond the rear wall of an 'original' (attached) house by more than 3m, or 4m from a detached house.
- Materials are similar in appearance to the existing house.
- Maximum eaves and ridge height of an extension is no higher than the existing house.
- Maximum eaves height of an extension within 2m of the boundary is 3m.
- Maximum height of a single-storey rear extension is no more than 4m.
- No extension is forward of the principal elevation or side elevation fronting a highway.
- No extension is higher than the highest part of the existing building's roof.
- No more than half the area of land around the 'original' house would be covered by additions (including decking) or other buildings.
- There are no verandas, balconies or raised platforms.
- Roof pitch of extensions higher than one storey should match the existing house.
- Side extensions should be single storey with a maximum height of 4m and width of no more than half that of the 'original' house.
- Single-storey rear extensions on designated land may not extend beyond the rear wall of the 'original' house. For other sites with prior approval this is increased to 8m and 6m.

Yes Most extensions of properties require approval under the Building Regulations.

 Note: Work that adds over 100m² floor space may attract a charge under the Community Infrastructure Levy.

 Note: Where prior approval applies the Local Authority will consult adjoining neighbours.

Table 2.17 (Continued)

Planning Permission	Building Regulations approval
• Any two-storey extension is no closer than 7m to the rear boundary. • Upper-floor, side-facing windows should be obscure-glazed	

On Designated Land
(Conservation Areas, National Parks and the Broads, Areas of Outstanding Natural Beauty and World Heritage Sites)

No	Provided that, in addition to the above, there is: • No cladding of the exterior. • No side extensions. • No rear extensions of more than one storey.	**Yes**	Most extensions of properties require approval under the Building Regulations.

Table 2.17A Buildings that are exempt (some partially and others completely) from the regulations

Class of work	Approved Documents A–K and M	Approved Document L	Approved Document P
Class 1 (Buildings controlled under other legislation)	Exempt	May apply	Exempt
Class 2 (Buildings not frequented by people)	Exempt	May apply	Exempt
Class 3 (Greenhouses)	Exempt	May apply	
Class 3 (Agricultural buildings)	Exempt	May apply	Exempt
Class 4 (Temporary buildings)	Exempt	May apply	Exempt
Class 5 (Ancillary buildings)	Exempt	May apply	Exempt
Class 6 (Small detached buildings)	Exempt	May apply	
Class 7 (Extensions)	Exempt	Exempt	

2.18 External walls

If, as part of a conversion project, you are converting an existing external wall, then the existing wall will need to be checked for its adequacy in terms of:

Table 2.18 Authorisation requirements for external walls

Planning Permission		Building Regulations approval	
Repairs, maintenance or minor improvements			
No	Provided it is for repairs, maintenance or minor improvements, e.g. painting your house, and it is not a Listed Building or in a Conservation or designated area.	Yes	• If more than 25% or more of an external wall is re-rendered, re-clad, re-plastered or re-lined internally. • If 25% or more of the external leaf of a wall is rebuilt. • If you want to insert insulation into a cavity wall.
Yes	• If you want to insert insulation into a cavity wall. • If you live in a Conservation Area, a National Park, an Area of Outstanding Natural Beauty or the Broads, and the work is significant.		
Cladding the outside of a house with stone, artificial stone, pebble dash, render, timber, plastic or tiles			
Yes	If you live in a Listed Building, or in a Conservation Area, a National Park, an Area of Outstanding Natural Beauty or the Broads.	Maybe	Depending on the extent of the work involved in re-cladding the building.
No	If you live outside the areas above and any cladding is of a similar appearance to those used in the construction of the house.		

- damp-proofing arrangements and protecting new timbers from damp;
- thermal resistance and changes to 'thermal elements';
- weather resistance;
- weight and structural stability.

2.19 Fascias

Table 2.19 Authorisation requirements for fascias

Planning Permission		Building Regulations approval	
No	Unless you live in a Listed Building or designated area.	**No**	Provided the ventilation for the roof void is not compromised, then the replacement of fascia and soffit boards does not normally require permission.

You are advised to check that the replacement work does not reduce the ventilation provided to the roof void as this could cause condensation to occur within the roof, which can then leave damp on the timbers. Any existing vents will need to be maintained.

2.20 Fences, gates and garden walls

You will **not** need to apply for Planning Permission to take down a fence, wall or gate, or to alter, maintain or improve an existing fence, wall or gate (no matter how high) if you don't increase its height. You will need to apply for Planning Permission if you put an ornament on your gate post and together with the post it stands higher than 1m tall.

Where a fence, wall or gate is classed as a 'party fence wall', you must notify the adjoining owner of the work planned.

Table 2.20 Authorisation requirements for fences, gates and garden walls

Planning Permission	Building Regulations approval
Yes • If it would be over 1m high and next to a highway used by vehicles (or the footpath of such a highway); or over 2m high elsewhere. • If your right to put up or alter fences, walls and gates has been removed by an Article 4 direction or a planning condition. • If your house is a Listed Building or in the curtilage of a Listed Building. • If it forms a boundary with a neighbouring Listed Building or its curtilage. • If you plan to put an ornament on your gate post and together with the post it stands higher than 1m tall. • Structures must be sound and suitably maintained.	**No** If the garden wall is a 'party fence wall', and depending on the type of building work you intend to carry out, then you must notify the adjoining owner of the work in respect of the Party Wall Act 1996. This does not include wooden fences.

2.21 Flats and maisonettes

There is a different planning regime for flats and maisonettes, and the permitted development rights which apply to many common projects for houses may not apply to flats. In particular, see the table below.

In addition to Planning Permission, you may also require Listed Building or Conservation Area Consent, as work on a building that affects its special historic character without consent is a criminal offence.

Table 2.21 Authorisation requirements for flats and maisonettes

Planning Permission		Building Regulations approval	

To extend a flat

Yes	You must apply for Planning Permission.	Yes	You may need to consult the Fire Service regarding issues relating to fire escapes.

To sub-divide a house or single flat

Yes	You must obtain Planning Permission to sub-divide a house into multiple units.	Yes	Conversions will require Building Regulations approval. You may also need to consult the Fire Service and the property will need to be licensed.

To build an extension to a ground-floor flat

Yes	You must obtain Planning Permission to add an extension. Also, in a Conservation Area where work involves demolition, separate permission may be required.	Yes	Conversions will require Building Regulations approval. You may also need to consult the Fire Service.

Loft conversion in a top floor flat

No	Provided it is internal works. However, you should get permission if you are a leaseholder.	Yes	Approval is required to convert a loft or attic into a liveable space.
Yes	If you intend to extend or alter the roof space.		

Converting a space above a shop into a flat

No	Planning Permission is not required if: • The space is in the same class of use as the shop to start with (either class A1 or A2). • The space is not in a separate planning unit.	Yes	This will class as a material change of use and therefore the whole (or at least part) of the building may need to be upgraded to comply with the requirements.

(continued)

Table 2.21 (Continued)

Planning Permission		Building Regulations approval
• The outside appearance of the building will not be changed.		
• None of the ground floor will be incorporated into the flat if there is a display window at ground floor level.		

Painting the exterior of a flat

No	Unless you live in an area where an Article 4 direction (i.e. the removal of specified permitted development rights related to operational development or change of use) applies.	**No**	

Erecting satellite dishes for a flat

Perhaps	Planning Permission may be required in certain circumstances – best to check with your Local Council.	**Yes**	

Replacing windows in a flat

Perhaps	Permission may be needed to fit new double-glazed windows.	**Yes**	In relation to thermal performance and other areas such as safety, air supply, means of escape and ventilation.
No	If you are replacing like with like or adding internal secondary glazing.	**Yes**	As stated above.

2.22 Flues, chimneys and soil and vent pipes

If the building is on designated land, the flue should not be installed on a wall or roof slope which fronts a highway.

Table 2.22 Authorisation requirements for flues, chimneys and soil and vent pipes

Planning Permission	Building Regulations approval
Domestic flues	
No Provided that: • Flues are not fitted on the principal or side elevation that fronts a highway in a designated area. • Flues on the rear or side elevation of the building are no more than 1m above the highest part of the roof. • The building is not listed or in a designated area.	**Yes** Particularly with regard to ventilation and general safety. Installation should be carried out by a suitably qualified installer.
Flues for biomass and combined heat and power systems (non-domestic)	
No Provided that: • It is the first installation of a flue as part of either a biomass heating system or a combined heat and power system (further installations will require Planning Permission). • The building is not listed or a Scheduled Monument. • The capacity of the system is no more than 45kW thermal. • The flue is no more than 1m higher than the highest part of the roof, or the height of an existing flue which is being replaced, whichever is the highest.	**Yes** Particularly with regard to ventilation and general safety. Installation should be carried out by a person who is registered with the Competent Person Scheme.

2.23 Fuel tanks

Storage of oil, or any other liquids, especially petrol, diesel and chemicals, is strictly controlled and would **not** be allowed on residential premises. If you are considering installing an external oil storage tank for central heating, the storage tanks, then the pipes connecting them to combustion appliances, should be constructed and protected in order to reduce the

Table 2.23 Authorisation requirements for fuel tanks

Planning Permission		Building Regulations approval	
No	Provided that: • The maximum area that will be covered by buildings, enclosures and containers more than 20m from a house is limited to 10m² in National Parks, the Broads, Areas of Outstanding Natural Beauty and World Heritage Sites. • The maximum height is 2.5m within 2m of a boundary. • The maximum overall height is 3m. • It is not located at the side of properties on designated land. • It is not forward of the principal elevation fronting a highway. • The capacity is not more than 3,500 litres. • No more than half the area of land around the 'original' house will be covered by additions or other buildings.	**No**	Provided the installation tank meets the necessary Building Regulations requirements: • Installations above ground should have adequate shielding of the tank from any surrounding fire and, in the case of an oil tank, containment of oil leakages so that groundwater is not contaminated. • Any new oil-connecting pipework will require a fire valve at the point where the pipe enters the building. If you are installing an oil tank and/or connecting pipework and you employ an Installer registered with one of the related Competent Person Schemes, you will not need to involve a Building Control Service.
Yes	Any container within the curtilage of a Listed Building will require Planning Permission.		

risk of the oil escaping and causing pollution. This **includes** liquid petroleum gas tanks as well as oil storage tanks.

2.24 Garages and carports

A typical floor for a garage generally consists of hard-core, sand blinding, DPM (damp-proof membrane) and concrete. It is considered to be good practice to place reinforced mesh within the concrete as this can reduce the chances of cracking from the load (weight) of a vehicle. The floor will not normally need to be insulated.

Table 2.24 Authorisation requirements for garages and carports

Planning Permission		Building Regulations approval	
Carport			
No	Provided it is open on at least two sides and less than 30m² in floor area.	No	If it is an attached carport less than 30m².
New garage – attached			
No	Provided the floor area is between 15m² and 30m².	Yes	If it is attached to an existing home.
New garage – detached			
No	The floor area must be either less than 15m², or it can be between 15m² and 30m² provided that: • It is at least 1m from any boundary. • It is constructed from substantially non-combustible materials.	No	If the floor area is no more than 15m² or the floor area is between 15m² and 30m² and is at least 1m from any boundary.
Yes	If the building is listed or in a Conservation Area.		

(continued)

Table 2.24 (Continued)

Planning Permission	Building Regulations approval
Garage conversion	
No Provided the work is internal **Yes** and permitted development rights have not been removed in the area.	Regulations apply in relation to the following: doors and windows; drainage; electrics; external walls; internal walls and roofs.
Infill garage door	
No	**Yes** As the foundation to the existing garage is not likely to be traditional, a new foundation may be needed for the new wall.

2.25 Hardstanding for cars, caravans and boats

You do **not** need to apply for Planning Permission to build a hardstanding for a car provided that it is within your boundary and is not used for a commercial vehicle. However, there are rules for commercial parking (e.g. taxis or commercial delivery vans), and a 'change of use' as a trade premises would probably need to be granted for this to be allowed.

Whilst access from a new hardstanding to an unclassified roadway does not require Planning Permission, if the access crosses a pedestrian thoroughfare, pavement or roadside verge you will need to gain approval from the Highways Department for the kerb to be dropped.

Table 2.25 Authorisation requirements for hardstanding for cars, caravans and boats

Planning Permission	Building Regulations approval
No Provided it is within your boundary at, or near, ground level, and does not require significant embanking or terracing work.	**No** Unless you introduce steps where none existed before.

Rules on siting of static caravans or mobile homes are also quite stringent. Caravans smaller than 65×22ft can be placed in your garden but they may **only** be used as additional living space and **not** as independent accommodation.

2.26 Heat pumps

Table 2.26 Authorisation requirements for heat pumps

Planning Permission		Building Regulations approval	
Domestic ground-source or water-source heat pumps			
No	Unless the building is listed or in a Conservation Area.	Yes	You should use an installer, preferably one who belongs to either the Microgeneration Certification Scheme or the relevant Competent Person Scheme.
Domestic air-source heat pumps			
No	All of the below limits and conditions must be met: • All parts of the air-source heat pump must be at least 1m from the property boundary. • The air-source heat pump installation must comply with the Microgeneration Certification Scheme Planning Standards (or equivalent standards). • Within a Conservation Area or World Heritage Site, the installation is not on a wall that is above the level of the ground storey. • On a flat roof, all parts of the pump must be at least 1m from the external edge of that roof.	Yes	You should use an installer, preferably one who belongs to either the Microgeneration Certification Scheme or the relevant Competent Person Scheme.

(*continued*)

Table 2.26 (Continued)

Planning Permission	Building Regulations approval
• It is the first installation of an air-source heat pump and there is no existing wind turbine on the building or within the curtilage of that property. • The installation is not on a pitched roof. • The installation is not within the curtilage of a Listed Building or within a site designated as a Scheduled Monument. • The installation is not on a wall or roof which fronts a highway. • Within a Conservation Area or World Heritage Site, the installation is not nearer to any highway which bounds the property than any part of the building. • The volume of the air-source heat pump's outdoor compressor unit (including housing) must not exceed 0.6m³. In addition, the following conditions must also be met: • The pump must be removed as soon as reasonably practicable when it is no longer needed for microgeneration. • The pump must be sited, so far as is practicable, to minimise its effect on the external appearance of the building and its effect on the amenity of the area. • The air-source heat pump must be used solely for heating purposes.	

Table 2.26 (Continued)

Planning Permission	Building Regulations approval
Non-domestic ground-source or air-source heat pumps	
No Provided that all the following conditions are met: • Only the first standalone installation will be permitted development. • When no longer needed for microgeneration, pumps should be removed as soon as reasonably practicable and the land restored to its condition before the development took place (or to the condition agreed in writing between the Local Planning Authority and the developer). • The total surface area covered by the water-source heat pump (including any pipes) must not exceed 0.5 hectares.	**Yes** You should use an installer, preferably one who belongs to either the Microgeneration Certification Scheme or the relevant Competent Person Scheme.

2.27 Hedges

You do **not** need Planning Permission for hedges but hedges should not be allowed to block out natural light and the positioning of fast-growing hedges (such as Leylandii) should be checked with your Local Authority.

Table 2.27 Authorisation requirements for hedges

Planning Permission	Building Regulations approval
No Unless it obscures the view of traffic at a junction or access to a main road.	**No**

2.28 Home energy generation

Table 2.28 Authorisation requirements for home energy generation

Planning Permission		Building Regulations approval	
Perhaps	Depending on the type of energy generation and the regulations for the local area.	**Yes**	Particularly for electrical installations.

The use of renewable energy has become increasingly popular as an alternative to fossil fuel. Further information is available from: www.energysavingtrust.org.uk/renewable-energy.

2.29 Hydroelectricity

Table 2.29 Authorisation requirements for hydroelectricity

Planning Permission		Building Regulations approval	
Yes	Some form of environmental assessment will be essential for this type of project.	**Yes**	Particularly for electrical installations.

This is a complex area and the Environment Agency **must** also be consulted about water extraction licences. More details can be found at: www.gov.uk/government/organisations/environment-agency.

2.30 Insulation

Table 2.30 Authorisation requirements for insulation

Planning Permission		Building Regulations approval	
No	Provided it is not a Listed Building or in a Conservation Area, and there is no external change in appearance.	**Yes**	Cavity wall insulation is notifiable work if you install insulation as part of a roof renovation, and more than 25% of the roof is being renewed.
		No	However, loft insulation and floor insulation will have to comply with the relevant Approved Documents.

2.31 Internal walls

Table 2.31 Authorisation requirements for internal walls

Planning Permission	Building Regulations approval
No	**Yes** • Care should be taken before removing any internal wall as they have a number of functions that could affect the building and the safety of its occupants. • There should be adequate separation – in terms of fire resistance and thermal insulation – between a new habitable space and the remaining space. • Doors in such a separation wall should have adequate fire resistance and be self-closing. • New separating walls may need sound insulation (depending on the use of the new habitable room).

2.32 Kitchens and bathrooms

Under the Greener Homes initiative, homeowners are encouraged to pay attention to appliances and fittings by looking for A-rated kitchen appliances and using aerated taps to reduce water use.

In bathrooms, homeowners are asked to reduce water consumption by buying low flush toilets, low flow showers and basin taps, and a smaller capacity bath.

Table 2.32 Authorisation requirements for kitchens and bathrooms

Planning Permission		Building Regulations approval	
No	Unless it is part of a house extension, or the property is a Listed Building.	**No**	Unless drainage or electrical works form part of the refit.
		Yes	If work includes installing additional fittings (or a new fitting such as a kitchen sink).

2.33 Loft conversions

Work on a loft conversion could affect bats and you will need to consider this protected species when planning work of this type.

Table 2.33 Authorisation requirements for loft conversions

Planning Permission	Building Regulations approval
Yes • For flats – even where construction works may not be intended. • For conversion to shops and offices.	**Yes** • Unless you are not proposing any building work to make the change. • Regulations apply in relation to sound insulation, fire safety and structure stability.
No Provided that you do not alter or extend the roof space and abide by the conditions shown in Schedule 2 Part 1 (section 2.17) of the Town and Country Planning (England) Order 2015 (as amended and listed below): • The volume of additional roof space is less than 40m^3 for terraced houses and 50m^3 for detached and semi-detached houses. • The extension is not beyond the plane of the existing roof slope of the principal elevation that fronts the highway. • Side-facing windows are obscure-glazed.	

2.34 Micro combined heat and power Microgeneration Certification Scheme

The Microgeneration Certification Scheme was developed to support the microgeneration industry and to drive the quality and reliability of installations. The main purpose of the scheme is to build consumer confidence so that the industry can move to a sustainable position. As part of the code there is certification for products and installation companies, and a code of practice. Where equipment is installed by a person who is certified through the scheme the installer is responsible for ensuring that the installation meets all the appropriate standards at the time of installation. For further details, see (https://mcscertified.com/).

2.35 New homes and self-build homes

Table 2.34 Authorisation requirements for new homes and self-build homes

Planning Permission		Building Regulations approval	
Yes	In all cases.	**Yes**	In all cases.

All new houses, or premises of any kind, require Planning Permission.

Self-build projects account for approximately 12,000 new homes each year (7–10% of new housing). In all cases, unless you are an architect or a builder, you **must** seek professional advice.

2.36 Outbuildings

Outbuildings intended to go in the garden of a house do not normally require any Planning Permission, as long as they are associated with the residential amenities of the house and a few requirements are adhered to, such as position and size. If, however, your new building exceeds 10m^2 (and/or comes within 5m of the house), it would be treated as an extension and would count against your overall volume entitlement.

Table 2.35 Authorisation requirements for outbuildings

Planning Permission	Building Regulations approval
No Provided that: • There is a maximum height of 2.5m in the case of a building, enclosure or container within 2m of a boundary of the curtilage of the dwelling house. • No more than half the area of land around the 'original' house would be covered by additions or other buildings. • No outbuilding is to be sited on land forward of a wall forming the principal elevation of the dwelling house. • The outbuilding does not include verandas, balconies or raised platforms. • Outbuildings and garages are single storey with a maximum eaves height of 2.5m and a maximum overall height of 4m with a dual pitched roof or 3m for any other roof. • In National Parks, the Broads, Areas of Outstanding Natural Beauty and World Heritage Sites, the maximum area to be covered by buildings, enclosures, containers and pools more than 20m^2 from the house is limited to 10m^2.	**No** If the floor area of the building is between 15m^2 and 30m^2 you will **not** normally be required to apply for Building Regulations Approval – **providing** that the building contains **NO** sleeping accommodation and is either at least one metre from any boundary or it is constructed of substantially non-combustible materials.

Table 2.35 (Continued)

Planning Permission	Building Regulations approval
Yes Planning Permission is required for: • Buildings, enclosures, containers and pools at the side of properties on designated land. • Any outbuilding within the curtilage of Listed Buildings.	

Polytunnels

Planning Permission	Building Regulations approval
Yes If it is a Listed Building or in a Conservation Area, National Park or Area of Outstanding Natural Beauty.	**Perhaps** Depending on work involved.
No Provided the polytunnel is no nearer to the road than the nearest part of the house (unless there is more than 20m between the tunnel and the road), and is less than 3m high.	

2.37 Patios, decking and driveways

You are strongly advised to seek advice from a builder, architect, drainage engineer or Local Authority Building Control before committing to, or commencing, work.

Table 2.36 Authorisation requirements for patios, decking and driveways

Planning Permission		Building Regulations approval	
No	Provided that: • No significant embanking or terracing works are required. • It is not a Listed Building.	**No**	Provided the alterations do not make access to the dwelling any less satisfactory than it was before.
Decking			
No	Provided that: • The decking is no more than 30cm above the ground. • Together with extensions, outbuildings, etc., the decking covers no more than 50% of the garden.	**No**	Provided the structure does not require Planning Permission.
Kerbs and paving over your front garden			
Perhaps	If you wish to cross a footpath to gain access to your garden, you will need to obtain permission from the highways department of your Local Council.	**No**	Provided the alterations do not make access to the dwelling any less satisfactory than it was before.
No	• Although you do not usually need to apply for Planning Permission to install a patio or driveway, you will need approval from the Local Council if you wish to cross a pavement. • New or replacement driveways should use permeable or porous surfacing. • Surface water from hardstandings must not be allowed to run onto the highway.		
Yes	If the surface to be covered is more than 5m².		

2.38 Paving your front garden

Table 2.37 Authorisation requirements for paving your front garden

Planning Permission		Building Regulations approval	
No	If the replacement driveway of any size uses permeable surfacing or if the rainwater is directed to a lawn or border to drain naturally.	**No**	Provided the alterations do not make access to the dwelling any less satisfactory than it was before.
Yes	If the surface to be covered is more than 5m² and you plan to lay a traditional impermeable driveway.		

In order to avoid flooding, it is preferable for any paved area to be sloped towards permeable ground or to be made of pervious materials. If you are making a new access into the garden across the footpath, you will need to obtain permission from the Local Council to drop the kerbs and the pavement may need strengthening. This is to protect any services buried in the ground such as water pipes.

2.39 Plumbing

Table 2.38 Authorisation requirements for plumbing

Planning Permission		Building Regulations approval	
No	Unless it is a Listed Building.	**No**	If it is installed by an approved person and complies with Approved Document J.
		Yes	If you use an unregistered installer or DIY, you will need to get approval from the Local Authority Building Control.

2.40 Porches

Table 2.39 Authorisation requirements for porches

Planning Permission		Building Regulations approval	
No	Unless: • Any part is less than 2m from a boundary adjoining a highway. • Any part is more than 3m high. • The floor area exceeds 3m².	**No**	Provided glazing and electrical installations comply with Building Regulations and: • The front entrance door between the existing house and the new porch remains in place. • If the house has ramped or level access for disabled people, the porch must not adversely affect access.

It is advisable to ensure that a porch is not constructed so that it restricts ladder access to windows serving a room in the roof or a loft conversion, particularly if that window is required as an emergency means of escape in the case of fire.

2.41 Roofs

After a period of time the roof on existing buildings will need to be replaced. In most situations, this work will need Building Regulations approval. However, if a roof containing integral insulation is to be replaced, then you may be required to upgrade this 'thermal element' of the structure and reduce the amount of heat that was originally lost by upgrading the insulation – in which case, approval under the Building Regulations is likely to be needed.

 Work on a roof may affect bats and you will need to consider this protected species when planning work of this type.

Table 2.40 Authorisation requirements for roofs

Planning Permission	Building Regulations approval
No Provided the alterations to any side-facing windows are obscure-glazed; and any opening is: • At least 1.7m above the floor. • Not projecting more than 150mm from the existing roof plane. • No higher than the highest part of the roof.	**Yes** You will need approval if: • The performance of the new covering will be significantly different to that of the existing covering in the event of a fire. • You are replacing/repairing more than 25% of the roof area, in which case, the roof thermal insulation would normally have to be improved. • You are carrying out structural alterations.
	No If you want to carry out repairs on or re-cover less than 25% of the area of a pitch or flat roof, you will not normally need to submit a Building Regulations application.
Rooflights	
No Provided any side-facing windows are obscure-glazed and openings are 1.7m above the floor.	**Yes** Approval will generally be needed for the installation of a new rooflight.

2.42 Security lighting

Beams from your security lights should **not** point directly at windows of other houses and security lights fitted with passive infra-red detectors (PIRs) and/or timing devices should be adjusted so that they minimise nuisance to neighbours and are set so that they are not triggered by traffic or pedestrians passing outside your property. External lighting should automatically turn off when there is enough daylight and have a capacity that does not exceed 150W per light fitting.

Table 2.41 Authorisation requirements for security lighting

Planning Permission	Building Regulations approval
No Unless the property is a Listed Building, but you should make sure that the intensity and direction of light does not disturb others.	**Yes** You should either use an installer who is registered with the Competent Person Scheme or make an application to your Local Authority's Building Control department or Approved Inspectors.

2.43 Shops

Shopping and commercial areas are usually defined by Local Authorities in their planning policies. It is therefore often difficult to get permission to use a property as a shop outside these areas, especially if the area is mainly or completely residential.

Changing from one use to another, especially from residential to retail, will be classed as a change of use.

Table 2.42 Authorisation requirements for shops – type A

Planning Permission	Building Regulations approval
To change residential use to a shop	
Yes	**Yes** Particularly with respect to escape and other fire precautions, hygiene, energy conservation, and access to and use of buildings.
To convert a shop to a I cafe, public house, or takeaway	
No A shop may change to a café as long as Prior Approval is sought and the premises is less than 150m².	**Perhaps** Depending on what work is involved.

Table 2.42 (Continued)

Planning Permission	Building Regulations approval
Yes To convert a shop to a public house or takeaway because the proposed new use is a different use class within the planning system.	**Yes** Particularly with respect to escape and other fire precautions, hygiene, energy conservation, and access to and use of buildings.

To convert a shop to an office, storage, storage or other uses

No You may convert a shop to an office use for a single period of up to two years.	**Perhaps** Particularly with respect to escape and other fire precautions, hygiene, energy conservation, and access to and use of buildings.
Yes For a material change of use (e.g. storage), and where the new use is a different use class.	**Yes** Particularly with respect to escape and other fire precautions, hygiene, energy conservation, and access to and use of buildings.

 Note: You may also require the consent of the landlord or landowner.

To erect/change/alter a shop's adverts, fascia or projecting signs

Yes	No

2.44 Solar panels

Installing solar panels on domestic and non-domestic land is likely to be considered permitted development with no need to apply to the Council for Planning Permission. There are, however, important limits and conditions and it is best to seek official advice before proceeding.

Table 2.43 Authorisation requirements for solar panels

Planning Permission		Building Regulations approval
Solar panels mounted on a house or on a building within the grounds of a house		
No	Provided all the following conditions are observed: • When no longer needed for microgeneration, panels should be removed as soon as reasonably practicable. • Panels should not be installed above the highest part of the roof (excluding the chimney) and should project no more than 200mm from the roof slope or wall surface. • Panels must not be installed on a building that is within the grounds of a Listed Building, or on a site designated as a Scheduled Monument.	**Yes** The ability of the roof to carry the load (weight) of the panels needs to be checked, and electrical work needs to be carried out by an approved contractor.
Standalone solar panel installations (i.e. panels not on a building)		
No	Provided all the following conditions are observed: • When no longer needed for microgeneration, panels should be removed as soon as reasonably practicable. • No part of the installation should be higher than 4m.	**Yes** The ability of the surface to carry the load (weight) of the panels will need to be checked and proven. Some strengthening work may be needed. Building Regulations also apply to other aspects of the work such as electrical installation.

- The installation should be at least 5m from the boundary of the property.
- The size of the array should be no more than 9m² or 3m wide × 3m deep.
- If your property is in a Conservation Area, or in a World Heritage Site, no part of the solar installation should be nearer to any highway than the part of the house that is nearest to that highway.
- Panels should not be installed within the boundary of a Listed Building or a Scheduled Monument.

Non-domestic solar panels mounted on a building

No Provided all the following conditions are observed:

- Panels should be sited to minimise the effect on the external appearance of the building and the amenity of the area.
- When no longer needed for microgeneration, panels should be removed as soon as reasonably practicable.
- Equipment mounted on a roof must not be within 1m of the external edge of the roof.
- Equipment mounted on a wall must not be within 1m of a junction of that wall with another wall or with the roof of the building.

Yes The ability of the existing roof to carry the load (weight) of the panels will need to be checked and proven. Some strengthening work may be needed.
Building Regulations also apply to other aspects of the work such as electrical installation.

(continued)

Table 2.43 (Continued)

Planning Permission	Building Regulations approval
<ul><li>Where panels are installed on a flat roof, the highest part of the equipment should not be more than one metre above the highest part of the roof (excluding the chimney).</li><li>Panels installed on a wall or a pitched roof should project no more than 200mm from the wall surface or roof slope.</li><li>If the building is on designated land, the equipment must not be installed on a wall or a roof slope which fronts a highway.</li><li>Panels must **not** be installed on a Listed Building or on a building that is within the grounds of a Listed Building, or on a site designated as a Scheduled Monument.</li></ul>	The ability of the surface to carry the load (weight) of the panels will need to be checked and proven. Some strengthening work may be needed. Building Regulations also apply to other aspects of the work such as electrical installation.

Non-domestic standalone solar panel installations

No	Provided all the following conditions are observed:
	<ul><li>Panels should be sited, as far as is practicable, to minimise the effect on the amenity of the area.</li><li>When no longer needed for microgeneration, panels should be removed as soon as reasonably practicable.</li><li>Only the first standalone solar installation will be permitted development. Further installations will require Planning Permission from the Local Authority.</li><li>No part of the installation should be higher than 4m.</li></ul>

- The installation should be at least 5m from the boundary of the property.
- The size of the array should be no more than 9m² or 3m wide × 3m deep.
- If the property is in a designated area, no part of the solar installation should be nearer to any highway bounding the grounds of the property than the part of the building that is nearest to that highway.
- Panels should not be installed within the boundary of a Listed Building or a Scheduled Monument.

2.45 Structural alterations – inside

You will need approval if:

* The alterations are to the structure (such as the removal or part removal of a joist, beam or chimney breast).
* The alterations would affect fire precautions of a structural nature either inside or outside your house.
* Work is required to the drainage system.
* Work is required to maintain the means of escape in case of fire.

Table 2.44 Authorisation requirements for structural alterations – inside

Planning Permission		Building Regulations approval	
No	Unless it is a Listed Building or within a Conservation Area.	**Possibly**	If it is a Listed Building or within a Conservation Area.
Yes	If the alterations are major, such as removing (or part removing) a load-bearing wall, or altering the drainage system.	**Yes**	If you wish to build or remove an internal wall or make openings in an internal wall.

2.46 Swimming pools

Installing swimming pools (sauna cabins and hot tubs) are subject to special requirements (for example: a covered swimming pool will involve regulations applicable to sheds and outbuildings as well as BS7671).

Table 2.45 Authorisation requirements for swimming pools

Planning Permission		Building Regulations approval
Possibly	Consult your local planning officer.	**Yes**

2.47 Trees

The use and nature of trees can be controlled through planning conditions and legal covenants. Many trees are protected by Tree Preservation Orders (TPOs), which mean that, in general, you need the Council's consent to prune or fell them. All trees in Conservation Areas are automatically protected. When planting or removing trees, or building new structures, you should be aware that certain tree species can affect foundations as much as 20m away!

Table 2.46 Authorisation requirements for trees

Planning Permission		Building Regulations approval
No	You can fell or lop trees on your property unless the trees are protected by a Tree Preservation Order or you live in a Conservation Area.	**No**

2.48 Underpinning

Table 2.47 Authorisation requirements for underpinning

Planning Permission		Building Regulations approval	
No	Unless it is a Listed Building or in a designated area (i.e. a Conservation Area, National Park or Area of Outstanding Natural Beauty).	**Yes**	The Regulations specifically define underpinning as 'building work' and appropriate measures must be applied to ensure the underpinning stabilises the movement of the building. Particular attention will need to be given to any sewers and drains near the work.

The foundations to a building may be increased by underpinning.

2.49 Warehouses and industrial buildings

Any new, extended or altered warehouse must relate to the current use of the building or the provision of staff facilities, and the development must be within the curtilage of an existing industrial building or warehouse.

There are stringent regulations relating to the development of warehouses and industrial buildings and you are strongly advised to

Table 2.48 Authorisation requirements for warehouses and industrial buildings

Planning Permission	Building Regulations approval
New industrial buildings and warehouses	
No Provided that: • No new building that is within 10m of the boundary is higher than 5m. • No new building is to exceed a gross floor space of 100m^2 on designated land and Sites of Special Scientific Interest. On all other sites, the gross floor space limit is 200m^2.	**Yes**
Extending or altering industrial buildings and warehouses	
No Provided that: • An extension/alteration that is within 10m of the boundary does not make the building higher than 5m. • Any other extension/alteration must not be higher than the building being extended or altered.	**Yes**

read all the interim guidance on the new permitted development rules and to seek guidance from your Local Planning Authority before you commence work.

2.50 Wind turbines

Planning Permission rules vary depending on the region of the UK and it is recommended that you check with your Local Authority to see if there are any limits in your local area.

Table 2.49 Authorisation requirements for wind turbines

Planning Permission	Building Regulations approval
Wind turbine: building mounted	
No Provided that:	**Yes** The size, weight and force exerted on fixed points would be considerable.
• When no longer needed, turbines are removed as soon as reasonably practicable.	You are advised to use an installer who is able to self-certify their work by being registered with the relevant Competent Person Scheme.
• Turbines are sited in a way that minimises its effect on the external appearance of the building and on the amenity of the area.	
• The installation complies with the Microgeneration Certification Scheme Planning Standards.	
• Turbines are mounted on a wall or roof slope which does not front a highway in a Conservation Area.	
• No part (including blades) protrudes more than 3m above the highest part of the roof (excluding the chimney) or exceeds an overall height (including building, hub and blade) of 15m, whichever is the lesser.	
• No part of the turbine (including blades) is within 5m of the property boundary.	
• Turbines can only be attached to detached houses (as opposed to blocks of flats) and other detached buildings within the boundaries of a house or block of flats.	
• It is the first installation of a wind turbine and there is no existing air-source heat pump at the property.	

(continued)

Table 2.49 (Continued)

Planning Permission	Building Regulations approval
• The installation is not within the curtilage of a Listed Building or within a site designated as a Scheduled Monument or on designated land (other than Conservation Areas – see above). • The distance between ground level and the lowest part of any wind turbine blade is not less than 5m. • The installation is not sited on safeguarded land. • The turbine blade has a swept area of no more than $3.8m^2$. • The turbine blades must use non-reflective materials.	

Wind turbine: standalone

	Planning Permission	Building Regulations approval
No	Provided that: • When no longer needed for microgeneration, turbines are removed as soon as reasonably practicable. • Turbines are sited in a way that minimises its effect on the amenity of the area. • The installation is not in a position which is less than a distance equivalent to the overall height of the turbine (including blades) plus 10% of its height when measured from any point along the property boundary. • The installation complies with the Microgeneration Certification Scheme (MCS020) Planning Standards, or equivalent standards.	**Yes** — The size, weight and force exerted on fixed points would be considerable. You are advised to use an installer who is able to self-certify their work by being registered with the relevant Competent Person Scheme.

- In a Conservation Area, the turbine is no nearer to any highway which bounds the curtilage (garden or grounds) of the dwelling or block of flats than the part of the house or block of flats which is nearest to that highway.
- It is the first installation of a wind turbine and there is no existing air-source heat pump at the property.
- The turbine is not within the curtilage of a Listed Building or within a site designated as a Scheduled Monument or on designated land (other than Conservation Areas – see above).
- The distance between ground level and the lowest part of any wind turbine blade is not less than 5m.
- The highest part of the turbine does not exceed 11.1m.
- The installation is not sited on safeguarded land (the Aviation Safeguarding Tool can be used to check whether the installation will be on safeguarded land).
- The turbine blade has a swept area of no more than 3.8m^2.
- The turbine blades must use non-reflective materials.

2.51 **Working from home**

Table 2.50 Authorisation requirements for working from home

Planning Permission		Building Regulations approval	
No	Provided the overall character of the dwelling does not change.	**No**	Provided the work does not affect the structure of the building, the means of escape and other fire precautions, or affect the access to and use of the building.

Part 3
Requirements of the Approved Documents

3.1 Introduction

Approved Documents (ADs) are made available by the Secretary of State and are aimed at providing practical guidance concerning the requirements of the Building Regulations. Part 3 provides a series of tables containing details of the ADs' requirements, together with an overview of what it really means 'in a nutshell'.

Table 3.1 lists the current editions of the ADs, including incorporated amendments. All of these ADs can be accessed on the UK government legislation website:

www.legislation.gov.uk

and the Planning Portal:

www.gov.uk/government/collections/approved-documents.

They can also be purchased from The Stationery Office or RIBA.

DOI: 10.1201/9781003173786-3

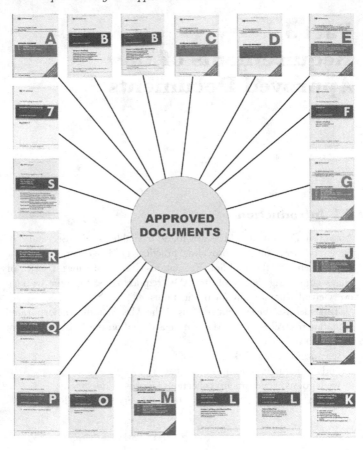

Figure 3.1 Building Regulations Approved Documents

Table 3.1 Current editions of Approved Documents (including incorporated amendments)

Section	Title	Edition
A	Structure	2013
B	Fire safety	2020
C	Site preparation and resistance to contaminants and moisture	2013
D	Toxic substances	2013
E	Resistance to the passage of sound	2015

Table 3.1 (Continued)

Section	Title	Edition
F	Ventilation	2021
G	Sanitation, hot water safety and water efficiency	2016
H	Drainage and waste disposal	2015
J	Combustion appliances and fuel storage systems	2013
K	Protection from falling, collision and impact	2013
L	Conservation of fuel and power	2021
M	Access to and use of buildings	2015
O	Overheating	2021
P	Electrical safety	2013
Q	Security in dwellings	2015
R	High-speed electronic communications networks	2016
S	Infrastructure for the charging of electric vehicles	2021
Regulation 7	Materials and workmanship	2018

3.2 A – Structure

Requirement	Title	Regulation	Requirement (in a nutshell)
A1	Loading	*The building shall be constructed so that the combined dead, imposed and wind loads are sustained and transmitted by it to the ground:* • *safely; and* • *without causing such deflection or deformation of any part of the building, or such movement of the ground, as will impair the stability of any part of another building.*	• Structural elements of a building shall safely carry the loads expected to be placed on them. This will include elements such as the: • type of loading • properties of materials • design analysis • details of construction • safety factors • workmanship.

Requirement	Title	Regulation	Requirement (in a nutshell)
A2	Ground movement	*The building shall be constructed so that ground movement caused by:* • *swelling, shrinkage or freezing of the sub-soil; or* • *land-slip or subsidence (other than subsidence arising from shrinkage), in so far as the risk can be reasonably foreseen, will not impair the stability of any part of the building.*	• Foundations shall be adequate for any movement of the ground (e.g. caused by landslip or subsidence). • Horizontal and vertical ties should be provided.
A3	Dispropor-tionate collapse	*The building shall be constructed so that in the event of an accident the building will not suffer collapse to an extent disproportionate to the cause.*	• Large buildings shall be strong enough to withstand an accident without collapsing. • Horizontal and vertical ties should be provided.

3.3 B – Fire safety: Volume 1: Dwellings and Volume 2: Buildings other than dwellings

The main changes to AD-B (which came into effect on 30 August 2019 in England) were as a result of the fire which virtually destroyed Grenfell Tower in June 2017, and these changes concerned guidance on:

- type of cladding used to make buildings warmer and drier;
- design of sprinkler systems;
- installation and use of external stairs;

- improved fire safety information;
- insulating core panels;
- fire dampers and ventilation systems.

Requirement	Title	Regulation	Requirement (in a nutshell)
B1 B1 does not apply to any prison provided under Section 33 of the Prisons Act 1952.	Means of warning and escape	*The building shall be designed and constructed so that there are appropriate provisions for the early warning of fire, and appropriate means of escape in case of fire to a place of safety outside the building capable of being safely and effectively used at all material times.**	• There shall be an early warning fire alarm system for persons in the building. • Emergency escape routes shall be available to enable persons to evacuate the building in the event of a fire. • All emergency escape routes shall be protected from the effects of fire. • In an emergency, the occupants from any part of the building shall be able to escape without any external assistance.
B2	Internal fire spread (linings)	*(1) To inhibit the spread of fire within the building, the internal linings (i.e. materials or products used in lining any partition, wall, ceiling or other internal structure) shall:*	• The spread of flame over the internal linings of the building shall be restricted. • During an internal fire, the heat released from the building fabric shall be restricted.

* The extent to which these arrangements and facilities apply varies with the size and use of building.

Requirement	Title	Regulation	Requirement (in a nutshell)
		(a) adequately resist the spread of flame over their surfaces; and *(b) have, if ignited, a rate of heat release or a rate of fire growth which is reasonable in the circumstances.* *(2) In this paragraph 'internal linings' means the materials or products used in lining any partition, wall, ceiling or other internal structure.*	
B3 B3(3) does not apply to any prison provided under Section 33 of the Prisons Act 1952.	Internal fire spread (structure)	*(1) The building shall be designed and constructed so that, in the event of fire, its stability will be maintained for a reasonable period.* *(2) A wall common to two or more buildings shall be designed and constructed so that it adequately resists the spread of fire between those buildings. (For the purposes of this sub-paragraph, a house in a terrace and a semi-detached house are each to be treated as a separate building).*	• All load-bearing elements of a building structure shall be capable of withstanding the effects of fire. • The spread of fire to (or from) semi-detached buildings shall be restricted. • Buildings shall be sub-divided into fire-resisting compartments. • Openings in fire-separating elements shall be suitably protected.

Requirement	Title	Regulation	Requirement (in a nutshell)
		(3) Where reasonably necessary to inhibit the spread of fire within the building, measures shall be taken, to an extent appropriate to the size and intended use of the building, comprising either or both of the following: *(a) sub-division of the building by means of fire-resisting construction;* *(b) installation of suitable automatic fire suppression systems.* *(4) The building shall be designed and constructed so that the unseen spread of fire and smoke within concealed spaces in its structure and fabric is inhibited.*	• Hidden voids in the construction of a building shall be sealed and sub-divided to prevent the unseen spread of fire.
B4	External fire spread	*(1) The external walls of the building shall adequately resist the spread of fire over the walls and from one building to another, having regard to the height, use and position of the building.*	• External walls shall have a low rate of heat release in order to reduce the risk of ignition from an external source and the potential spread of fire over their surfaces.

Requirement	Title	Regulation	Requirement (in a nutshell)
		(2) The roof of the building shall adequately resist the spread of fire over the roof and from one building to another, having regard to the use and position of the building.	• Unprotected sides of buildings shall be restricted in order to limit the amount of thermal radiation that can pass through the wall. • Roofs shall be constructed so that the risk of spread of flame and/or fire penetration from an external fire source is restricted. • The risk of fire spreading from one building to another beyond the boundary shall be limited.
B5	Access and facilities for the fire service	*(1) The building shall be designed and constructed to provide reasonable facilities to assist firefighters in the protection of life.* *(2) Reasonable provision shall be made within the site of the building to enable fire appliances to gain access to the building.*	The building shall have: • Sufficient external access for fire appliances to be brought near to the building. • Sufficient access into and within the building for firefighting personnel to search, rescue and fight fire.

Requirement	Title	Regulation	Requirement (in a nutshell)
			• Sufficient internal fire mains and other facilities to assist firefighters in their tasks. • Adequate means to vent heat and smoke from a fire in a basement.

3.4 C – Site preparation and resistance to contaminants and moisture

Requirement	Title	Regulation	Requirement (in a nutshell)
C1	Preparation of the site and resistance to contaminants	*(1) The ground to be covered by the building shall be reasonably free from any material that might damage the building or affect its stability, including vegetable matter, top-soil and pre-existing foundations.* *(2) Reasonable precautions shall be taken to avoid danger to health and safety caused by contaminants in, or on, the ground covered, or to be covered by the building and any land associated with the building.*	• Buildings should be safeguarded from the adverse effects of: • vegetable matter • contaminants in or on the ground covered by the building • groundwater.

Requirement	Title	Regulation	Requirement (in a nutshell)
		(3) Adequate sub-soil drainage shall be provided if it is needed to avoid: *(a) the passage of the ground moisture to the interior of the building;* *(b) damage to the building, including damage through the transport of waterborne contaminants to the foundations of the building.* *(4) For the purpose of this requirement, 'contaminant' means: any substance that is, or may become, harmful to persons or buildings, including substances which are corrosive, explosive, flammable, radioactive or toxic.*	
C2	Resistance to moisture	*The floors, walls and roof of the building shall adequately protect the building and people who use the building from harmful effects caused by:*	All floors next to the ground, walls and roof shall not be damaged by moisture from the ground, rain or snow and

Requirement	Title	Regulation	Requirement (in a nutshell)
		(a) ground moisture; *(b) precipitation including wind-driven spray* *(c) interstitial and surface condensation; and* *(d) spillage of water from or associated with sanitary fittings or fixed appliances.*	shall not carry that moisture to any part of the building which it would damage. Methods of achieving this ambition include: • Building a solid or suspended floor next to the ground to prevent undue moisture from reaching the upper surface of the floor. • Erecting a wall to prevent undue moisture from the ground reaching the inside of the building, and (if it is an outside wall) the penetration of rain and snow to the inside of the building. • Ensuring that the roof is resistant to the penetration of moisture from rain or snow to inside the building.

3.5 D – Toxic substances

Requirement	Title	Regulation	Requirement (in a nutshell)
D1	Cavity insulation	*If insulating material is inserted into a cavity in a cavity wall, reasonable precautions shall be taken to prevent the subsequent permeation of any toxic fumes from that material into any part of the building occupied by people.*	Fumes given off by insulating materials should not be allowed to penetrate occupied parts of buildings where they could become a health risk to persons in the building.

3.6 E – Resistance to the passage of sound

Requirement	Title	Regulation	Requirement (in a nutshell)
E1	Protection against sound from other parts of the building and adjoining buildings	*Dwelling houses, flats and rooms for residential purposes shall be designed and constructed in such a way that they provide reasonable resistance to sound from other parts of the same building and from adjoining buildings.*	Dwellings shall be designed so that any domestic noise that is generated internally is kept to a level that: • Does not affect the health of the occupants of the dwelling. • Will allow them to sleep, rest and engage in their normal activities in satisfactory conditions.

Requirement	Title	Regulation	Requirement (in a nutshell)
E2	Protection against sound within a dwelling house, etc.	*Dwelling houses, flats and rooms for residential purposes shall be designed and constructed in such a way that:* *(a) internal walls between a bedroom or a room containing a water closet, and other rooms; and* *(b) internal floors provide reasonable resistance to sound.* **Note:** E2 does not apply to: • An internal wall which contains a door. • An internal wall which separates an ensuite toilet from the associated bedroom. • Existing walls and floors in a building which is subject to a material change of use.	Suitable sound-absorbing material shall be used in domestic buildings to restrict the transmission of noise from other rooms.

Requirement	Title	Regulation	Requirement (in a nutshell)
E3	Reverberation in the common internal parts of buildings containing flats or rooms for residential purposes	*The common internal parts of buildings which contain flats or rooms for residential purposes shall be designed and constructed in such a way as to prevent more reverberation around the common parts than is reasonable.* **Note:** E3 only applies to corridors, stairwells, hallways and entrance halls which give access to the flats or rooms for residential purposes.	Suitable sound-absorbing material shall be used in domestic buildings to restrict the transmission of echoes.
E4	Acoustic conditions in schools	*(1) Each room or other space in a school building shall be designed and constructed in such a way that it has the acoustic conditions and the insulation*	Suitable sound-insulation materials shall be used within school buildings to reduce the level of ambient noise (particularly echoing in corridors, etc.).

Requirement	Title	Regulation	Requirement (in a nutshell)
		against disturbance by noise appropriate to its intended use. *(2) For the purposes of this Approved Document, 'school' has the same meaning as in Section 4 of the Education Act 1996(4); and 'school building' means any building forming a school or part of a school.*	

3.7 F – Ventilation

Requirement	Title	Regulation	Requirement (in a nutshell)
F1	Means of ventilation	*(1) There shall be adequate means of ventilation provided for people in the building.* *(2) Fixed systems for mechanical ventilation and any associated controls must be commissioned by testing and adjusting as necessary to secure that the objective referred to in sub-paragraph (1) is met.*	Ventilation (mechanical and/or air-conditioning systems designed for domestic buildings) shall be capable of restricting the accumulation of moisture and pollutants originating from within a building.

Requirement	Title	Regulation	Requirement (in a nutshell)
		Note: F1 does not apply to a building or space within a building into which people do not normally go, or one that is used solely for storage, or to a garage used solely by a single dwelling.	

3.8 G – Sanitation, hot water safety and water efficiency

Requirement	Title	Regulation	Requirement (in a nutshell)
G1	Cold water supply	*(1) There must be a suitable installation for the provision of:* *(a) wholesome water to any place where drinking water is drawn off;* *(b) wholesome water or softened wholesome water to any washbasin or bidet provided in or adjacent to a room containing a sanitary convenience*	The cold water supply must: • Be reliable. • Be wholesome. • Have sufficient pressure and flow rate for the operation of all appliances and locations planned in the building.

Requirement	Title	Regulation	Requirement (in a nutshell)
		(c) wholesome water or softened wholesome water to any washbasin, bidet, fixed bath or shower in a bathroom; and *(d) wholesome water to any sink provided in any area where food is prepared.* *(2) There must be a suitable installation for the provision of water of suitable quality to any sanitary convenience fitted with a flushing device.* **Note:** For sanitary conveniences, the water supplied may be either wholesome, softened wholesome or of a suitable quality.	• Be capable of conveying wholesome water or softened wholesome water without waste, misuse, undue consumption or contamination of water.
G2	Water efficiency	*Reasonable provision must be made by the installation of fittings and fixed appliances that use water efficiently for the prevention of undue consumption of water.* **Note:** G2 applies only when a dwelling is erected or formed by a material change of use of a building within the meaning of Regulation 5 (a) or (b).	White goods shall be installed so that they use water efficiently and without undue consumption.

Requirement	Title	Regulation	Requirement (in a nutshell)
G3	Hot water supply and systems	*(1) There must be a suitable installation for the provision of heated wholesome water or heated softened wholesome water to:* *(a) any washbasin or bidet provided in or adjacent to a room containing a sanitary convenience;* *(b) any washbasin, bidet, fixed bath or shower in a bathIm; and* *(c) any sink provided in any area where food is prepared.* *(2) A hot water system, including any cistern or other vessel that supplies water to or receives expansion water from a hot water system, shall be designed, constructed and installed to resist the effects of temperature and pressure that may occur either in normal use, or in the event of such malfunctions as may reasonably be anticipated, and must be adequately supported.* *(3) A hot water system that has a hot water storage vessel shall incorporate precautions to:*	• The water supplied can either be heated wholesome water or heated softened water. • The installation shall convey hot water without waste, misuse or undue consumption of water. • Hot water systems (including cisterns) shall safely contain the hot water during normal operation, as well as following failure of any thermostat used to control temperature, and during the operation of any of the safety devices.

Requirement	Title	Regulation	Requirement (in a nutshell)
		(a) prevent the temperature of the water stored in the vessel at any time exceeding 100°C; and *(b) ensure that any discharge from safety devices is safely conveyed to where it is visible but will not cause a danger to persons in or about the building.* **Note:** G3(3) does not apply to a system which heats or stores water only for the purposes of an industrial process. *(4) The hot water supply to any fixed bath must be so designed and installed as to incorporate measures to ensure that the temperature of the water that can be delivered to that bath does not exceed 48°C.* **Note:** G3(4) applies only when a dwelling is erected or formed by a material change of use within the meaning of Regulation 5 (a) or (b).	• Storage vessels shall have a suitable vent pipe connecting the top to a point open to the atmosphere above the level of the water in the cold water storage cistern. • The hot water system shall have pipework that allows hot water from the safety devices to discharge into a place open to the atmosphere where it will cause no danger to persons in or about the building. • Storage vessels shall have at least two independent safety devices that release pressure and in doing so, prevent the temperature of the stored water at any time exceeding 100°C.

Requirement	Title	Regulation	Requirement (in a nutshell)
G4	Sanitary conveniences and washing facilities	*(1) Adequate and suitable sanitary conveniences must be provided in rooms provided to accommodate them or in bathrooms.* *(2) Adequate hand-washing facilities must be provided in:* *(a) rooms containing sanitary conveniences; or* *(b) rooms or spaces adjacent to rooms containing sanitary conveniences.* *(3) Any room containing a sanitary convenience, a bidet, or any facility for washing hands provided in accordance with paragraph (2) (b) must be separated from any kitchen or any area where food is prepared.*	• There should be appropriate sanitary conveniences for the building in sufficient numbers, taking into account the nature of the building and the age and sex of the users. • Hand-washing facilities should be provided in, or adjacent to, rooms containing sanitary conveniences.
G5	Bathrooms	*A bathroom must be provided with a washbasin and either a fixed bath or a shower.* **Note:** G5 applies only to dwellings and to buildings containing one or more rooms for residential purposes.	If a dwelling or a building contains one or more rooms for residential purposes, it shall be provided with a washbasin and either a fixed bath or a shower.

Requirement	Title	Regulation	Requirement (in a nutshell)
G6	Food preparation areas	*A suitable sink must be provided in any area where food is prepared.*	A sink should be provided in a room that is used to prepare food. However, if a dishwasher is provided in a separate room, there is no need for a separate sink in that room.

3.9 H – Drainage and waste disposal

Requirement	Title	Regulation	Requirement (in a nutshell)
H1	Foul water drainage	*(1) An adequate system of drainage shall be provided to carry foul water from appliances within the building to one of the following, listed in order of priority:* *(a) a public sewer; or* *(b) a private sewer communicating with a public sewer or,* *(c) a septic tank which has an appropriate form of secondary treatment (or another wastewater treatment system); or* *(d) a cesspool.*	The foul water drainage system shall: • Convey the flow of foul water to a foul water outfall (i.e. sewer, cesspool, septic tank or holding tank). • Minimise the risk of blockage or leakage.

Requirement	Title	Regulation	Requirement (in a nutshell)
		(2) In this Approved Document, 'foul water' means wastewater which comprises or includes: *(a) waste from a sanitary convenience, bidet or appliance used for washing receptacles for foul waste; or* *(b) water which has been used for food preparation, cooking or washing.*	• Prevent foul air from the drainage system entering the building. • Be ventilated and accessible for clearing blockages. • Not increase the vulnerability of the building to flooding.
H2	Wastewater treatment systems and cesspools	*(1) Any septic tank and its chosen form of secondary treatment, other wastewater treatment system or cesspool, shall be so sited and constructed that:* *(a) it is not prejudicial to the health of any person;* *(b) it will not contaminate any watercourse, underground water oIater supply;* *(c) there are adequate means of access for emptying and maintenance; and* *(d) where relevant, it will function to a sufficient standard for the protection of health in the event of a power failure.*	• Wastewater treatment systems shall have sufficient capacity to enable breakdown and settlement of solid matter in the wastewater from the buildings. • Cesspools shall have sufficient capacity to store the foul water from the building until they are emptied.

Requirement	Title	Regulation	Requirement (in a nutshell)
		(2) Any septic tank or holding tank which is part of a wastewater treatment system or cesspool shall be: *(a) of adequate capacity;* *(b) so constructed that it is impermeable to liquids; and* *(c) adequately ventilated.* *(3) Where a foul water drainage system from a building discharges to a septic tank, wastewater treatment system or cesspool, a durable notice shall be affixed in a suitable place in the building containing information on any continuing maintenance requirement to avoid risks to health.*	• Wastewater treatment systems and cesspools shall be sited and constructed so as not to: • Be prejudicial to health or be a nuisance. • Adversely affect water sources or resource. • Pollute controlled waters. • Be in an area where there is a risk of flooding. • Septic tanks and wastewater treatment systems and cesspools shall: • Have adequate ventilation. • Prevent leakage of the contents and ingress of sub-soil water.

Requirement	Title	Regulation	Requirement (in a nutshell)

			• Take into consideration water table levels at any time of the year and rising groundwater levels.
			• Drainage fields shall be sited and constructed to:
			  • Avoid overloading of the soakage capacity.
			  • Provide the availability of an aerated layer in the soil at all times.

H3	Rainwater drainage	*(1) Adequate provision shall be made for rainwater to be carried from the roof of the building.* *(2) Paved areas around the building shall be so constructed as to be adequately drained.* *(3) Rainwater from a system provided pursuant to sub-paragraphs (1) or (2) shall discharge to one of the following, listed in order of priority:*	Rainwater drainage systems shall: • Minimise the risk of blockage or leakage. • Be accessible for clearing blockages.

Requirement	Title	Regulation	Requirement (in a nutshell)
		(a) an adequate soakaway or some other adequate infiltration system; or, where that is not reasonably practicable, *(b) a watercourse; or, where that is not reInably practicable,* *(c) a sewer.* 📝 ***Note:*** H3(2) applies **only** to paved areas which provide access to: • The building or a place of storage. • Any passage giving access to the building, where this is intended to be used in common by the occupiers of one or more other buildings. H3(3) does **not** apply to the gathering of rainwater for re-use.	• Ensure that rainwater soaking into the ground is distributed sufficiently so that it does not damage the foundations of a building or any adjacent structure. • Ensure that rainwater from roofs and paved areas is carried away from the surface either by a drainage system or by other means. • Ensure that the rainwater drainage system carries the flow of rainwater from the roof to a soakaway, watercourse, or surface water or combined sewer.

Requirement	Title	Regulation	Requirement (in a nutshell)
H4	Building over sewers	(1) *The erection or extension of a building or work involving the underpinning of a building shall be carried out in a way that is not detrimental to the building or building extension or to the continued maintenance of the drain, sewer or disposal main.* (2) *In this paragraph 'disposal main' means any pipe, tunnel or conduit used for the conveyance of effluent to or from a sewage disposal works, which is not a public sewer.* (3) *In this paragraph and paragraph H5 'map of sewers' means any records kept by a sewerage undertaker under Section 97 of the Water Act 2003.* **Note:** H4 applies **only** to work carried out: • Over a drain, sewer or disposal main which is shown on any map of sewers. • On any site or in such a manner as may result in interference with the use of, or obstruction of the access of any person to, any drain, sewer or disposal main which is shown on any map of sewers.	Building or extensions or work involving underpinning shall be carried out in a manner which will not: • Overload or otherwise cause damage to the drain, sewer or disposal main. • Obstruct reasonable access to any manhole or inspection chamber on the drain, sewer or disposal main. • Unduly obstruct work to replace a drain, sewer or disposal main if required. • Increase the risk of damage to the building as a result of failure of the drain, sewer or disposal main.

Requirement	Title	Regulation	Requirement (in a nutshell)
H5	Separate systems of drainage	*Any system for discharging water to a sewer which is provided in accordance with paragraph H3 shall be separate from that provided for the conveyance of foul water from the building.* **Note:** H5 applies only to a system provided in connection with the erection or extension of a building where it is reasonably practicable for the system to discharge directly or indirectly to a sewer for the separate conveyance of surface water which is: • Shown on a map of sewers. • Under construction either by the sewerage undertaker or by some other person (where the sewer is the subject of an agreement to make a declaration of vesting pursuant to Section 96 of the Water Act 2003).	Separate systems of drains and sewers shall be provided for foul water and rainwater where: • The rainwater is not contaminated. • The drainage is to be connected, either directly or indirectly, to the public sewer system that includes separate systems for foul water and surface water, or to a system of sewers which are provided only for the passage of the construction's surface water.

Requirement	Title	Regulation	Requirement (in a nutshell)
H6	Solid waste storage	*(1) Adequate provision shall be made for storage of solid waste.* *(2) Adequate means of access shall be provided:* *(a) for people in the building to the place of storage; and* *(b) from the place of storage to a collection point (where one has been specified by the waste collection authority under Section 46 (household waste) or Section 47 (commercial waste) of the Environmental Protection Act 1990, or to a street [where no collection point has been specified]).*	Solid waste storage shall be: • Designed and sited so as not to be prejudicial to health. • Capable of containing the quantity of solid waste to be removed at the envisaged frequency of removal. • Accessible for use by people in the building as well as the street for emptying and removal.

3.10 J – Combustion appliances and fuel storage systems

Requirement	Title	Regulation	Requirement (in a nutshell)
J1	Air supply	*Combustion appliances shall be so installed that there is an adequate supply of air to them for combustion, to prevent overheating and for the efficient working of any flue.* **Note:** J1 only applies to fixed combustion appliances (including incinerators).	The building shall allow the admission of sufficient air for the proper combustion of fuel and the operation of flues and cooling appliances where necessary so that they: • Enable normal operation. • Have been inspected and tested to establish suitability for the purpose intended. • Have been suitably performance labelled.
J2	Discharge of products of combustion	*Combustion appliances shall have adequate provision for the discharge of products of combustion to the outside air.* **Note:** J2 only applies to fixed combustion appliances (including incinerators).	Combustion appliances shall discharge combustion products to the outside air efficiently.

Requirement	Title	Regulation	Requirement (in a nutshell)
J3	Warning of release of carbon monoxide	*Where a fixed combustion appliance is provided, appropriate provision shall be made to detect and give warning of the release of carbon monoxide.* **Note:** J3 only applies to fixed combustion appliances located in dwellings.	All fixed combustion appliances shall be capable of detecting and giving warning of the release of carbon monoxide.
J4	Protection of building	*Combustion appliances and flue pipes shall be so installed, and fireplaces and chimneys shall be so constructed and installed, as to reduce to a reasonable level the risk of people suffering burns or the building catching fire in consequence of their use.* **Note:** J4 only applies to fixed combustion appliances located in dwellings.	Combustion appliances, flue pipes, fireplaces and chimneys shall be constructed and installed so that the risk of people suffering burns or the building catching fire is minimised.

Requirement	Title	Regulation	Requirement *(in a nutshell)*
J5	Provision of information	*Where a hearth, fireplace, flue or chimney is provided or extended, a durable notice containing information on the performance capabilities of the hearth, fireplace, flue or chimney shall be affixed in a suitable place in the building for the purpose of enabling combustion appliances to be safely installed.*	Notices containing information about performance (and the capabilities of hearths, fireplaces, flues and chimneys) shall be located in a suitable place.
J6	Protection of liquid fuel storage systems	*Liquid fuel storage systems and the pipes connecting them to combustion appliances shall be so constructed and separated from buildings and the boundary of the premises as to reduce to a reasonable level the risk of the fuel igniting in the event of fire in adjacent buildings or premises.* ***Note:** J6 applies only to:* • Fixed oil storage tanks with capacities greater than 90 litres and their connecting pipes.	• Oil and liquid petroleum gas (LPG) fuel storage installations shall be located and constructed so that they are reasonably protected from fires that may occur in buildings or beyond boundaries. • Oil storage tanks used wholly or mainly for private dwellings shall be: • Reasonably resistant to physical damage and corrosion.

Requirement	Title	Regulation	Requirement (in a nutshell)
		• Fixed liquefied petroleum gas storage installations with capacities greater than 150 litres and connecting pipes, which are located outside the building and which serve fixed combustion appliances (including incinerators) in the building.	• Designed and installed so that it minimises the risk of oil escaping during the filling or maintenance of the tank.
J7	Protection against pollution	*Oil storage tanks and the pipes connecting them to combustion appliances shall:* *(a) be so constructed and protected as to reduce to a reasonable level the risk of the oil escaping and causing pollution; and* *(b) have affixed in a prominent position a durable notice containing information on how to respond to an oil escape to reduce to a reasonable level the risk of pollution.*	Oil storage tanks with capacities of 3,500 litres or less shall be constructed and protected to reduce the level of risk of the oil escaping and causing pollution to a reasonable level.

Requirement	Title	Regulation	Requirement (in a nutshell)
		Note: J7 applies only to fixed oil storage tanks with capacities of 3,500 litres or less, and connecting pipes, which are: • Located outside the building. • Serving fixed combustion appliances (including incinerators) in a building used wholly or mainly as a private dwelling, but does not apply to buried systems.	

3.11 K – Protection from falling, collision and impact

Requirement	Title	Regulation	Requirement (in a nutshell)
K1	Stairs, ladders and ramps	*Stairs, ladders and ramps shall be so designed, constructed and installed as to be safe for people moving between different levels in or about the building.*	Stairs, steps and ladders shall provide reasonable safety between all of the building's levels.

Requirement	Title	Regulation	Requirement (in a nutshell)
		Note: K1 applies **only** to stairs, ladders and ramps which form part of the building.	**Note:** Stairs, etc., in public buildings may need to be of a higher standard than those in a dwelling.
K2	Protection from falling	(a) *Any stairs, ramps, floors and balconies, and any roof to which people have access, and* (b) *any light well, basement area or similar sunken area connected to a building, shall be provided with barriers where it is necessary to protect people in or about the building from falling.* **Note:** K2(a) applies **only** to stairs and ramps which form part of the building.	Pedestrian guarding should be provided for any part of a floor, gallery, balcony, roof, or any other place to which people have access; and any light well, basement area or similar sunken area next to a building.

Requirement	Title	Regulation	Requirement (in a nutshell)
K3	Vehicle barriers and loading bays	*(1) Vehicle ramps and any levels in a building to which vehicles have access shall be provided with barriers where it is necessary to protect people in or about the building.* *(2) Vehicle loading bays shall be constructed in such a way, or be provided with such features, as may be necessary to protect people in them from collision with vehicles.*	• Vehicle barriers that are capable of resisting or deflecting the impact of vehicles shall be provided. • Loading bays shall include exits (or refuges) to enable people to avoid being crushed by vehicles.
K4	Protection against impact with glazing	*Glazing with which people are likely to come into contact whilst moving in or about the building shall:* *(a) if broken on impact, break in a way which is unlikely to cause injury; or* *(b) resist impact without breaking; or* *(c) be shielded or protected from impact.*	The risk of injury from glazing shall be minimised by ensuring that: • If a breakage occurs, glass particles would be relatively harmless. • The glazing is sufficiently robust.

Requirement	Title	Regulation	Requirement (in a nutshell)
K5.1	Protection from collision with open windows, etc.	*Provision shall be made to prevent people moving in or about the building from colliding with open windows, skylights or ventilators.* **Note:** K5.1 does **not** apply to dwellings.	All windows, skylights and ventilators shall be capable of being left open without danger of people colliding with them.
K5.2	Manifestation of glazing	*Transparent glazing with which people are likely to come into contact while moving in or about the building, shall incorporate features which make it apparent.* **Note:** K5.2 does **not** apply to dwellings.	There should be a permanent means of indicating the presence of large uninterrupted areas of clear glazing.
K5.3	Safe opening and closing of windows, etc.	*Windows, skylights and ventilators which can be opened by people in or about the building shall be so constructed or equipped so that they may be opened, closed or adjusted safely.* **Note:** K5.3 does **not** apply to dwellings.	People should be able to operate the windows safely.

Requirement	Title	Regulation	Requirement (in a nutshell)
K5.4	Safe access for cleaning windows, etc.	*Provision shall be made for any windows, skylights, or any transparent or translucent walls, ceilings or roofs to be safely accessible for cleaning.* **Note:** K5.4 does **not** apply to dwellings; or to any transparent or translucent elements whose surfaces are not intended to be cleaned.	There should be a safe means of access to clean both sides of all glass windows (and any transparent or translucent element) in buildings.
K6	Protection against impact from and trapping by doors	*(1) Provision shall be made to prevent any door or gate:* *(a) which slides or opens upwards, from falling onto any person; and* *(b) which is powered, from trapping any person.*	In buildings other than dwellings, measures should be taken to prevent the opening and closing of doors and gates that present a safety hazard.

Requirement	Title	Regulation	Requirement (in a nutshell)
		(2) Provision shall be made for powered doors and gates to be opened in the event of a power failure. (3) Provision shall be made to ensure a clear view of the space on either side of a swing door or gate. **Note:** K6 does **not** apply to dwellings; or to any door or gate that is part of a lift.	

3.12 L – Conservation of fuel and power: Volume 1: Dwellings and Volume 2: Buildings other than dwellings

Requirement	Title	Regulation	Requirement (in a nutshell)
L (Volume I)	Conservation of fuel and power (dwellings)	(1) Reasonable provision shall be made for the conservation of fuel and power in buildings by:	Energy-efficient methods shall be utilised to:

Requirement	Title	Regulation	Requirement (in a nutshell)
		(a) limiting heat gains and losses: *(i) through thermal elements and other parts of the building fabric; and* *(ii) from pipes, ducts and vessels used for space heating, space cooling and hot water services;* *(b) providing fixed building services which:* *(i) are energy efficient;* *(ii) have effective controls; and* *(iii) are commissioned by testing and adjusting as necessary to ensure that they use no more fuel and power than is reasonable in the circumstances.*	• Limit heat gains and losses. • Provide fixed building services which are energy efficient.

Requirement	Title	Regulation	Requirement (in a nutshell)
L (Volume 2)	Conservation of fuel and power (buildings other than dwellings)	(2) Where a system for on-site electricity generation is installed: (a) reasonable provision must be made to ensure that: (i) the system and its electrical output are appropriately sized for the site and are available infrastructure; (ii) the system has effective controls; and (iii) it must be commissioned by testing and adjusting as necessary to ensure that it produces the maximum electricity that is reasonable in the circumstances.	The installation should: • Be appropriate for the particular site in question. • Have effective controls. • Be officially commissioned and tested.

Note: Regulation 40 provides the owner with sufficient information about how the building, the fixed building services and the maintenance requirements of the building can be operated in such a manner as to use no more fuel and power than is reasonable in the circumstances.

3.13 M – Access to and use of buildings: Volume 1: Dwellings and Volume 2: Buildings other than dwellings

Requirement	Title	Regulation	Requirement (in a nutshell)
Volume 1: Dwellings			
M4(1)	Category 1 – Visitable dwellings: Access and use	*Reasonable provision shall be made for people to:* *(a) gain access to; and* *(b) use, the building and its facilities.* ✏️ **Note:** M4(1) does **not** apply to: • An extension to a dwelling. • Any part of a building which is used solely to enable the building or any service or fitting in the building to be inspected, repaired or maintained.	Dwellings should make reasonable provision for most people (including wheelchair users) to approach and enter the dwelling and gain access to habitable rooms and sanitary facilities on the entrance storey.
M4(2)	Category 2 – Accessible and adaptable dwellings: Optional requirement	*(1) Reasonable provision shall be made for people to:* *(a) gain access to; and* *(b) use, the building and its facilities.* *(2) The provision made must be sufficient to:* *(a) meet the needs of occupants with differing needs, including some older or disabled people; and* *(b) allow adaptation of the dwelling to meet the changing needs of occupants over time.*	The dwelling should include features that make it potentially suitable for a wide range of occupants (including older people, those with reduced mobility and wheelchair users).

Requirement	Title	Regulation	Requirement (in a nutshell)
		📝 ***Note:*** M4(2) is **optional** and it: • May apply only in relation to a dwelling that is erected. • Will apply in substitution for requirement M4(1). • Does not apply where optional requirement M4(3) applies. • Does not apply to any part of a building that is used solely to enable the building or any service or fitting in the building to be inspected, repaired or maintained.	
M4(3)	Category 3 – Wheelchair-user dwellings: Optional requirement	*(1) Reasonable provision shall be made for people to:* *(a) gain access to; and* *(b) use, the building and its facilities.* *(2) The provision made must be sufficient to:* *(a) meet the needs of occupants who use wheelchairs; or* *(b) allow the simple adaptation of the dwelling to meet the needs of occupants who use wheelchairs.*	New dwellings shall make reasonable provision for a wheelchair user to live in the dwelling and use any associated private outdoor space, parking, and communal facilities provided for the occupants of the dwelling.

Requirement	Title	Regulation	Requirement (in a nutshell)
		Note: M4(3) is **optional** and it: • May apply only in relation to a dwelling that is erected. • Will apply in substitution for requirement M4(1). • Does not apply where optional requirement M4(2) applies. • Does not apply to any part of a building that is used solely to enable the building or any service or fitting in the building to be inspected, repaired or maintained.	
Volume 2: Buildings other than dwellings			
M1	Access to and use of buildings other than dwellings	*Reasonable provision shall be made for people to:* *(a) gain access to; and* *(b) use, the building and its facilities.* **Note:** M1 does **not** apply to any part of a building which is used solely to enable the building to be inspected, repaired or maintained.	People, regardless of disability, age or gender, should be able to gain access to buildings and use their facilities both as visitors and as people who live or work in them.

Requirement	Title	Regulation	Requirement (in a nutshell)
M2	Access to extensions to buildings other than dwellings	*Suitable independent access must be provided to the extension where reasonably practicable.* **Note:** M2 does **not** apply where suitable access to the extension is provided through the building that is extended.	People, regardless of disability, age or gender, should be able to gain access to extensions to buildings other than dwellings.
M3	Sanitary conveniences in extensions to buildings other than dwellings	*If sanitary conveniences are provided in any building that is to be extended, reasonable provision shall be made within the extension for sanitary conveniences.* **Note:** M3 does **not** apply where there is reasonable provision for sanitary conveniences elsewhere in the building, such that people occupied, or otherwise having occasion to enter the extension, can gain access to and use those sanitary conveniences.	Where the requirement exists, sanitary conveniences shall be provided in extensions to buildings other than dwellings.

3.14 O – Overheating

Requirement	Title	Regulation	Requirement (in a nutshell)
O1	Overheating	*Reasonable provision must be made in respect of a dwelling, institution, any other building containing one or more rooms for residential purposes, other than a room in a hotel ('residences') to:* *(a) limit unwanted solar gains in summer;* *(b) provide an adequate means to remove heat from the indoor environment.* **Note:** *In meeting the obligations in paragraph (1):* *(a) account must be taken of the safety of any occupant, and their reasonable enjoyment of the residence; and* *(b) mechanical cooling may only be used where insufficient heat is capable of being removed from the indoor environment without it.*	• Overheating is reduced by: • Limiting unwanted solar gains in summer. • Removing excess heat from the indoor environment. • In a new residential building, the overheating mitigation strategy must take account of: • noise at night • pollution • security • protection from falling • protection from entrapment.

3.15 P – Electrical safety: Design and installation of electrical installations

Requirement	Title	Regulation	Requirement (in a nutshell)
P1	Design and installation of electrical installations	*Reasonable provision shall be made in the design and installation of electrical installations in order to protect persons operating, maintaining or altering the installations from fire or injury.* **Note:** These requirements only apply to electrical installations that are intended to operate at low or extra-low voltage in: • A dwelling (or attached to a dwelling). • Common parts of a building serving one or more dwellings (but excluding power supplies to lifts).	The design and connection of electrical installations shall be capable of protecting persons from fire or injury when operating, maintaining or altering an installation.

Requirement	Title	Regulation	Requirement (in a nutshell)
		• A building that receives its electricity from a source located within (or shared with) a dwelling. • A garden or in or on land associated with a building where the electricity is from a source located within (or shared with) a dwelling.	

All electrical work shall be carried out in accordance with BS 7671:2018.

3.16 Q – Security in dwellings

Requirement	Title	Regulation	Requirement (in a nutshell)
Q1	Unauthorised access	*Reasonable provision must be made to resist unauthorised access to:* *(a) any dwelling; and* *(b) any part of a building from which access can be gained to a flat within the building.*	To resist unauthorised access, easily accessible doors and windows shall be sufficiently robust and fitted with appropriate hardware particularly where they provide access: • To a dwelling from outside.

Requirement	Title	Regulation	Requirement (in a nutshell)
		Note: Q1 only applies in relation to **new** dwellings.	• To parts of a building containing flats from outside. • To a flat from the common parts of a building.

3.17 R – High-speed electronic communications networks

Requirement	Title	Regulation	Requirement (in a nutshell)
R1	Physical infrastructure for high-speed electronic communication networks	*(1) Building work must be carried out so as to ensure that the building is equipped with high-speed-ready in-building physical infrastructure, up to a network termination point for high-speed electronic communications networks.*	Buildings shall be designed and constructed so that high-speed electronic communications networks can be installed in the future.

Requirement	Title	Regulation	Requirement (in a nutshell)
		(2) Where the work concerns a building containing more than one dwelling, the work must be carried out to ensure that the building is equipped in addition with a common access point for high-speed electronic communications networks. **Note:** R1 applies to **all** building work that consists of: • The erection of a building. • Major renovation works to a building.	

3.18 S – Infrastructure for the charging of electric vehicles

Requirement	Title	Regulation	Requirement (in a nutshell)
S1	The erection of new residential buildings	*(1) A new residential building with associated parking must have access to electric vehicle charging points as provided for in paragraph (2).* *(2) The number of associated parking spaces which have access to electric vehicle charging points must be:* *(a) the total number of associated parking spaces, where there are fewer associated parking spaces than there are dwellings contained in the residential building; or* *(b) the number of associated parking spaces that is equal to the total number of dwellings contained in the residential building, where there are the*	All new residential buildings with associated parking must have access to electric vehicle charging points.

Requirement	Title	Regulation	Requirement (in a nutshell)

		same number of associated parking spaces as, or more associated parking spaces than, there are dwellings.
		 (3) *Cable routes for electric vehicle charging points must be installed in any associated parking spaces which do not, in accordance with paragraph (2), have an electric vehicle charging point where:*
		   (a) *a new residential building has more than 10 associated parking spaces; and*
		   (b) *there are more associated parking spaces than there are dwellings contained in the residential building.*	

Requirement	Title	Regulation	Requirement (in a nutshell)
S2	Dwellings resulting from a material change of use	*Where one or more dwellings with associated parking result from a building, or a part of a building, undergoing a material change of use, at least one associated parking space for the use of each such dwelling must have access to an electric vehicle charging point.*	Each dwelling resulting from a material change of use should have its own vehicle charging point.
S3	Residential buildings undergoing major renovation	*Where a residential building undergoing major renovation will have more than 10 associated parking spaces after the major renovation is completed:* *(a) at least one associated parking space for the use of each dwelling must have access to an electric vehicle charging point; and* *(b) cable routes for electric vehicle charging points must be installed in all additional associated parking spaces.*	Buildings undergoing major renovation that have more than 10 parking spaces must have access to one charging point for each dwelling, and cable routes for the remainder of the spaces.

Requirement	Title	Regulation	Requirement (in a nutshell)
S4	Erection of new buildings which are not residential buildings or mixed-use buildings	*Where a new building which is not a residential building or a mixed-use building has more than 10 parking spaces:* *(a) one of those parking spaces must have access to an electric vehicle charging point; and* *(b) cable routes for electric vehicle charging points must be installed in a minimum of one-fifth of the total number of remaining parking spaces.*	New non-residential buildings with more than 10 parking spaces must have at least one accessible electric charging point and cable routes for one-fifth of the remaining spaces.
S5	Buildings undergoing major renovation which are not residential buildings or mixed-use buildings	*Where a building undergoing major renovation, which is not a residential building or a mixed-use building, will have more than 10 parking spaces after the major renovation is completed:* *(a) one of those parking spaces must have access to an electric vehicle charging point; and* *(b) cable routes for electric vehicle charging points must be installed in a minimum of one-fifth of the total number of remaining parking spaces.*	Major underground renovation to a non-residential building which will have more than 10 parking spaces must have at least one accessible electric charging point and cable routes for one-fifth of the remaining spaces.

Requirement	Title	Regulation	Requirement (in a nutshell)
S6	The erection of new mixed-use buildings and mixed-use buildings undergoing major renovation	(1) *The requirements of paragraph S1 apply with respect to the part of the new mixed-use building that contains one or more dwellings and the associated parking spaces that are assigned to those dwellings.* (2) *The requirements of paragraph S3 apply with respect to the part of the mixed-use building that is undergoing major renovation that contains one or more dwellings and the associated parking spaces that are assigned to those dwellings.* (3) *The requirements of paragraph S4 apply with respect to the part of the new mixed-use building that contains one or more new premises that are not dwellings and the parking spaces that are assigned to those premises.*	

Requirement	Title	Regulation	Requirement (in a nutshell)
		(4) The requirements of paragraph S5 apply with respect to the part of the mixed-use building that is undergoing major renovation that contains one or more premises that are not dwellings and the parking spaces that are assigned to those premises.	

3.19 Regulation 7 – Materials and workmanship

Regulation 7 in a nutshell:

(1) Building work shall be carried out with adequate and proper materials which are:
  * Appropriate for the circumstances in which they are used.
  * Adequately mixed or prepared.
  * Applied, used or fixed (in a workmanlike manner) so as to adequately perform the functions for which they were designed.

And (2) materials other than:

  * cavity trays (when used between two leaves of masonry);
  * any part of a roof connected to an external wall;
  * door frames and doors;
  * electrical installations;
  * insulation and waterproofing materials used below ground level;
  * instrument and fire-stopping materials that have been included in accordance with the requirements of part b of schedule 1;
  * membranes;
  * seals, gaskets, fixings, sealants and backer rods;
  * thermal break materials that have been included in order to meet the thermal bridging requirements of part l of schedule 1;
  * window frames and glass;

and which are used for building work that becomes part of an external wall (or specified attachment) of a relevant building, shall be in accordance with 'Reaction to fire tests for building products' (BS EN 13823:2020).

Notes:

1. Changes which came into effect in 2018 and 2019 included a major revision of Approved Document B and Regulation 7 (as a result of the Grenfell Tower inquiry, which concluded that the external cladding did not comply with the Building Regulation aimed at preventing the external spread of fire, but instead 'actively promoted it').

2. New editions of Approved Documents F and L were released in December 2021 and two new documents (O and S) have also been introduced. All of these will come into effect on 15 June 2022.

3. It should be noted that Approved Documents F and L have been completely rewritten and the numbering and order of guidance has changed (F is now in 2 volumes and L has been reduced from 4 volumes to 2 volumes).

4. Approved Document O only applies with respect to new residential buildings (including dwellings, institutional and student accommodation where people sleep on the premises).

Part 4
Meeting the requirements of the Building Regulations

4.1 Introduction

The **Building Regulations** set standards for the design and construction of buildings to ensure the safety and health for people, including those with disabilities, in or about those buildings and to help conserve fuel and power.

Approved Documents describe how the requirements of Schedule 1 and Regulation 7 of the Building Act 1984 can be met.

Each Approved Document (AD) reproduces the **actual requirements** contained in the Building Regulations relevant to the subject area (e.g. AD-P deals with electrical safety).

In this Part, **practical and technical guidance** (together with examples) shows how the requirements can be met in some of the more common building situations.

4.1.1 How to use Part 4

Part 4 of the book has been designed to help you meet the requirements of the Building Regulations in dwelling houses, flats and rooms for residential purposes (hereafter referred to as 'dwellings'). However, as a high percentage of the requirements, information and explanations about Building Regulations in dwellings is also applicable to buildings other than dwellings, rather than writing an additional part purely for 'Buildings other than dwellings', we have included any differences that might occur at the end of an associated subsection of Part 4.

In compiling Part 4, we have adopted a foundations-up approach and put all of the relevant AD requirements that relate to each specific area (e.g. windows) into one subsection.

DOI: 10.1201/9781003173786-4

Each subsection of Part 4 is then arranged to discuss the precise requirement together with notes on how to meet the requirement. Subsections start with a table showing the essential requirements of which part of each AD applies to a particular aspect of the building.

Very often a number of ADs will apply; therefore each subsection follows the alphanumeric order of the AD numbering system. Where a specific AD does not apply, it is omitted from that particular subsection.

Owing to the huge amount of information contained in ADs (and in an effort to help you 'sort the wood from the trees'!) the following format has been used in this Part of the book:

Shaded boxes are used in the subsections to show **either** the full text of the Building Regulation's legal requirements **or** a paraphrased version of these requirements.	The numbers in the right-hand column relate to the appropriate Approved Document and the paragraph number in the relevant guidance section of the Approved Document.

For example:

• A protected shaft containing a protected stairway and/or a lift should not also contain a ventilating duct unless:
  • ducts have been provided in order to keep it smoke free;
  • ducts are only used to ventilate the protected stairway.
	B(V1) 7.26

4.1.2 Cross references to AD diagrams, tables and other publications

In order to retain the pocket-book format of this publication, rather than reproducing a whole host of illustrations and tables extracted from ADs, you will find that throughout Part 4 we provide you with a link to a relevant diagram or table that is contained in one of the ADs. In a similar manner, rather than repeating word for word what is contained in a British Standard or Government publication, we have provided you with a reference point. For example:

• Non-engineered fill (see BRE Digest 427) should be avoided and there should **not** be a wide variation in ground conditions within the loaded area.	A 2E1a
• The **minimum** thickness of a concrete foundation should be 150mm or P, whichever is the greater (where P is derived using AD-A Diagram 23 and Table 10).	A 2E2c

Copies of the full text of these ADs can, of course, either be viewed or downloaded onto your iPhone or similar device from the Governments Planning Portal – for free! – by simply logging on to www.gov.uk/gov ernment/collections/apfo/200135/approved_documents.

4.1.3 Associated publication

If you need a little more advice on a particular aspect of the Building Regulations, then you should cross refer to our other publication (*Building Regulations in Brief*, 10th edition), which should provide you with the infor- mation that you require, or advise you where this material can be found in another publication, organisation or elsewhere on the web.

4.2 Foundations

A foundation refers to the lower part of a building which is designed to not only distribute the weight of the new building evenly but also to provide a firm footing for it. Even though the foundations are not visible when the building has been completed, their design and structure are probably the most important parts of any building project.

Reminder: The subsections include details for buildings other than dwellings only where they differ from the requirements for dwellings.

Subsection	Description	Page
Requirements for dwellings		
4.2.1	Structure	146
Requirements for buildings other than dwellings – Additional requirements		
4.2.2	Structure	158
4.2.3	Site preparation and resistance to contaminants and water	159

Reminder: Free downloads of all the diagrams and tables referenced in the subsections are available on-line at the Go https://www.gov.uk/government/collections/approved-documents.

Requirements for dwellings

4.2.1 Structure

Laying the foundation of a house is completed in five main steps:

1   **Ground cleaning:**
    - agree soil conditions;
    - clear area of boulders, sticks, etc.
2   **Area preparation:**
    - dig area twice the depth of your intended foundation;
    - level and compact the soil;
    - add gravel/sand to keep moisture at bay;
    - lay polyethylene on top as waterproofing.
3   **Building a frame:**
    - create a wooden outline of your foundation's dimensions;
    - put screed rails on both sides;
    - lay a straightedge (e.g. 2×4) across the frame, resting on the rails;
    - add supports to the rails so they stay in place after concrete pouring;
    - place the straightedge so that its bottom is level to where the slab's top will be.
4   **Mixing and pouring the concrete:**
    - mix the aggregate with water and concrete;
    - pour into the foundation frame;
    - as it fills, slide the straightedge across to level it out;
    - once the concrete is level and excess is removed, take off the rails and straightedge.
5   **Finishing touches:**
    - remove defects using a trowel while the concrete is still wet;
    - allow the foundation to dry and cure.

4.2.1.1 General requirements

The safety of a structure depends on the successful combination of design and completed construction.

- The design should: A 0.2a
  - identify the hazards (to which the structure is likely   A 0.2b
    to be subjected) And assess the risks;
  - reflect conditions reasonably foreseen during   A 0.2
    future use;
  - ensure that the dead load, imposed load and wind
    loads are in accordance with current codes of
    practice;
  - guarantee loads used in calculations to allow for
    possible dynamic, concentrated and peak load
    effects that may occur;
  - include safety factors, workmanship and
    requirements for materials used in the assembly of
    the structure.

4.2.1.2 Stability

- Known or recorded conditions of ground instability   A 1.11
  (from landslides, disused mines, etc.) should be
  taken into account when designing and building
  foundations.
- Adequate provision shall be made to ensure that the   A 2A2
  building is stable under all wind-loading conditions.
  The size and position of the building should be
  limited in accordance with the guidance for that type
  of structure.
- The layout of walls (internal and external) should   A 2A2b
  form a robust three-dimensional box constructed
  according to the specific guidance for each form of
  construction.
- The internal and external walls shall be connected by A 2A2c
  either masonry bonding or mechanical connections.
- The intermediate floors and roof shall be constructed A 2A2d
  so that they:
  - provide local support to the walls;
  - act as horizontal diaphragms capable of
    transferring the wind forces to buttressing elements
    of the building.

A map showing wind speed experienced in England and Wales is given
in Figure 1 of AD-A Diagram 6.

4.2.1.2.1 HEIGHTS OF BUILDINGS

• The maximum height of a dwelling is less than 15m.	A 2C4a(i)
• The height of the building should not exceed twice the least width of the building.	A 2C4a(ii)
• The height of a wing should not exceed twice the least width of a wing (see AD-A Diagram 1).	A 2C4a(iii)
• The maximum heights of buildings should be in accordance with BS EN 1991-1-4 (and as shown in Table C of AD-A Diagram 7) correlating to various site exposure conditions and wind speeds.	A 2 C16
• The height of an annex (as shown in AD-A Diagram 2) should not exceed 3m.	A 2C4c

4.2.1.2.2 HEIGHTS OF WALLS AND STOREYS

• The height of a wall or storey should be measured in accordance with AD-A Diagram 8.	A 2C18
• Differences in the level of ground of other solid constructions between one side of a wall and another should be less than four times the thickness of the wall.	
• The combined dead and imposed load at the base of A wall should not exceed 70kN/m (see AD-A Diagram 9).	A 2C21

4.2.1.2.3 MAXIMUM FLOOR AREA

• No floor enclosed by structural walls on all sides shall exceed 70m².	A 2C14
• No floor with a structural wall on one side shall exceed 36m² (see AD-A Diagram 5).	A 2C14
• The imposed loads on roofs, floors and ceilings shall not exceed those shown in AD-A Table 4.	A 2C15

4.2.1.3 *Types of foundations*

The foundation of any building serves two main purposes. The first is to evenly distribute the weight from load-bearing walls to the soil or bedrock beneath the building and the second is to keep groundwater or soil moisture out.

There are many different types of foundation – for example shallow (strip, raft or mat foundation) or deep (pile or drilled foundation) – but the main ones covered in the Building Regulations primarily deal with plain concrete or strip foundations.

4.2.1.3.1 FOUNDATIONS – PLAIN CONCRETE

* Non-engineered fill (see BRE Digest 427) should be avoided.   A 2E1a
* There should be no wide variation in ground conditions within the loaded area.

* The ground below the foundations should ***not*** be weaker or more compressible as it could impair the stability of the structure.   A 2E1b
* The foundations should be situated centrally under the wall.   A 2E2a
* If the soil is non-aggressive, the concrete should be composed of Portland cement (BS EN 197-1 and 197-2) and a fine, coarse aggregate (conforming to BS EN 12620).   A 2E2b

The mix should be either:
* 50kg of Portland cement to not more than 200kg (0.1m³) of fine aggregate and 400kg (0.2m³) of coarse aggregate; or
* Grade ST2 or Grade GEN I concrete to BS 8500-2.

* For foundations in chemically aggressive soil conditions, guidance in BS 8500-1: Part 1 (and BRE Special Digest 1) should be followed.   A 2E2b
* The ***minimum*** thickness of a concrete foundation should be 150mm or P, whichever is the greater. (Where P is derived using AD-A Diagram 23 and AD-A Table 10.)   A 2E2c

Note: The depth might need to be increased in areas subject to long periods of frost or in order to transfer the loading onto satisfactory ground.

For foundations of piers, buttresses and chimneys, the projection X (shown in AD-A Diagram 22) should never be less than the value of P where there is no local thickening of the wall.

• Stepped foundations should overlap the step, by:
    • twice the height of the step;
    • the thickness of the foundation;
    • or 300mm;
 whichever is greater (see AD-A Diagram 21).
	A 2E2d

• Trench fill foundations may be used as an acceptable alternative to strip foundations. • The overlap for trench fill foundations should be twice the height of the step, or 1m – whichever is greater.	A 2E2c A 2E2d A 2E2f

Note: Foundations for piers, buttresses and chimneys should project as shown in AD-A Diagram 22 which indicates that the projection X should never be less than the value of P where there is no local thickening of the wall.

4.2.1.3.2 STRIP FOUNDATIONS

• The recommended minimum widths of strip foundations shall be as indicated in AD-A Table 10.	A 2E3
• Except where strip foundations are founded on rock, the strip foundations should have a minimum depth of 0.45m to their underside to avoid the action of frost.	A 2E4
• In clay soils that are subject to volume change on drying (i.e. 'shrinkable clays'), strip foundations should be taken to a depth where anticipated ground movements (caused by vegetation and trees on the ground) will not weaken the stability of any part of the building.	
• The depth to the underside of foundations on clay soils should not be less than 1.0m on high-shrinkage clay soils.	

Table 4.1 Types of sub-soil

Type	Applicable field test
Rock	Requires at least a pneumatic or other mechanically operated pick for excavation.
Compact gravel and/or sand	Requires a pick for excavation.
Stiff clay or sandy clay	Can be indented slightly by thumb.
Firm clay or sandy clay	Thumb makes impression easily.
Loose sand, silty sand or clayey sand	Can be excavated with a spade.
Soft silt, clay, sandy clay or silty clay	Finger can (perhaps with a little difficulty) be pushed up to 10mm.
Very soft silt, clay, sandy clay or silty clay	Finger pushed up to 10mm.

4.2.1.4 Foundations and types of soil

Table 4.1 provides guidance on determining the type of soil on which it is intended to lay a foundation.

4.2.1.5 Contaminated ground

Building sites that have previously been used as a factory, mine, steel mill or landfill, etc., are quite likely to contain possible contaminants. These need to be identified at an early stage (see AD-C Table 3).

• The underlying geology of a potential site has to be considered, as natural contaminants may be present – such as:	C 2.3 C 2.4
• naturally occurring heavy metals (e.g. cadmium and arsenic) originating in mining areas;	
• gasses (e.g. methane and carbon dioxide) originating in coal-mining areas;	C 2.5
• organic-rich soils and sediments (e.g. peat and river silts);	
• radioactive radon gas – in certain parts of the country;	
• sulphate attack from concrete floor slabs and oversite concrete also have to be considered.	

4.2.1.6 Corrective measures

- When building work is undertaken on sites affected by contaminants and which have control measures already in place, care must be taken not to compromise these measures.   C 2.15
  - If the risks posed by the gas are unacceptable then these should be managed by appropriate remedial measures.   C 2.36
  - Site-wide gas-control measures may be required (if the risks are deemed unacceptable on land associated with it).   C 2.37
  - The design and layout of buildings should maximise the availability of natural ventilation.   C 2.38
- Continued maintenance and calibration of mechanical (as opposed to passive) gas-control systems is required.
- Sub-floor systems should be carefully designed.

Depending on the contaminant, three generic types of corrective measure can be considered: treatment, containment and/or removal.

4.2.1.7 Flood resistance

Although flood resistance is **not** currently a requirement of the Building Regulations, as part of its aim to improve the energy efficiency of buildings and protect the environment, one of the policies set out in the Government's revised 2019 *National Planning Policy Framework* proposes that Local Planning Authorities should steer new development to areas with the lowest risk of flooding (see https://www.gov.uk/guidance/national-planning-policy-framework/10-meeting-the-challenge-of-climate-change-flooding-and-coastal-change paragraphs 155-156).

4.2.1.8 Gaseous contaminants – radon

Radon is a colourless, odourless, naturally occurring radioactive gas formed by decaying uranium that happens naturally in all rocks and soils. Normally the gas escapes from rock or soil and is immediately diluted by the atmosphere and thus poses no harm to humans. However, when radon is trapped in an enclosed space, it can seep out of the ground and build up in houses, buildings and indoor spaces, and precautions against radon may be necessary.

- All new buildings, extensions and conversions   C 2.39
  (whether residential or non-domestic), which are built
  in areas where there may be high radon emissions,
  may need to incorporate precautions against radon.

4.2.1.9 Gaseous contaminants – landfill gas

Landfill gas generally consists of methane and carbon dioxide together with small quantities of volatile organic compounds (VOCs), which give the gas its characteristic odour. Landfill gas can migrate under pressure through the sub-soil and through cracks and fissures into buildings.

- Methane is an asphyxiant which will burn and can   C 2.27
  explode in air; whilst
- carbon dioxide is non-flammable and toxic.

Note: Many of the other components of landfill gas are flammable and some are toxic: all will require careful analysis.

4.2.1.10 Gaseous risk assessment

- A risk assessment should be completed for methane   C2.28
  and other gases, particularly:
  - on a landfill site or within 250m of the boundary
    of a landfill site;
  - on a site subject to the wide-scale deposition of
    biodegradable substances (including made ground
    or fill);
  - on a site that has been subject to a use that could
    give rise to petrol, oil or solvent spillages;
  - in an area subject to naturally occurring methane,
    carbon dioxide and other hazardous gases.
- During a site investigation for methane and other   C 2.30
  gases:
  - measurements should be taken over a sufficiently
    long period of time to fully characterise gas
    emissions;
  - measurements should also include periods when
    gas emissions are likely to be higher.

- Gas risks (i.e. to human receptors) should be considered for: C2.32
  - gas entering the dwelling through the substructure (and building up to hazardous levels);
  - subsequent householder exposure in garden areas including outbuildings, extensions and garden features.

- When land affected by contaminants is being developed and 'receptors' (i.e. buildings, building materials, building services and people) are introduced onto the site, it is necessary to break the pollutant linkages. This can be achieved by: C 2.7
  - treating the or removing the contaminant (e.g. use of physical, chemical or biological processes to eliminate or reduce the contaminant's toxicity or harmful properties);
  - blocking or removing the pathway (e.g. isolating the contaminant beneath protective layers or installing barriers to prevent migration);
  - protecting or removing the receptor (e.g. changing the form or layout of the development and using appropriately designed building materials, etc.).

A risk assessment based on the concept of a 'source–pathway–receptor' relationship, or pollutant linkage of a potential site should be carried out to ensure the safe development of land that is affected by contaminants.

4.2.1.11 Ground investigation

The detailed ground investigation:

- Is likely to involve collection and analysis of soil, soil gas, and surface and groundwater samples by the use of invasive and/or non-invasive techniques.   C 2.13
- Must provide sufficient information to confirm the risk assessment, design and specification of any corrective works.

During the development of land affected by contaminants, the health and safety of both the public and workers should be considered.

4.2.1.12 *Hazards associated with the ground*

- Hazards may include: C 0.4
  - the effects of vegetable matter including tree roots;
  - health hazards associated with chemical and biological contaminants;
  - gas generation from biodegradation of organic matter.
- Hazards to the built environment can be physical, chemical or biological and include:
  - items such as underground storage tanks or foundations that may create hazards to both health and the building;
  - physical hazards such as unstable fill or unsuitable hardcore containing sulphate.

Naturally occurring radioactive gas (e.g. radon) and gases produced by some soils and minerals can also be a potential hazard.

- Problems from rain and moisture include: C 0.6
  - driving rain or wind-driven spray from the sea or other water bodies adjacent to the building penetrating walls or roofs and damaging the structure, internal fittings or equipment;
  - condensation that may cause damage to the structure;
  - moisture rising from the ground to damage floors and the base of walls;
  - surface condensation from the water vapour generated within the building causing moulds to grow;
  - spills and leaks from sanitary fittings or fixed appliances that use water (e.g. bathrooms and kitchens) causing damage to floor decking or other parts of the structure.

- More severe problems can arise in sites that are liable   C 1.6
  to flooding.
- Building services such as below-ground drainage should be sufficiently robust or flexible to accommodate the presence of any tree roots.
- Joints should be made so that roots will not penetrate them.

Where roots could pose a hazard to building services, consideration should be given to their removal.

> • On sites previously used for buildings, consideration   C 1.7
>   should be given to the presence of other infrastructure
>   (such as existing foundations, services and buried
>   tanks, etc.) that could endanger persons in and about
>   the building and/or land associated with the building.
>
> • If the site contains fill or made ground, consideration   C 1.8
>   should be given to its compressibility and its potential
>   to collapse when wet.

Construction activities undertaken as part of building development can alter the gas regime on the site (e.g. a site strip can increase surface gas emissions, as can piling and excavation for foundations, and dynamic compaction which can push dry biodegradable waste into moist, gas-active zones). General excavation work for foundations and services can also alter groundwater flows through the site. Potential problems include:

> • Aggressive substances – including inorganic and   C 2.23
>   organic acids, alkalis, organic solvents and inorganic
>   chemicals such as sulphates and chlorides.
> • Combustible fill – including domestic waste, colliery
>   spoil, coal, plastics, petrol-soaked ground, etc.
> • Expansive slags – e.g. blast furnace and steel-making
>   slag.
> • Floodwater affected by contaminants – e.g.
>   substances in the ground, waste matter or sewage.

4.2.1.13 Site preparation

> • The effects of roots close to the building needs to   C 1.4
>   be assessed and any vegetable matter such as turf
>   and roots should be removed from the ground that
>   is going to be covered by the building, sufficient to
>   prevent later growth.

Where mature trees are present (particularly on sites with shrinkable clays), the potential damage arising from ground-heave to services, floor slabs and oversite concrete should be assessed.

4.2.1.14 Risk assessment

Following a site investigation, a risk assessment may be required. There are three types of assessment:

- Preliminary (once the need for a risk assessment has been identified).
- Generic Quantitative Risk Assessment (GQRA).
- Detailed Quantitative Risk Assessment (DQRA).

Where the site is potentially affected by contaminants, a combined geo-technical and geo-environmental investigation should be considered.

4.2.1.15 Sub-soil drainage

• Where the water table can rise to within 0.25m of the lowest floor of the building, or where surface water could enter or adversely affect the building, either the ground to be covered by the building should be drained by gravity, or other effective means of safeguarding the building should be taken.	C 3.2

• If an active sub-soil drain is cut during excavation and if it passes under the building it should be either:
  • re-laid in pipes with sealed joints and have access points outside the building;
  • re-routed around the building; or
  • re-run to another outfall (see AD-C Diagram 3).
	C 3.3

• Where contaminants are present in the ground, consideration should be given to sub-soil drainage to prevent the transportation of waterborne contaminants to the foundations or into the building or its services.	C 3.7

4.2.1.15.1 TREATMENT

The choice of the most appropriate treatment process for a particular site is a highly site-specific decision and specialist advice should be sought.

4.2.1.15.2 CONTAINMENT

• In-ground, vertical barriers may also be required to control lateral migration of contaminants.	C 2.17

- Cover systems involving the placement of one or   C 2.18
  more layers of materials over the site may be used to:
  - break the pollutant linkage between receptors and
    contaminants;
  - sustain vegetation;
  - improve geotechnical properties;
  - reduce exposure to an acceptable level.

- Imported fill and soil for cover systems should be   C 2.20
  assessed at source to ensure that it is suitable for use.
- The size and design of cover systems (particularly
  soil-based ones used for gardens) should take account
  of their long-term performance.
- Gradual intermixing due to natural effects and
  activities such as burrowing animals or gardening
  should be considered.

4.2.1.15.3 REMOVAL

- Imported fill should be assessed at source to   C 2.21
  ensure that there are no materials that will pose
  unacceptable risks to potential receptors.

Requirements for buildings other than dwellings – Additional requirements

This section provides the details for the foundations of buildings other than dwellings, only where they differ from the requirements already given above.

4.2.2 Structure

The basic requirements for foundations are the same as for dwellings (see above) with the addition of the following requirements.

- As shown in AD-A Diagram 2, in a small single-   A 2C4b
  storey non-residential building:
  - the height (H) should be less than 3m;
  - the width (W) should be less than 9m (the greatest
    length or width of the building).

4.2.3 Site preparation and resistance to contaminants and water

In single-storey industrial and commercial buildings, where there is no need for the structure to be fire resistant, portal frames are often used.

Portal frames are frequently used in buildings other than dwellings if there is no real need for the structure to be fire resistant. However, if the building is near a relevant boundary, the external wall near the boundary may need fire resistance to restrict the spread of fire.

• If a portal frame is made of reinforced concrete, it should be capable of supporting external walls and will not need any additional base protection to resist overturning.	B2 13.15
• Unless there is a sprinkler system in place, the rafter and columns of a portal frame may need fire protection.	

4.3 Ventilation

Ventilation is defined in the Building Regulations as 'The supply and removal of air (by natural and/or mechanical means) to and from a space or spaces in a building'.

Note: Removing 'stale' air and replacing it with 'fresh' air helps to control internal temperatures as well as ensuring a reduced accumulation of moisture, odours and other gases that can build up during occupied periods: however, ventilation systems should **not** prejudice the use or character of a historic or Listed Building.

Reminder: The subsections include details for buildings other than dwellings only where they differ from the requirements for dwellings.

Subsection	Description	Page
Requirements for dwellings		
4.3.1	Ventilation – General requirements	161
4.3.2	The purpose of ventilation	163
4.3.3	Ventilation system performance	163
4.3.4	Control of the ventilation systems	163

Reminder: Free downloads of all the diagrams and tables referenced in the subsections are available on-line at the Governments https://www. gov.uk/government/collections/approved-documents.

Requirements for dwellings

4.3.1 General requirements

In most cases where it is proposed to carry out ventilation work on a building, it will be necessary to notify the work to a Building Control Body (BCB) in advance via a Full Plans Application or an Initial Notice given jointly with the Approved Inspector.

> • Some means of ventilation (natural or mechanical) is required in the common corridors/lobbies to control smoke from a possible fire in a flat.   B(V1) 3.49

Table 4.2 is a handy checklist to confirm if a dwelling's ventilation systems comply with the official requirements (it can also be found in AD-F(V1) Table D1).

Table 4.2 Checklist for ventilation in existing dwellings

Natural ventilation		
What is the total equivalent area of background ventilators currently in the dwelling?		mm²
Does each habitable room satisfy the minimum equivalent area standards in Table 1.7?	Yes	No
Have all background ventilators been left in the open position?	Yes	No
Are fans and background ventilators in the same room at least 0.5m apart?	Yes	No
Are there working intermittent extract fans in all wet rooms?	Yes	No
Is there the correct number of intermittent extract fans to satisfy the standards in Table 1.1?	Yes	No
Does the location of fans satisfy the standards in paragraph 1.20?	Yes	No
Do all automatic controls have a manual override?	Yes	No
Does each room have a system for purge ventilation (e.g. windows)?	Yes	No
Do the openings in the rooms satisfy the minimum opening area standards in Table 1.4?	Yes	No
Do all internal doors have sufficient undercut to allow air transfer between rooms as detailed in paragraph 1.25 (i.e. 10mm above the floor finish or 20mm above the floor surface)?	Yes	No

(continued)

Table 4.2 (Continued)

Continuous mechanical extract ventilation

Does the system have a central extract fan, individual room extract fans, or both?	Yes	No
Does the total combined continuous rate of mechanical extract ventilation satisfy the standards in Table 1.3?	Yes	No
Does each minimum mechanical extract ventilation high rate satisfy the standards in Table 1.2?	Yes	No
Is it certain that there are no background ventilators in wet rooms?	Yes	No
Do all habitable rooms have a minimum equivalent area of 5,000mm^2?	Yes	No
Does each room have a system for purge ventilation (e.g. windows)?	Yes	No
Do the openings in the rooms satisfy the minimum opening area standards in Table 1.4?	Yes	No
Do all internal doors have sufficient undercut to allow air transfer between rooms as detailed in paragraph 1.25 (i.e. 10mm above the floor finish or 20mm above the floor surface)?	Yes	No

---

Mechanical ventilation with heat recovery

Does each habitable room have mechanical supply ventilation?	Yes	No
Does the total continuous rate of mechanical ventilation with heat recovery satisfy the standards in Table 1.3?	Yes	No
Does each minimum mechanical extract ventilation high rate satisfy the standards in Table 1.2?	Yes	No
Have all background ventilators been removed or sealed shut?	Yes	No
Does each room have a system for purge ventilation (e.g. windows)?		
Do the openings in the rooms satisfy the minimum opening area standards in Table 1.4?	Yes	No
Do all internal doors have sufficient undercut to allow air transfer between rooms as detailed in paragraph 1.25 (i.e. 10mm above the floor finish or 20mm above the floor surface?	Yes	No

4.3.2 The purpose of ventilation

In urban areas, buildings are exposed to a large number of pollution sources causing internal contamination which can have a detrimental effect on the buildings' occupants, and so it is very important to ensure that the ventilation system provided is sufficient and, above all, that the air intake cannot be contaminated.

Ventilation, therefore, is required for the:

- Provision of outside air for breathing.
- Dilution and removal of airborne pollutants, including odours.
- Control of excess humidity (arising from water vapour in the indoor air).
- Provision of air for fuel-burning appliances.

Note: Ventilation may also provide a means to control thermal comfort.

4.3.3 Ventilation system performance

The key aim of the requirement is that a ventilation system is provided which is capable of limiting the accumulation of moisture which would otherwise become a hazard to the health of the people in the building. This requirement may be achieved by providing a ventilation system which:

- Extracts water vapour from areas where it is produced in significant quantities (e.g. kitchens and bathrooms).
- Rapidly dilutes pollutants and water vapour produced in habitable rooms, occupiable rooms and sanitary accommodation.
- Makes available over long periods a minimum supply of outdoor air for occupants, and disperses, where necessary, residual pollutants and water vapour.

4.3.4 Control of the ventilation systems

Ventilation should be controllable so that it can maintain reasonable indoor air quality and avoid waste of energy. These controls can be either **manual** or **automatic**.

4.3.4.1 Control of ventilation intakes and exhausts

• Ventilation intakes should be located away from the direct impact of the sources of local pollution.	F(V1) 2.2–2.3
• Ventilation intakes should not be located in courtyards or enclosed urban spaces.	F(V1) 2.4
• Air intakes should point in the opposite direction to the exhaust outlets.	F(V1) 2.5

• Exhaust outlets should:
  • be located so that exhaust air does not re-enter the building and nor does it have a harmful effect on the surrounding area;
  • be downwind of intakes;
  • not discharge into courtyards, enclosures or architectural screens.
	F(V1) 2.7–2.9

4.3.5 Ventilation ducts

Ventilation ducts or vents that are installed to supply air to combustion appliances should not penetrate any airtight membranes in the floors in a way that might render them ineffective.

• Ventilation ducts supplying or extracting air directly to or from a protected stairway should **not** serve any other areas.	B(V1) 2.9
• Ventilation ducts serving an enclosure should not serve other areas.	B(V1) 3.23c
• If a ventilation duct passes through a compartment wall or compartment floor (or is built into a compartment wall), each wall of the duct should have a fire resistance of at least half that of the wall or floor (see AD-B(V1) Diagram 9.2).	B(V1) 9.5
• Openings in compartment walls or floors should be limited to those for pipes ventilation ducts and chimneys.	B(V1) 7.20

Note: Ductwork in mechanical ventilation systems should not transfer fire and smoke through the building.

Rigid ducts should be used wherever possible.	F(V1) 1.76

• Flexible ductwork should:
  • only be used for final connections;
  • not be any longer than 1.5m;
  • be installed so that the full internal diameter is maintained and flow resistance is minimised;
  • be pulled taut so as to avoid peaks and troughs;
  • be both mechanically secured and adequately sealed to prevent leaks.
	F(V1) 1.77 &
	1.78
	
	F(V1) 1.81

4.3.5.1 Ventilation ducts and flues passing through fire-separating elements

The ventilation of heat and smoke from a fire in a basement is one of the provisions required to safeguard the health and safety of people (and of importance, firefighters) in and around the building.

• Under fire conditions, ventilation and air-conditioning systems should be compatible with smoke control.	B(V1) 9.11
• Only services associated with the firefighting shaft, such as ventilation systems and the lighting of firefighting shafts, should pass through, or be contained within, the firefighting shaft.	B(V1) 15.8
• The walls of a flue or ventilation duct should have a fire resistance of at least half of any compartment wall or floor that it is built into, or passes through (see AD-B(V1) Diagram 9.5).	B(V1) 9.23

4.3.6 Types of ventilation

Buildings may be ventilated through a combination of infiltration and purpose-provided ventilation which can be delivered by either a natural ventilation system, a mechanical ventilation system or a combination of both (i.e. a 'mixed-mode' or 'hybrid' ventilation system).

The air-flow resistance of all components should be considered when specifying ventilation systems and they should comply with the following standards.

4.3.6.1 Background ventilators

• All rooms with external walls (except wet rooms) should have background ventilators:	
• to avoid draughts (e.g. typically 1.7m above floor level), with a minimum total area as shown in AD-F(V1) Table 1.7;	F(V1) 1.34
• if fans and background ventilators are fitted in the same room, they should be at least 500mm apart;	F(V1) 1.54
• with at least 5,000mm^2 equivalent area in each habitable room and 2,500mm^2 equivalent area in each wet room;	
• background ventilators with automatic controls should *also* have manual override.	F(V1) 1.37
• Background ventilators should:	F(V1) 1.64
• not be in wet rooms;	
• provide a minimum equivalent area of 4,000mm^2 for each habitable room;	
• provide a minimum total number of ventilators that is the same as the number of bedrooms *plus* two additional ventilators.	

To ensure good transfer of air throughout the dwelling, there should be an undercut of minimum area 7,600mm^2 in all internal doors above the floor finish.

4.3.6.2 Mechanical ventilation

• A mechanical ventilation system which serves both the stairway and other areas should be designed to shut down on the detection of smoke within the system.	B(V1) 2.9.d B(V1) 3.23d
• In mixed-use buildings, non-domestic kitchens should have a separate and independent extraction system. *Extracted air should not be recirculated*.	B(V1) 9.10
• Thermally activated fire dampers and automatically activated fire and smoke dampers should not be used for extract ductwork serving kitchens.	B(V1) 9.15
• Mechanical ventilation systems must be commissioned in accordance with an approved procedure as laid out in AD-F(V1) Appendix C.	F(V1) 1.83

4.3.6.2.1 CONTINUOUS MECHANICAL EXTRACT

- A continuous mechanical extract system may consist   F(V1) 1.60
  of either a central extract system or individual room
  fans, or a combination of both.

Note: Where ducts, etc., are provided in a dwelling with a protected
stairway, precautions may be necessary to avoid the possibility of the
system allowing smoke or fire to spread into the stairway.

4.3.6.2.2 MECHANICAL INTERMITTENT EXTRACT

- Ventilators equipped with intermittent extract
  shall be capable of being operated manually and/
  or automatically by a sensor (e.g. humidity sensor,
  occupancy/usage sensor, or a moisture/pollutant
  release detector, etc.).
- Automatic ventilator controls must be provided with
  a manual override to allow the occupant to turn the
  extract on.
- Automatic controls for ventilators used in kitchens
  must be capable of providing sufficient flow during
  cooking with fossil fuels so as to avoid the build-up of
  combustion products.
- If a fan is installed in an internal room ***without*** an
  openable window, then the fan should have a
  15-minute overrun.
- In rooms with ***no*** natural light, fans could be
  controlled by the operation of the main room light
  switch.

In dwellings, humidistat controls should be available to regulate the
humidity of the indoor air and hence minimise the risk of condensation
and mould growth. Ideally, they are best installed as part of an extract
ventilation system in moisture-generating rooms such as a kitchen or a
bathroom.

Controls based on humidity sensors should **not** be used for intermittent
extract in sanitary accommodation where odour is the main pollutant.

• In kitchens, any automatic control must provide sufficient flow during cooking with fossil fuels to avoid build-up of combustion products.	F(V1) Table 5.2c
• Where manual controls are provided, they should be within reasonable reach of the occupants.	F(V1) 0.20 Table 5.2c
• Any system of mechanical ventilation which recirculates air and which serves both a stairway and other areas should be designed to shut down on the detection of smoke within the system.	B(V1) 2.9d
• A ducted warm-air heating system's room thermostat should be mounted on the living room wall 1,370mm to 1,830mm above the floor.	B(V1) 3.23e

4.3.6.2.3 MECHANICAL VENTILATION AND AIR-CONDITIONING SYSTEMS

Exhaust points should be sited away from final exits, combustible cladding or roofing materials, and openings into the building.

• A separate ventilation system should be provided for each protected stairway.	B(V1) 9.7
• Ventilation ducts supplying or extracting air directly to, or from, a protected stairway should **not** serve other areas.	
• In a system that recirculates air, smoke detectors should be fitted in the extract ductwork. When smoke is detected, detectors should either cause the system to immediately shut down or divert smoke to outside the building.	B(V1) 9.11

Note: All ventilation and air-conditioning systems should be compatible with smoke-control systems under fire conditions.

4.3.6.3 *Passive stack ventilation*

Passive stack ventilation (PSV) is effectively a natural ventilation system which uses a combination of cross ventilation, buoyancy (warm air rising) and the venturi effect (i.e. wind passing over the terminals causing suction).

- If and where used, passive stack ventilation terminals   F2 1.30b
  should be located in the ceiling.

Placing the outlet terminal at the ridge of the roof is the preferred option, as it is then not prone to wind gusts and/or certain wind directions.

 Precautions must be taken where ducting passes through a fire-resisting wall, floor or fire compartment: openings in compartment walls or floors should be limited to pipes containing service cables or flue pipes – or being used as ventilation ducts and chimneys.

 Note: Instead of a PSV, an open-flued appliance may provide sufficient extract ventilation.

4.3.6.4 Purge ventilation

 Note: Purge ventilation may be provided by natural means (e.g. an openable window) or mechanical means (e.g. a fan).

- Purge ventilation is required in each habitable room   F(V1)
  and should be capable of extracting a minimum of   1.26-1.30
  four air changes per hour (4 ach) per room directly
  to outside. (Normally, openable windows can provide
  this function). (See AD-F(V1) Table 1.4 concerning
  purge ventilation openings.)

4.3.6.5 Trickle ventilators

Trickle ventilators are devices that enable fresh air to circulate naturally through a room and allow polluted air out. They should either be fitted to a window frame or, if they are fitted between the glass and the frame, they should be temporarily sealed rather than just closed.

 Note: Manually controlled trickle ventilators can be located over the window frames, in window frames, just above the glass or directly through the wall. They are normally left open in occupied rooms.

4.3.6.6 Ventilation ducts

Ventilation ducts supplying or extracting air directly to or from a protected stairway should not also serve other areas.

> * If a ventilation duct passes through a compartment   B(V1) 9.2
>   wall or compartment floor (or is built into a
>   compartment wall), then each wall of the duct should
>   have a fire resistance of at least half that of the wall
>   or floor. (See AD-B(V1) Diagram 9.2 and Table 9.1.)
> * A protected shaft containing a protected stairway   B(V1) 7.26
>   and/or a lift should not also contain a ventilating
>   duct unless:
>   * ducts have been provided in order to keep it smoke
>   free; and
>   * ducts are only used to ventilate the protected
>   stairway.

4.3.6.7 Refuse chute

Rooms being used to store refuse or containing a refuse chute should either be accessed directly from the open air or through a protected lobby with a minimum of $0.2m^2$ of permanent ventilation.

4.3.6.8 Transfer grilles

Transfer grilles should not be fitted in any wall of an individual dwelling's stair enclosure if the floor is more than 4.5m above ground level and the dwelling already has an air-circulation system.

4.3.6.9 Protection from falling, collision and impact

> * If parts of ventilators project outside (or inside) a   K 6.1
>   building, then a barrier (as shown in AD-K Diagram
>   6.1) should be used.

- Controls to operated windows, skylights and   K 8.1
  ventilators should be provided and these can be one
  of the following:
  - controls as shown in AD-K Diagram 8.1;
  - manual or electrical controls that are within safe
    reach of a permanent stable surface.

4.3.7 *Fire safety*

4.3.7.1 *Potentially fire hazardous locations*

- Where a stair serves an enclosed car park or place of   B(V1) 3.75
  special fire hazard (such as a boiler room), the lobby or
  corridor should either have permanent ventilation or
  be protected by a mechanical smoke-control system.

4.3.7.2 *Fire escape routes*

In dwellings with three or more storeys and blocks of flats, PSV ducts
should not impede fire escape routes.

- Escape routes should have a minimum clear   B(V1) 3.38
  headroom of 2m.
- If the mains-electricity power supply fails, escape   B(V1) 3.41
  lighting should illuminate the route (including
  external escape routes).
- The maximum escape route travel distance from flat   B(V1) 3.32
  entrance door to a common stair should be less than:
  - 7.5m in one direction;
  - 30m in more than one direction.
- Escape route floor finishes should minimise their   B(V1) 3.39
  slipperiness when wet. (Finishes include the treads of
  steps and surfaces of ramps and landings, and any
  sloping floor or tier should have a pitch of not more
  than 35°.)
- Common corridors should be protected corridors.   B(V1) 3.34

- A common corridor connecting two or more storey exits should be divided with a fire doorset fitted with a self-closing device as shown in AD-B(V1) Diagram 3.8.   B(V1) 3.35
- All associated screens should be fire resisting.
- Doors should be fitted so that smoke does not affect access to more than one stair.

- A fire doorset fitted with a self-closing device should separate the dead-end portion of a common corridor from the rest of the corridor.   B(V1) 3.36

Note: Doors should be sited so that smoke does not affect access to more than one stair.

- The wall between each flat and the corridor should be a compartment wall.   B(V1) 3.34

In divided corridors with cavities, fire-stopping systems should be provided to prevent alternative escape routes being affected by fire and/or smoke.

4.3.7.2.1 SMOKE CONTROL FOR COMMON ESCAPE ROUTES

- Ventilation in common corridors and lobbies can be natural or mechanical.   B(V1) 3.49

4.3.7.2.2 FIREFIGHTING FACILITIES

- Floor identification signs should be located on every landing of a protected stairway and every protected corridor/lobby (or open-access balcony) into which a firefighting lift opens.   B(V1) 15.14

4.3.8 Control of smoke

4.3.8.1 Smoke detectors

• In single-stair buildings, smoke vents on the storey where the fire started should be activated by smoke detectors in the common parts.	B(V1) 3.53
• In buildings with more than one stair, smoke vents may be activated manually.	

4.3.8.2 Smoke vents

• Smoke vents should be provided next to each stair in corridors and lobbies of all buildings that have more than one stair.	B(V1) 3.50

Note: The vent should be positioned as high as the top of the door to the stair.

• If smoke is detected in the common corridor or lobby of a building, then:
   • vents at the top of the smoke shaft and to the stairs of the storey affected should simultaneously open; whilst
   • the vents on all other storeys should remain closed.
	B(V1) 3.51

• Smoke vents should be a minimum of 1m² from the corridor or lobby into the shaft.	B(V1) 3.52
• Corridors and/or lobbies next to a stair should have a smoke vent positioned as high as possible.	B(V1) 3.50

Note: Away from the fire, vents from the corridors or lobbies on all other storeys should remain closed, even if smoke is subsequently detected on storeys other than where the fire is located.

• Smoke vents (with a minimum free area of 1.5m²) should be located on an external wall and should discharge into a vertical smoke shaft that is closed at the base but open at roof level.	B(V1) 3.51

Note: The vent should be positioned as high as the top of the door to the stair.

- A vent to the outside with a minimum free area of $1m^2$ B(V1) 3.52
  should be provided from the top storey of the stair.

- In single-stair buildings, the smoke vents on the storey   B(V1) 3.53
  where the fire is initiated and the vent at the head
  of the stair should be activated by smoke detectors
  in the common parts. Whereas, in buildings with
  more than one stair, smoke vents may be activated
  manually and the smoke detection is not required for
  ventilation purposes in this instance.

 For more details concerning these openings (for example, the total width
of all openings and recesses in a wall), refer to AD-A Diagram 14; and for
further guidance on the design of smoke-control systems using pressure
differentials, see BS EN 12101-6.

4.3.9 Ventilation operation

Acceptable levels for moisture and pollutants are listed in AD-F1
Appendix B and these tables show ventilation rates which are designed
to meet the performance criteria set out in Appendix B.

4.3.9.1 Structure

- No openings should be provided in walls below   A 2C29
  ground floor except for small holes for services
  and ventilation, etc., which should be limited to a
  maximum area of $0.1m^2$ at not less than 2m centers.

4.3.9.2 Gas-control measures

Gas-control measures for dwellings consist of a gas-resistant barrier
across the whole footprint above an extraction (or ventilation) layer from
which gases can be dispersed and vented to the atmosphere. They are
normally passive.

- A shaft conveying piped flammable gas should be   B(V1) 7.28
  ventilated directly to the outside air. (See BS 8313
  for further details.)

Note: The free movement of air should not be compromised.

4.3.9.3 Safety

- Air should normally be supplied to each habitable   F(V1)
  room.   1.17–1.32
- The supply of air flow should be distributed in
  proportion to the habitable room volumes.
- Recirculation by the system of moist air from the
  wet rooms to the habitable rooms should be avoided.
- Extraction should be from each wet room.
- Cooker hoods should be 650mm to 750mm above
  the hob surface (or follow the manufacturer's
  instructions).
- Mechanical extract terminals and fans should be
  installed as high as is practical and preferably less
  than 400mm below the ceiling.
- Mechanical supply terminals should be located and
  directed to avoid draughts.
- Where ducts, etc., are provided in a dwelling with a
  protected stairway, precautions may be necessary to
  avoid the possibility of the system allowing smoke or
  fire to spread into the stairway.

- The maximum ('boost') rate should be the greater   F(V1) 1.7
  than the whole building ventilation rate and the
  whole dwelling air extract rate.

- The minimum air supply rates for purge ventilation   F(V1)
  should be at least four air changes per hour.   1.27 and
  Performance based ventilation is designed to control   Appendix B
  moisture levels, indoor air pollutants and bio
  effluents.

Extract terminals located on the prevailing windward façade should be
protected against the effects of wind by using ducting to another façade,
using a constant-volume flow rate unit or a central extract system.

- All fans should operate quietly at their minimum   F(V1) 4.11
  (i.e. normal) rate so as not to disturb the occupants
  of the building.

- Ventilation devices designed to work continuously: F(V1) 1.22
  - shall be set up to operate without occupant
    intervention;
  - may have automatic controls such as humidity
    control, occupancy/usage sensor, moisture/
    pollutant release detector, etc.;
  - may have a manual control to select maximum
    'boost'.

- Automatic controls for ventilators that are designed   F(V1) 1.22
  to work continuously in kitchens must be capable
  of providing sufficient flow during cooking with
  fossil fuels so as to avoid the build-up of combustion
  products.

4.3.9.3.1 COMMISSIONING

- Fixed mechanical ventilation systems (when capable   F(V1)
  of being tested and adjusted) shall be commissioned   4.1–4.3
  and a Commissioning Notice given to the BCB.
- For mechanical ventilation systems installed in new
  dwellings, the air-flow rates shall be measured on-
  site and a notice given to the BCB. This shall apply
  to intermittently used extract fans and cooker hoods,
  as well as continuously running systems.

The owner needs to be given sufficient information about the ventilation
system and its maintenance requirements so that the ventilation system
can be operated to provide adequate air flow.

4.3.9.3.2 VENTILATION OF ROOMS CONTAINING OPENABLE
  WINDOWS

- Manually controlled trickle ventilators can be located   F(V1) 1.15
  over the window frames, in window frames, just
  above the glass or directly through the wall.

A window with a night latch position is not recommended due to the likelihood of draughts and the potential increased security risk in some locations.

4.3.9.4 Stairs

Note: In buildings with more than one stair, smoke vents may be activated manually. Smoke detection is not required for ventilation purposes in this instance.

• In single-stair buildings:	
• electricity meters should be installed in securely locked cupboards;	B(V1) 3.79
• these cupboards should be separated from the escape route by fire-resisting construction;	B(V1) 3.53
• the corridor or lobby next to each stair should have a smoke vent.	B(V1) 3.50

4.3.9.5 Work on existing buildings – windows

Under the Building Regulations, **windows are a controlled fitting**. The following clauses, therefore, make it mandatory that, when windows in an existing building are replaced, the replacement work shall comply with the following requirements.

If a unit contains both living accommodation and space to be used for commercial purposes (e.g. a workshop or office), the whole unit should be treated for dwelling replacement windows.

• All replacement windows should include trickle ventilators or have an equivalent background ventilation opening in the same room.	F(V1) 3.14
• Ventilation openings should not be smaller than the original opening and should be controllable.	
• The opening part of hinged or pivot windows that are designed to open more than 30° (and/or sliding sash windows) should be at least 1/20 of the floor area of the room.	F(V1) Table 1.4

If the opening is designed to open less than 30°, then it should be at least 1/10 of the floor area of the room.

4.3.9.6 Maintenance

• There should be reasonable access to ventilation systems to enable changing of filters, replacing of defective components, cleaning of ductwork and other maintenance activities.	F(V1) 1.8

4.3.9.7 Refurbishing a kitchen or bathroom in an existing dwelling

• If any of the work being carried out in the kitchen or bathroom of an existing building is classified as 'building work', you should retain (always assuming that it is in good working order!) or replace an existing extract fan or passive stack ventilator.	F(V1) 7.23

Note: This has now become a **mandatory** requirement!

4.3.9.8 The addition of a conservatory to an existing building

If a conservatory is added to an existing habitable room which does not have windows opening to outside, then the habitable room may be ventilated through the conservatory provided that there is background ventilation of at least 8,000mm² equivalent area between the habitable room and the conservatory and between the conservatory and outside.

4.3.9.9 The addition of a habitable room

• If an additional room is connected to an existing habitable room which still has windows opening to outside, but with: • a total background ventilator equivalent area **less than** 5,000mm² equivalent area, then the ventilation opening (or openings) shall be greater than 8,000mm² equivalent area; • a total background ventilator equivalent area of **at least** 5,000mm² equivalent area, then there should be background ventilators of at least 8,000mm² equivalent area between the two rooms and between the additional room and outside.	F(V1) 3.17–3.20

4.3.9.10 Installation of fans

Ensure that the free area of the grill opening of a room terminal extract grille and/or discharge terminal has a minimum of 85% of the free area of the ducting being used.

• Intermittent extract fans may be operated manually and/or automatically by a sensor (e.g. humidity, occupancy/usage, pollutant release).	F(V1) Appendix B
• Humidity controls should not be used for sanitary accommodation, as odour is the main pollutant.	
• In kitchens, any automatic control must provide sufficient flow during cooking with fossil fuels to avoid build-up of combustion products.	
• Any automatic control should have a manual override to allow the occupant to turn the extract on.	
• In a room with no openable window (i.e. an internal room) an intermittent extract fan should have a 15-minute overrun.	

In rooms with no natural light, the fans could be controlled by the operation of the main room light switch.

• Minimum extract air-flow rates for intermittent extract fans should be greater than those shown in AD-F Table 5.2b.	F(V1) App A
• Background ventilators may be either manually adjustable or automatically controlled.	F(V1) 0.20, 4.12 & 4.16
• Manual controls should be within reasonable reach of the occupants.	
• Fans should be quiet so as not to discourage their use by occupants.	
• Background ventilators for dwellings with a single exposed façade should be located at both high (typically 1.7m above floor level) and low (i.e. at least 1.0m below the high ventilators) positions in the façade.	
• Dwellings with only a single exposed façade should be designed so that the habitable rooms are on the exposed façade in order to achieve cross-ventilation.	F(V1) 1.53
• Background ventilators should be at least 0.5m from an extract fan so as to avoid draughts.	

- Controllable background ventilators with a minimum equivalent area of 2,500mm² shall be fitted in each room (except wet rooms from which air is extracted).   F(V1) Table 1.7

- Windows with night latches should **not** be used, as they are more liable to draughts as well as being a potential security risk.

4.3.9.10.1 NOISE GENERATED BY VENTILATION FANS

Noise generated by ventilation fans (which may travel through ducts) and noise from a fan unit may disturb the occupants of the building and so discourage their use in some circumstances.

- Continuous and intermittent mechanical ventilation systems should be designed and installed to minimise noise.   F(V1) 1.5

- Fan units should be appropriately sized for the room they serve.   F(V1) 1.6

- The average A-weighted sound pressure level in noise sensitive rooms, such as bedrooms and living rooms, should not exceed 30 dB Laeq,T.   F(V1) 1.7

- In less sensitive rooms, such as kitchens and bathrooms, a higher level would be acceptable (e.g. 45 dB Laeq,T).

Noise from a continuously running mechanical ventilation system on its minimum low rate should not normally exceed the above levels, and should preferably be lower in order to minimise the impact of the ventilation system.

4.3.9.11 *The addition of a wet room to an existing building*

- Internal doors between the wet room and the existing building should have an undercut of at least 10mm above the floor finish for a standard 760mm width door.   F(V1) 3.25–3.29

- Whole building and extract ventilation can be provided by:
  - intermittent extract and a background ventilator; or
  - single room heat recovery ventilator; or
  - passive stack ventilator; or
  - continuous extract fan.

4.3.9.12 Historic buildings

Ventilation systems should **not** introduce a new or increased technical risk, or in any other way prejudice the use or character of the building – particularly historic buildings that are:

- Listed Buildings.
- Buildings in Conservation Areas.
- Buildings which are of architectural and historical interest.
- Buildings of traditional construction with permeable fabric that both absorbs and readily allows the evaporation of moisture.

• In general, new extensions to historic or traditional dwellings should comply with the standards of ventilation as set out in AD-F.	F(V1) 0.7

However, in **all** cases, the overall aim should be:

• To improve ventilation of a historic building without:
  • having a detrimental influence on the character of the building;
  • increasing the risk of long-term deterioration of the building's fabric or fittings.
	F(V1) 0.7

The guidance given by English Heritage (and in BS 7913) should be taken into account when determining appropriate ventilation strategies for building work in historic buildings.

4.3.9.13 New dwellings

4.3.9.13.1 EXTRACT VENTILATION REQUIREMENTS

• All bathrooms and sanitary accommodation, kitchens and utility rooms shall be provided with extract ventilation to the outside, which is capable of operating either intermittently or continuously.	F(V1) 1.17 & 1.18
• The minimum extract air-flow rates for intermittent extract systems are shown in AD-F Tables 1.1 and 1.2. • Minimum extract ventilation rates for continuous operation extract systems are given in AD-F Table 1.11.	F(V1) 1.19

- Extract ventilation terminals and fans, not including cooker extract hoods, should be installed at a maximum of 400mm below the ceiling – but as high as is practicable.

Note: Where a cooker hood extracts to the outside, the height of the extract hood above the hob surface should be between 650 and 750mm.

4.3.9.13.2 VENTILATION RATES

- Purge ventilation is required in each habitable room and should be capable of extracting a minimum of four air changes per hour (4 ach) per room directly to outside.   F(V1) 1.27

Normally, openable windows or doors can provide this function; otherwise, a mechanical extract system should be provided. In other rooms (e.g. kitchens and bathrooms), mechanical or passive stack extract should be sufficient.

- The whole building ventilation rate for habitable rooms in a dwelling should be greater than that shown in AD-F1 Table 1.3.   F(V1) 1.24
- The minimum ventilation rate (based on two occupants in the main bedroom and a single occupant in all other bedrooms) should be not less than 0.3l/s per m^2.
- Internal doors should be undercut (10mm undercut in a 760mm wide door) to allow air to flow throughout the dwelling.   F(V1) 1.25

4.3.9.13.3 KITCHENS

All kitchens and utility rooms shall be provided with intermittent or continuous extract ventilation to the outside, which is capable of operating either:	F(V1) 1.17 & 1.18
• intermittently at a minimum extract rate of 30l/s (adjacent to the hob) and 60l/s elsewhere; or	F(V1) 1.19
• continuously with a minimum extract rate of 13l/s.	

Manual boost controls should also be provided in kitchens to guard against the possibility of a single centrally located switch being left in an incorrect mode of operation.

4.3.9.13.4 BASEMENTS

When ventilating a basement, you should select one of the following ventilation systems:

- Background ventilators and intermittent extract fans.
- Passive stack ventilation (PSV).
- Continuous mechanical extract ventilation (MEV).
- Mechanical ventilation with heat recovery (MVHR).

If the basement has a single exposed façade, while the rest of the dwelling above ground has more than one exposed façade, then PSV or MVHR should be used.

- Habitable rooms without an openable window shall be either ventilated through another habitable room or through a conservatory.   F(V1) 1.38–1.40

- If the basement is not connected to the rest of the dwelling by a large permanent opening, then:
  - the part of the dwelling above ground should be considered separately; and
  - the basement should be treated as a single-storey dwelling, as if it were above ground.

- If the part of the dwelling above ground has no bedrooms, then for the purpose of ventilation requirements:
  - assume that the dwelling has one bedroom; and
  - treat the basement as a single-storey dwelling (with one bedroom) as if it were above ground.

- If a dwelling only comprises a basement, then it should be treated as if it were a single-storey dwelling (with one bedroom) above ground.   F(V1) 1.41

4.3.9.13.5 VENTILATION OF A HABITABLE ROOM THROUGH

ANOTHER ROOM OR A CONSERVATORY

The general ventilation rate for conservatories (and adjoining rooms) with a floor area greater than 30m² can be achieved by the use of background ventilators (e.g. air bricks).

• A habitable room not containing openable windows may be ventilated through another habitable room or a conservatory (see F(V1) Diagram 1.1).	F(V1) 1.42
• A habitable room not containing openable windows may be ventilated through a conservatory (see Diagram 1.1).	F(V1) 1.43
• Openable doors can provide purge ventilation in habitable rooms.	F(V1) 1.43

Note: Approved Document O outlines strategies to reduce overhearing risks and limit solar gains. These may require a larger amount of ventilation than AD F(V1). Where more than one AD applies the most stringent of the regulations should be followed.

4.3.10 Drainage and waste disposal

• The discharge pipe should not be connected to a soil discharge stack unless it is capable of safely withstanding temperatures of the water discharged, in which case, it should contain a mechanical seal which allows water into the branch pipe without letting foul air from the drain being ventilated through the tundish.	H 3.60a

Note: Composting toilets should not be connected to an energy source other than for purposes of ventilation or sustaining the composting process.

4.3.11 Car parks

4.3.11.1 Car parks and shopping complexes

There is a low probability of fire spread between storeys **provided** that the car park and shopping complex is well ventilated by:

- open sides (high level of natural ventilation);
- natural ventilation;
- mechanical ventilation.

• Car parks are not normally expected to be fitted with sprinklers.	B(V1) 16.11
• Car parks and plant rooms should have separate and independent extraction systems and extracted air should not be recirculated.	B(V1) 9.10
• Where there is likely to be leakage or spillage of oil, drainage systems should be provided with oil interceptors.	H3 3.22 H3 App A

4.3.11.2 Basements and enclosed car parks

• In basements and enclosed car parks, the lift should be within the enclosure of a protected stairway.	B(V1) 3.102
• Mechanically ventilated basement car parks shall be capable of at least six air changes per hour (6 ach).	F(V2) 1.39bii

4.3.11.3 Sprinklers

Sprinklers should be provided within the individual flats. However, they do not need to be provided in the common areas such as stairs, corridors or landings when these areas are fire sterile.

4.3.12 Combustion appliances and fuel storage systems

Combustion appliances require ventilation to supply them with air for combustion.

4.3.12.1 Ensuring adequate ventilation

• Ventilation is to ensure the proper operation of flues or, in the case of flueless appliances, to ensure that the products of combustion are safely dispersed to the outside air.	J 1.2
• If an appliance is room sealed but takes its combustion air from another space in the building (e.g. the roof void), that space should have ventilation openings directly to the outside.	J 1.8

4.3.12.2 Extract ventilation

• Avoid installing extract ventilation for solid-fuel appliances in the same room.	J 1.20c
• When checking for spillage in an appliance, ensure all external doors, and other adjustable ventilators to outside, are closed.	J 1.21
• For oil-fired appliances, the effects of fans can be checked and, where spillage or flue draught interference is identified, it may be necessary to add additional ventilation to the room or space.	J 1.23

If mechanical extraction is unavoidable, then seek specialist advice.

4.3.12.3 Flueless appliances

• Spaces which contain flueless appliances may need permanent ventilation and purge ventilation as well as adjustable ventilation and rapid ventilation.	J 1.18

Openable elements installed for the rapid ventilation of rooms, and other provisions made for the rapid ventilation of kitchens, may be acceptable for flueless appliances in those locations.

• For some flueless appliances, it may be necessary to provide permanently open air vents and/or make provision for rapid ventilation.	J 3.15
• A room containing a gas point intended for use with a flueless appliance (e.g. for a cooker) should have the ventilation provision required for the installation of that appliance.	J 3.16

Ways of meeting the requirement when installing flueless cookers (including ovens, grills or hotplates), flueless water heaters and flueless space heaters are shown in **AD-J Diagram 32**.

4.3.12.4 *Permanently open air vents*

• A room containing an open-flued appliance may need permanently open air vents.	J 1.4 & J 1.18

• Permanently open air vents should be:
  • non-adjustable;
  • sized to admit sufficient air for the purpose intended;
  • positioned where they are unlikely to become blocked.
	J 1.10

• Ventilators should be installed so that building occupants are not provoked into sealing them against draughts or noise.	
• Ventilation openings should not be made in fire-resisting walls other than external walls (unless that particular wall shields an LPG tank).	

Unless otherwise advised, you should **not** locate air vents within a fire-place recess.

• Where ventilation is to be provided via a single proprietary assembly, the equivalent area of the ventilator should be taken as that declared by the manufacturer.	J 1.12
• Where two or more components are to be used to provide a non-proprietary assembly, the assembly should be kept as simple and smooth as possible.	J 1.13
• For an airbrick, grille or louvre with apertures no smaller than 5mm, the aggregate free area is shown in AD-J Diagram 9.	J 1.14

4.3.12.5 Ventilation of LPG and oil storage tanks

- Cylinders **shall** be stood upright, secured by straps   J 5.20
  or chains against a wall outside the building in a well-
  ventilated position at ground level, where they are
  readily accessible, reasonably protected from physical
  damage and where they do not obstruct exit routes
  from the building.
- A firm level base for the cylinder valves (such as
  concrete at least 50mm thick or paving slabs bedded
  on mortar) **shall** be provided so that cylinder valves
  will be:
  - at least 1m horizontally and 300mm vertically
    from openings in the building or heat sources such
    as flue terminals and tumble-dryer vents; and
  - at least 2m horizontally from drains without
    traps, unsealed gullies or cellar hatches unless
    an intervening wall not less than 250mm high is
    provided.
- To ensure good ventilation, firewalls should not   J 5.17
  normally be built on more than one side of an LPG
  tank.
- Oil storage tanks within a building should be directly   J Table 10
  ventilated to the outside.

AD-J Diagram 43d provides a very good description of these requirements.

4.3.13 Conservation of fuel and power

Buildings shall conserve fuel and power in buildings by:

- Limiting heat gains and losses through thermal elements, building
  fabric and pipes, ducts and vessels, etc., that are used for space
  heating, space cooling and hot water services.
- Providing fixed building services which are energy efficient, have
  effective controls and have been commissioned correctly.
- Limiting the effects of solar gain in summer by a combination of
  window size and orientation, solar-control measures, ventilation
  (day and night) and high thermal capacity.

 If ventilation is provided using a balanced mechanical system, consider
providing a summer bypass function to use during warm weather.

Requirements for buildings other than dwellings – Additional requirements

In buildings other than dwellings, more sophisticated automatic control systems such as occupancy sensors (using local passive infrared detectors) or indoor carbon dioxide concentration sensors (using electronic carbon dioxide detectors) can be used as an indicator of occupancy level and, therefore, body odour.

Note: This section provides the details for ventilation in buildings other than dwellings, only where they differ from the requirements already given above.

4.3.14 Fire safety

The basic requirements for ventilation are the same as for dwellings (see above), with the addition of the following requirements.

• For most buildings, basic information on the location of fire-protection measures should be provided, showing any ventilation system with a smoke-control function, including mode of operation and control system.	B(V2)19.3

4.3.14.1 Access and facilities for the fire service

• Depending on the size and use of the building (and as most of the firefighting will be carried out within the building), access and facilities for the fire service will need to be provided, including internal fire facilities and ventilation of heat and smoke from a fire in the basement.	B(V2) B5
• The stairs and lobbies within a firefighting shaft should have a means of venting smoke and heat. • Only services associated with the firefighting shaft (e.g. ventilation systems and lighting) should pass through or be contained within the shaft.	B(V2) 17.9
• External access for fire appliances that will be used near the building should also be taken into consideration.	B(V2) B5

It is advisable to obtain advice from the fire and rescue service as early as possible if you plan to deviate from the guidance above.

4.3.15 *Escape routes*

4.3.15.1 *Protected escape route*

• Ventilation ducts supplying or extracting air directly to or from a protected escape route, should not serve other areas.	B(V2) 10.7
• A separate ventilation system should be provided for each protected stairway.	
• A fire and smoke damper should be provided where ductwork enters or leaves each section of a protected escape route.	B(V2) 10.8
• Dampers should be operated by a smoke detector or suitable fire detection system and should close when smoke is detected.	
• Ducts passing through the enclosure of a protected escape route should be fire resisting.	B(V2) 10.15
• Ventilation ducts serving the enclosure should not serve any other areas.	B(V1) 3.23c

4.3.16 *Control of smoke*

4.3.16.1 *Smoke detectors*

• Under fire conditions, ventilation and air-conditioning systems should be compatible with smoke-control systems.	B(V2) 10.11
• An as-built plan of the building should be provided showing any smoke-control systems, or ventilation systems with a smoke-control function, including mode of operation and control systems.	B(V2) 19.3
• Smoke detectors should be fitted in the extract ductwork of a system which recirculates air before the point of separation of the recirculated air, and then discharges that air to the open air, and before any filters or other air-cleaning equipment. Such detector(s) should: • cause the system to immediately shut down on detection of smoke; • switch the ventilation system from recirculating mode to extraction to open air, so as to divert any smoke to the outside of the building.	B(V2) 10.9

4.3.16.2 Smoke control of common escape routes

• A corridor or lobby that is next to a stair should have a smoke vent located as high up as practicable, with the top edge at least as high as the top of the door to the stair.	B(V1) 3.50
• Smoke vents (with a minimum free area of 1.5m²) should be located on an external wall.	B(V1) 3.51
• Smoke vents should discharge into a vertical smoke shaft, that is closed at the base but open at roof level.	
• Smoke vents from the corridors or lobbies on all other storeys should remain closed, even if smoke is subsequently detected on storeys other than where the fire is located.	
• A vent to the outside with a minimum free area of 1m² should be provided from the top storey of the stair.	B(V1) 3.52
• In single-stair buildings, the smoke vents on the storey where the fire is initiated and the vent at the head of the stair should be activated by smoke detectors in the common parts.	B(V1) 3.53
• In buildings with more than one stair, smoke vents may be activated manually. Smoke detection is not required for ventilation purposes in this instance.	

Note: Further guidance on the design of smoke-control systems using pressure differentials is in BS EN 12101-6.

• If escape stairs are protected by a smoke-control system or they are approached on each storey through a protected lobby, then a sprinkler system is not required.	B(V2) 3.15
• A protected lobby should have a minimum 0.4m² of permanent ventilation, or be protected by a mechanical smoke-control system.	B(V2) 3.35
• Under fire conditions, ventilation and air-conditioning systems should be compatible with smoke-control systems.	B(V2) 10.11

4.3.16.3 Venting of heat and smoke from basements

• Smoke outlets should be evenly sited at high level in either the ceiling or wall of basements, and discharge heat and smoke to the open air.	B(V2) 18.5

4.3.17 Flues and ducts

• The walls of a flue or duct that contains an appliance's ventilation duct(s) should have a fire resistance that is at least half of any compartment wall or compartment floor it passes through or is built into.	B(V2) 10.23
• Proprietary, tested fire-stopping and sealing systems may be used. • Different materials suit different situations and not all are suitable in every situation.	B(V2) 10.26
• Joints and openings between fire-separating elements should be fire-stopped. 　• Openings through a fire-resisting element for pipes, ducts, conduits or cable should be as few as possible and as small as practicable.	B(V2) 10.24

4.3.18 Conservation of fuel and power

The basic requirements for ventilation are the same as for dwellings (see above) with the addition of the following requirements.

The approved procedure for pressure testing is given in the Air Tightness Testing and Measurement Association's (ATTMA) publication 'Measuring air permeability of building envelopes'.

4.3.18.1 Controls

• Systems should have controls to enable the achievement of reasonable standards of energy efficiency in use.	L(V2) 6.53

- Ventilation and air-conditioning system controls:
  - should be sub-divided into separate control zones to correspond to each area of the building that has a significantly different solar exposure, or pattern or type of use;
  - should be capable of enabling independent timing and temperature control for each separate control zone and, where appropriate, ventilation and air-recirculation rate.
- Central plant should operate only as and when the zone systems require it. The default condition should be OFF.
- Central mechanical ventilation systems should have both time control (at room level) and on/off time control (at air-handler level).   L(V2) 6.55
- System controls should be wired so that when there is no demand for space heating or hot water, the heating appliance are switched off.   L(V2) 6.54
- Supply temperature control should be provided via a variable set point with outdoor temperature compensation.   L(V2) 6.56

4.3.19 Single-stair buildings

- A single escape route is acceptable from a flat entrance door if either the flat is on a storey served by a single common stair or the flat is in a dead end of a common corridor served by two (or more) common stairs.   B(V1) 3.27

In a small single-stair building, not every flat needs access to an alternative escape route **if**:

- The stair does not also serve ancillary accommodation and if there is a protected lobby or corridor with permanent ventilation or a mechanical smoke-control system between them.
- Each floor level has a high-level openable vent with a minimum free area of $1m^2$.
- The head of the stairs has a single openable vent which is remotely operated from fire and rescue service access level.

4.3.20 Car parks and shopping complexes

There is a low probability of fire spread in car parks and high-rise shopping complexes because the fire load is well defined and the probability of fire spreading from one storey to another in a well-ventilated car park is low.

• Ventilation should be either natural, mechanical or an open-sided, high level of natural ventilation which is independent of other extraction systems.	B(V2) 11.2
• Each storey should have permanent openings (ideally at ceiling level) at each car parking height, and so they will be naturally ventilated.	B(V2) 11.4
• If the minimum amount of natural ventilation is not possible, then mechanical ventilation should be provided. Provided that it is both independent of any other ventilating system and designed to operate at 10 air changes per hour (10 ach) during a fire.	B(V2) 11.5 B(V2) 10.10

Note: For more information on equipment for removing hot smoke, and alternative methods of ventilating smoke, see BS EN 12101-3 and BS 7346-7.

4.3.21 Air-circulation systems in flats with a protected stairway or entrance hall

• The approach to rooms that are used to store refuse or contain a refuse chute come either directly from the open air or through a protected lobby with a minimum $0.2m^2$ of permanent ventilation.	B(V2) 5.44
• Openings may be made in a protected lobby for pipes, ventilation or flue pipe ducts, service cables or chimneys.	B(V2) 8.31b
• A protected shaft conveying piped flammable gas should be ventilated direct to the outside air.	B(V2) 8.38
• Any extension of a storey floor into a protected shaft should not compromise the free movement of air throughout the entire length of the shaft (see BS 8313).	

• A protected shaft that contains a protected stairway and/or a lift should not also contain an oil pipe or a ventilating duct (unless it ventilates or pressurises the stairway).	B(V2) 8.36 & 8.33

4.3.22 Non-domestic kitchens, car parks and plant rooms

• Non-domestic kitchens, car parks and plant rooms should have separate and independent extraction systems, and the extracted air should **not** be recirculated.	B(V2) 10.10

Note: Guidance on the use of mechanical ventilation in a place of assembly is in BS 5588-6.

• Where a pressure differential system is installed, ventilation and air-conditioning systems in the building should be compatible with it when operating under fire conditions.	B(V2) 5.52

BS 5720 provides guidance on the design and installation of mechanical ventilation and air-conditioning plants. Guidance on the provision of smoke detectors in ventilation ductwork can be found in BS 5839-1.

4.3.23 Openings in compartment walls or in compartment floors

Openings in compartment walls or floors should be limited to ventilation ducts, chimneys or pipes which contain service cables or flue pipes – or are being used as a refuse shaft.

• Ventilation ducts may pass through openings in a compartment wall or compartment floor.	B(V2) 8.31

Note: This requirement is provided that each wall of the duct has a fire resistance of at least half that of the wall or floor. Guidance on such shafts is provided in BS 8313.

- Where air-handling ducts pass through fire-   B(V2)
  separating elements, the reliability of those elements   10.12
  should be maintained.
- There are three basic methods of protection:
  - Method 1 – thermally activated fire dampers;
  - Method 2 – fire-resisting enclosures;
  - Method 3 – fire-resisting ductwork;
  - Method 4 – automatically activated fire and smoke
    dampers triggered by smoke detectors.

Note: As the build-up of grease within the duct can adversely affect dampers, these methods should **not** be used for extract ductwork serving kitchens.

4.4 Drainage

Sub-surface building drainage is used to remove underground water in order to keep the area under buildings dry and to protect buildings against damage from humidity or moisture from the surrounding soil. It preserves the value of the building fabric and provides a healthy living environment.

Reminder: The subsections include details for buildings other than dwellings only where they differ from the requirements for dwellings.

Subsection	Description	Page
Requirements for dwellings		
4.4.1	Site preparation and resistance to moisture	197
4.4.2	Sanitation, drainage and waste disposal	201
4.4.3	Wastewater treatment systems	207
4.4.4	Pumping installations	212
4.4.5	Combustion appliances and fuel storage systems	212
4.4.6	Resistance to the passage of sound	213
4.4.7	Power to examine and test drains and sewers	220
4.4.8	Fire safety	222
Requirements for buildings other than dwellings – Additional requirements		
4.4.9	Fire safety	222

Reminder: Free downloads of all the diagrams and tables referenced in the subsections are available on-line at the Governments' Planning Portal: www.gov.uk/government/collections/approved-documents.

Requirements for dwellings

4.4.1 Site preparation and resistance to moisture

4.4.1.1 Flood risk

• When building in flood-prone areas, there is always a possibility of a sewer flooding owing to the backflow or surcharging of sewers or drains. To counteract this, non-return valves and anti-flooding devices should be used.	C 0.8
• Below-ground drainage should be sufficiently robust or flexible to accommodate the presence of any tree roots. Joints should be made so that roots will not penetrate them, and where tree roots could pose a hazard to building services, they should ideally be removed.	C 1.6
• Low-lying buildings or basements should be protected from localised flooding – particularly where foul water drainage also receives rainwater.	C 3.5

Note: A dwelling that is only a basement should be treated as a single-storey dwelling above ground level.

4.4.1.2 Sub-soil drainage

• If the water table is within 0.25m of the lowest floor of the building, or surface water could enter the ground to be covered by the building, it should ideally be drained by gravity.	C 2.24
• If an active sub-soil drain that passes under the building is cut during excavation, it should be: • re-laid in pipes with sealed joints, together with access points outside the building; • re-routed around the building; • re-run to another outfall.	

• Consideration should be given to site drainage if there is a risk that groundwater beneath or around the building could affect the stability and properties of the ground.	C 3.4
• Where contaminants are present in the ground, consideration should be given to sub-soil drainage to prevent the transportation of waterborne contaminants to the foundations.	C 3.7

4.4.1.3 Protection against groundwater

• To prevent water collecting on the ground covering:
  • the top should be entirely above the highest level of the adjoining ground; or
  • on sloping sites, drainage should be installed on the outside of the up-slope side of the building (see AD-C Diagram 6).
	C 4.14

• A suspended concrete floor should include a damp-proof membrane.	C 4.19
• Damp-proof courses, cavity trays and closers should ensure that water drains outwards.	C 5.9

4.4.1.4 Greywater and rainwater tanks

• Greywater and rainwater tanks should:
  • prevent leakage of the contents and ingress of sub-soil water;
  • be ventilated;
  • have an anti-backflow device;
  • be provided with access for emptying and cleaning.
	H2 1.70

4.4.1.5 External walls

• The outer leaf of an external cavity wall should be separated from the inner leaf by a drained air space.	C 5.12
• Cladding for framed external walls should be separated from the insulation or sheathing by a vented and drained cavity with a membrane that is vapour open, but resists the passage of liquid water, on the inside of the cavity (see AD-C Diagram 11).	C 5.17

4.4.1.6 Roofs

- Roofing systems should be designed so that any precipitation (rain, drizzle, snow, hail, sleet) which enters the joints will be drained away without penetrating beyond the back of the roofing system.

C 6.7

4.4.1.7 Discharge to drains

- The discharge pipe should not be connected to a soil discharge stack unless that stack is capable of safely withstanding the temperature of the water being discharged, in which case, it should:
  - contain a mechanical seal which allows water into the branch pipe without allowing foul air from the drain to be ventilated through the tundish;
  - be a separate branch pipe with no sanitary appliances connected to it;
  - if plastic pipes are used as branch pipes, they should be either polybutylene (PB) or cross-linked polyethylene (PE-X).

G 3.60

- Any WC fitted with flushing apparatus should discharge to a properly designed and installed drainage system.

G 4.22

- A urinal fitted with flushing apparatus should discharge through a grating, a trap or mechanical seal and a branch pipe to a discharge stack or a drain.

G 4.23

- A WC fitted with a macerator and pump may be connected to a small-bore drainage system discharging to a discharge stack if:
  - there is also access to a WC discharging directly to a gravity system; and
  - the macerator and pump meet the requirements of BS EN 12050-1 or BS EN 12050-3.

G 4.24

- Any sanitary appliance used for personal washing should discharge through a grating, a trap and a branch discharge pipe to a drainage system.

G 5.9

- A sanitary appliance used for personal washing fitted   G 5.10
  with a macerator and pump may be connected to a
  small-bore drainage system connected to a discharge
  stack if:
  - there is also access to washing facilities discharging
    directly to a gravity system; and
  - the macerator and pump meet the requirements of
    BS EN 12050-2.
- Any sink should discharge through a grating, a trap   G 6.5
  and a branch discharge pipe to a drainage system.

4.4.1.8 Rodent control

If the site has been previously developed, the Local Authority should be
consulted to determine whether any special measures are necessary for
control of rodents. Special measures which may be taken include the
following:

- **Sealed drainage** – pipework access covers to be   H1 2.22a
  located in the inspection chamber instead of an open
  channel.
- **Intercepting traps** – of the locking type that can   H1 2.22b
  be easily removed from the chamber surface and
  securely replaced.
- **Rodent barriers** – including enlarged sections on   H1 2.22c
  discharge stacks to prevent rats from climbing.
- **Flexible downward-facing fins** – in the discharge   H1 2.22d
  stack; or one-way valves in underground drainage.
- **Metal cages on ventilator stack terminals** – to   H1 1.31
  discourage rats from leaving the drainage system.
- **Covers and gratings to gullies** – used to   H1 2.22e
  discourage rats from leaving the system.

4.4.1.8.1 RODENT CONTROL – SPECIAL MEASURES

- During construction, drains and sewers that are left   H1 2.56
  open should be covered when work is not in progress
  to prevent entry by rats.

- Disused drains or sewers less than 1.5m deep that are in open ground should, where practicable, be removed. H1 AppB.18
- Other pipes should be sealed at both ends (and at any point of connection) and grout filled to ensure that rats cannot gain access.

Sub-soil drainage shall be provided if it is required.

4.4.2 Sanitation, drainage and waste disposal

4.4.2.1 Bedding and backfill

Pipes need to be protected from damage, particularly pipes which could be damaged by the weight of backfilling.

- The choice of bedding and backfill depends on the depth at which the pipes are to be laid and the size and strength of the pipes. H1 2.41

Special precautions should be taken to take into account the effects of settlement where pipes run under or near buildings.

- Rigid pipes should be laid in a trench as shown in AD-H1 Diagram 10. H1 2.42
- Flexible pipes shall be supported to limit deformation under load (see AD-H1 Diagram 11). H1 2.44
- Flexible pipes with very little cover shall be protected from damage by a reinforced cover slab with a flexible filler and at least 75mm of granular material between the top of the pipe and the underside of the flexible filler below the slab. H1 2.42– 2.44
- Trenches may be backfilled with concrete to protect nearby foundations. In these cases, a movement joint (as shown in AD-H1 Diagram 12), formed with a compressible board, should be provided at each socket or sleeve joint. H1 2.45

4.4.2.2 Foul drainage

'Foul drainage' includes all underground drains and sewers from buildings to the point of connection to an existing sewer or a cesspool or wastewater treatment system plus any drains or sewers outside the curtilage of the building.

You should make sure pipework is well maintained at all times, as defective pipework is known to harbour rats.

• Foul drainage should be connected to:	
• a public foul or combined sewer; or	H1 2.3
• an existing private sewer that connects with a public sewer; or	H1 2.6
• a wastewater treatment system or cesspool.	H1 2.7
• Combined foul and rainwater sewers shall be designed to surcharge (i.e. if the water level in the manhole rises above the top of the pipe) during heavy rainfall.	H1 2.8
• Basements containing sanitary appliances, where the risk of flooding due to surcharge of the sewer is possible, should either use an anti-flooding valve (if the risk is low) or be pumped.	H1 2.9

Anti-flooding valves should preferably be of the double valve type, and should be suitable for foul water and have a manual closure device.

• For other low-lying sites (i.e. not basements) where the risk is considered low, a gully (at least 75mm below the floor level) can be dug outside the building.	H1 2.11
• Drainage unaffected by surcharge should discharge by gravity (by-passing protective measures).	H1 2.10
• The layout of the drainage system should be kept simple.	H1 2.12

Pipes should (wherever possible) be laid in straight lines. Changes of direction and gradient should be minimised.

• Access points should be provided only if blockages could not be cleared without them.	H1 2.13
• Connections should be made using prefabricated components.	H1 2.15

- Connection of drains to other drains or private or public sewers, and of private sewers to public sewers, should be made obliquely, or in the direction of flow. — H1 2.14

- Repair couplings and packing should be used on connections to existing drains or sewers which involve removal of pipes and insertion of a junction. — H1 2.16

- The system should be ventilated by a flow of air. — H1 2.18

- Ventilating pipes should not finish near openings in buildings. — H1 2.19

- Pipes should be laid to even gradients and any change of gradient should be combined with an access point. — H1 2.20

Drainage serving kitchens in commercial hot food premises should be fitted with a grease separator that complies with BS EN 1825-1.

4.4.2.3 Sanitary pipework

4.4.2.3.1 ACCESS TO AND USE OF BUILDINGS

Where the dwelling provides an accessible bathroom:

- Stack and drainage positions should be clear of access zones. (See AD-M Diagrams 2.6 and 2.7.) — M 2.6 / M 2.7

- Stacks or soil and vent pipes should only be positioned adjacent to a WC where there is no practical alternative and should always be on the wall side of the WC. — M 3.36ii

4.4.2.3.2 DISCHARGE STACKS

- All stacks should discharge to a drain. — H1 1.26
- The bend at the foot of the stack should have as large a radius (i.e. at least 200mm) as possible.

- Discharge stacks should be ventilated. — H1 1.29

- Offsets in the 'wet' portion of a discharge stack should be avoided. — H1 1.27

• Stacks serving urinals should be not less than 50mm.	H1 1.28
• Stacks serving closets with outlets less than 80mm should be not less than 75mm.	
• Stacks serving closets with outlets greater than 80mm should be not less than 100mm.	
• The internal diameter of the stack should be not less than that of the largest trap or branch discharge pipe.	
• Ventilating pipes open to outside air should finish at least 900mm above any opening into the building within 3m and should be fitted with a perforated cover or cage (see AD-H1 Diagram 6), and this should be metal if rodent control is a problem.	H1 1.31
• Air admittance valves complying with BS EN 12380 should be located in areas that have adequate ventilation.	H1 1.33
• Air admittance valves should not be used outside buildings or in dust-laden atmospheres.	
• Rodding points should be provided in discharge stacks.	H1 1.34

4.4.2.3.3 BRANCH DISCHARGE PIPES

• Branch pipes should discharge into either another branch pipe or a main discharge stack.	H1 1.7
• Appliances on the ground floor may discharge to a stub stack or discharge stack, directly to a drain or (if the pipe carries only wastewater) to a gully.	H1 1.8 H1 App A.5
• In a multi-storey building (up to five storeys), a branch pipe should **not** discharge into a stack less than 750mm above the invert of the tail of the bend at the foot of the stack.	H1 App A.5
• If the building has more than five storeys, ground-floor appliances (unless discharging to a gully or drain) should discharge into their own stack.	H1 App A.6
• If the building has more than 20 storeys, ground-floor **and** first-floor appliances should discharge into their own stack, unless discharging to a gully or drain.	

- A branch pipe from a ground-floor closet should only discharge direct to a drain if the depth from the floor to the drain is 1.3m or less (see AD-H1 Diagram 1).   H1 1.9

- A branch pipe should not discharge into a stack in a way which could cause cross-flow into any other branch pipe (see AD-H1 Diagram 2).   H1 1.10

- A branch pipe discharging to a gully should terminate between the grating or sealing plate and the top of the water seal.   H1 1.13

- Condensate drainage from boilers may be connected to sanitary pipework provided that the connection is to an internal stack with a 75mm condensate trap.   H1 1.14
- If an additional trap is provided externally to the boiler to provide the 75mm seal, an air gap should be provided between the boiler and:
  - if the connection is made to a branch pipe, the connection should be made downstream of any sink waste connection;   H1 1.14b
  - all sanitary pipework receiving condensate should be made from materials resistant to a pH value of 6.5 or lower and be installed in accordance with BS 6798.   H1 1.14c

- Pipes serving a single appliance should have at least the same diameter as the appliance trap (seeAD-H1 Table 2).   H1 1.15

- Bends in branch pipes should be avoided if possible.   H1 1.16

- Junctions on branch pipes should be made with a sweep of 25mm radius or at 45°.   H1 1.17

- Branch pipes up to 40mm diameter joining branch pipes of 100mm diameter or greater should connect to the upper part of the pipe wall of the larger branch.   H1 1.18

- A separate ventilating stack is preferred where the numbers of ventilating pipes and their distance to a discharge stack are large (see AD-H1 Table 2 or Diagram 3).   H1 1.19

- If the figures in AD-H1 Table 2 or Diagram 3 are exceeded, the branch pipe should be ventilated by a branch ventilating pipe to external air, to: H1 1.20
  - a ventilating stack (ventilated branch system); or
  - internally, by use of an air admittance valve.

A separate ventilating stack is only preferred where the numbers of sanitary appliances and their distance to a discharge stack are large. Ventilation stacks serving buildings with not more than 10 storeys and containing only dwellings should be at least 32mm diameter.

• Branch ventilation pipes should be connected to the discharge pipe within 750mm of the trap and should connect to the ventilating stack above the highest 'spill over' level.	H1 1.22

The ventilating pipe should have a continuous incline from the discharge pipe to the point of connection to the ventilating stack or stack vent. (See AD-H1 Diagram 4.)

• Branch ventilating pipes which run direct to outside air should finish at least 900mm above any opening into the building nearer than 3m.	H1 1.23
• Branch ventilating pipes to branch pipes serving one appliance should have a minimum diameter of 25mm (32mm where the branch is longer than 15m or has more than five bends).	H1 1.24
• Rodding points should be provided.	H1 1.25 & 1.6

4.4.2.3.4 TRAPS

All points of discharge into the system should be fitted with a trap (e.g. a water seal) to prevent foul air from the system entering the building.

• Traps should retain a minimum seal of 25mm of water.	H1 1.3
• Minimum trap sizes are given in AD-H1 Table 1.	H1 1.4
• Branch discharge pipes should prevent the water seal from being broken by pressure in the system.	H1 1.5
• Traps should be fitted directly over an appliance.	H1 1.6
• Traps should be removable or be fitted with a cleaning eye.	

4.4.2.3.5 NON-NOTIFIABLE WORK

• The replacement of a sanitary convenience is non-notifiable as long as it is replaced with one that uses no more water than the one it replaces, but only where the work does not include any work to:
  • underground drainage;
  • the hot or cold water system or above-ground drainage which could prejudice the health and safety of any person on completion of the work.
	G iii

4.4.3 Wastewater treatment systems

The use of non-mains foul drainage, such as wastewater treatment systems, septic tanks or cesspools, should only be considered where connection to mains drainage is **not** practicable.

• Constructed wetlands discharging to a suitable watercourse may be used to treat septic tank effluent where drainage fields are not practical.	H2 1.10
• Where a foul water drainage system from a building discharges to a septic tank, wastewater treatment system or cesspool, a durable notice shall be affixed in a suitable place in the building containing information on any continuing maintenance required (and particularly what can and cannot be flushed via a toilet!) to avoid risks to health.	H2 (3)

4.4.3.1 Sewers

If you are connected to the public sewer, then the public sewer and any shared sewer pipes or pipes beyond the boundary of the property are the responsibility of the Local Utility.

• Sewers should be laid at an appropriate distance from buildings so as to avoid damage to the foundations.	H4 Table C1
• Manholes and chambers should be located so that they are easily accessible.	
• The last access point on the house drain should be sized to allow man entry.	

- House 'collector' drains serving each property
  should normally discharge into the sewer via a single
  junction or a manhole.
- Sewers should not be laid deeper than necessary.
- Manholes on or near highways or other roads need
  to be of robust construction.
- Sewers should be laid in straight lines in both vertical
  and horizontal alignments.

4.4.3.1.1 BUILDING OVER EXISTING SEWERS

• Where it is proposed to construct a building over or near a drain or sewer shown on any map of sewers, the developer should consult the owner of the drain or sewer.	H4 0.3
• A building constructed over or within 3m of any rising main drain or sewer constructed from brick or masonry (or in poor condition) shall not be constructed in such a position unless special measures are taken.	H4 1.2
• Buildings or extensions should not be constructed over a manhole or inspection chamber or other access fitting on any sewer (serving more than one property).	H4 1.3
• A satisfactory diversionary route should be available so that the drain or sewer could be reconstructed without affecting the building.	H4 1.4
• The length of drain or sewer under a building should not exceed 6m except with the permission of the owners of the drain or sewer.	H4 1.5
• Buildings or extensions should not be constructed over or within 3m of any drain or sewer more than 3m deep, or greater than 225mm in diameter, except with the permission of the owners of the drain or sewer.	H4 1.6
• Where a drain or sewer runs under a building, at least 100mm of granular or other suitable flexible filling should be provided round the pipe.	H4 1.9
• Where a drain or sewer running below a building is less than 2m deep, the foundation should be extended locally so that the drain or sewer passes through the wall.	H4 1.10

• Where the drain or sewer is more than 2m deep to invert and passes beneath the foundations, the foundations should be designed with a lintel spanning over the line of the drain or sewer.	H4 1.12
• A drain trench should not be excavated lower than the foundations of any building nearby.	H4 1.13

4.4.3.2 *Cesspools*

A cesspool (i.e. a watertight tank, installed underground, for the storage of sewage) requires no treatment when a filling rate of 150 litres per person per day is assumed.

• Cesspools should be:	H2
• sited at least 7m from any habitable parts of buildings, preferably downslope;	
• provided with access for emptying and cleaning;	
• inspected fortnightly for overflow, and emptied as required;	
• emptied ideally on a monthly basis by a licensed contractor;	
• covered (with heavy concrete slabs) and ventilated;	
• without openings except for the inlet, access for emptying inspection and ventilation.	
• Cesspools and settlement tanks should:	
• prevent leakage of the contents and ingress of sub-soil water;	H2 1.63
• be sited within 30m of vehicle access;	H2 1.64
• have a capacity below the level of the inlet of at least 18,000 litres (18m³) for two users, increased by 6,800 litres (6.8m³) for each additional user;	H2 1.61
• be constructed in brickwork, concrete or glass-reinforced concrete.	H2 1.65 & 1.66

Brickwork should consist of engineering bricks at least 220mm thick. The mortar should be a mix of 1:3 cement/sand ratio, and *in situ* concrete should be at least 150mm thick of C/25/P mix (see BS 5328).

4.4.3.3 Drainage fields and mounds

Biological treatment takes place naturally in the aerated layers of soil.

• Drainage fields may be used to provide secondary treatment in conjunction with septic tanks, in which case:
   • they may be used where the sub-soil is sufficiently free draining and the site is not prone to flooding or waterlogging at any time of year;
   • they typically consist of a system of sub-surface irrigation pipes which allow the effluent to percolate into the surrounding soil.
	H2 1.4
	 H2 1.5

• Drainage fields should be set out as a continuous loop fed from the inspection chamber (see AD-H2 Diagram 1).	H2 1.44

Constructed wetlands discharging to a suitable watercourse may be used to treat septic tank effluent where drainage fields are not practical.

• Drainage fields (or mounds) serving a wastewater treatment plant or septic tank should be located:
   • at least 10m from any watercourse or permeable drain;
   • at least 50m from the point of abstraction of any groundwater supply;
   • at least 15m from any building;
   • sufficiently far from any other drainage fields, drainage mounds or soakaways so that the overall soakage capacity of the ground is not exceeded.
	H2 1.27

• No water supply pipes or underground services other than those required by the disposal system itself should be located within the disposal area.	H2 1.29
• No access roads, driveways or paved areas should be located within the disposal area.	H2 1.30
• The groundwater table should not rise to within 1m of the invert level of the proposed effluent distribution pipes.	H2 1.33
• An inspection chamber should be installed between the septic tank and the drainage field.	H2 1.43
• Constructed wetlands should **not** be located in the shade of trees or buildings.	H2 1.47

• The drainage field/mound should be checked on a monthly basis.	H2 (A.15)
• Drainage fields: • should only be used when percolation tests indicate average values of Vp of between 12 and 100 and the preliminary site assessment report and trial hole tests have been favourable;	H2 1.38
• should be designed and constructed to ensure aerobic contact between the liquid effluent and the sub-soil;	H2 1.39
• should be constructed using perforated pipe, laid in trenches of a uniform gradient which should be not steeper than 1:200.	H2 1.40
• Pipes should be laid on a 300mm layer of clean shingle or broken stone graded between 20mm and 50mm.	H2 1.41
• Drainage trenches should be from 300mm to 900mm wide, with areas of undisturbed ground 2m wide being maintained.	H2 1.42

The two main designs of constructed wetland system are:

- Horizontal flow systems – continuously fed with wastewater from one end.
- Vertical flow systems – intermittently fed with wastewater from the top flooding the surface followed by a period of rest.

4.4.3.4 Packaged treatment works

Packaged treatment works are used where it is either unacceptable or impractical to install a septic tank and a connection to a mains sewer is impossible. They are ideal for campsites and caravan parks and pubs.

• The discharge from the wastewater treatment plant should be sited at least 10m away from watercourses and any other buildings.	H2 1.54
• Regular maintenance and inspection should be carried out in accordance with the manufacturer's instructions.	H2 AppA.17

4.4.4 Pumping installations

Where gravity drainage is impracticable, or protection against flooding due to surcharge in downstream sewers is required, a pumping installation will be needed.

Package pumping installations suitable for installation within buildings are available, and floor-mounted units may be particularly suited for installing in basements.

> • Where foul water drainage from a building is to   H1 2.39
>   be pumped, the effluent receiving chamber should
>   be sized to contain 24-hour inflow to allow for
>   disruption in service.

The minimum daily discharge of foul drainage should be taken as 150 litres per head per day for domestic use.

> • To minimise the effects of any differential settlement,   H1 2.40
>   pipes should have flexible joints.
> • All joints should remain watertight under working
>   and test conditions.
> • Nothing in the pipes, joints or fittings should project
>   into the pipeline or cause an obstruction.
> • Different metals should be separated by non-metallic
>   materials to prevent electrolytic corrosion.

4.4.5 Combustion appliances and fuel storage systems

4.4.5.1 Location of flues

> • Balanced flues serving gas appliances should not be   J 1.52
>   located closer than:
>   • 300mm below any gutter soil pipe or drainpipe
>   where there is a natural draught;
>   • 75mm below any gutter soil pipe or drainpipe
>   where there is a fanned draught (as shown in AD-J
>   Diagram 34).

- Open flues serving gas appliances should not be: J 1.52
  - located closer than 75mm below any gutter soil pipe or drainpipe where there is a fanned draught (as shown in AD-J Diagram 34 and its associated Table);
  - used below any gutter soil pipe or drainpipe where there is a natural draught.

- Outlets from oil-fired appliances with jet burners (see   J 4.7 AD-J Diagram 41) should not be located closer than:
  - 75mm below a plastic/painted gutter, drainage pipe or eaves if combustible material protected;
  - 600mm below a plastic/painted gutter, drainage pipe or eaves if combustible material is not protected;
  - 300mm from any vertical sanitary pipework.

- Open flues serving oil-fired appliances with   J 4.7 vaporising burners should not be used near gutters, drainage pipes or sanitary pipework.

4.4.5.2 Location and support of cylinders

- Cylinders should be stored at least 2m horizontally   J 5.20b from drains without traps, unsealed gullies or cellar hatches, unless an intervening wall not less than 250mm high is provided.

4.4.6 Resistance to the passage of sound

4.4.6.1 Rainwater drainage

- Adequate provision shall be made for rainwater to be   H3 (1) carried from the roof of the building.

- Paved areas around the building shall be so   H3 (2) constructed as to be adequately drained.

- Methods of drainage other than connection to a   H3 0.2 public surface water sewer should be used where technically feasible.

- Rainwater or surface water should **not** be discharged   H3 0.6 to a cesspool or septic tank.

4.4.6.2 Surface water drainage

The following guidance applies to surface water drainage systems for small catchments with water resistant areas up to 2 hectares (approximately 5 acres).

• Gully gratings should be set approximately 5mm below the level of a surrounding paved area.	H3 2.15
• Provision should be made to prevent silt and grit entering the system.	H3 2.16
• Drainage from large paved areas should be designed in accordance with BS EN 752-4.	H3 2.17
• Surface water drainage should discharge to a soakaway or other infiltration system.	H3 3.2
• Surface water drainage connected to combined sewers should have traps on all inlets.	H3 3.7
• Drains should be at least 75mm diameter.	H3 3.14
• 75mm and 100mm rainwater drains should be laid at a gradient not less than 1:100, 150mm drains and sewers should be laid at gradients not less than 1:150 and 225mm drains should be laid at gradients not less than 1:225.	H3 3.15
• Where any materials that could cause pollution are stored or used, separate drainage systems should be provided.	H3 3.21
• On car parks, petrol filling stations or other areas where there is likely to be leakage or spillage of oil, drainage systems should be provided with oil interceptors.	H3 3.22 H3 App A
• Separators should be leak-tight and comply with the requirements of the Environment Agency and BS EN 858.	H3 A.9 H3 A.10
• Infiltration devices (including soakaways, swales, infiltration basins and filter drains) should not be built: • within 5m of a building or road or in areas of unstable land; • in ground where the water table reaches the bottom of the device at any time of the year;	H3 3.23 H3 3.24 H3 3.25 H3 3.26

- near any drainage fields, drainage mounds or other soakaways;
- where the presence of any contamination in the run-off could result in pollution of groundwater source or resource.

- Soakaways should be designed to a return period of once in 10 years. H3 3.27

- Soakaways for areas less than 100m² shall consist of square or circular pits filled with rubble or lined with dry-jointed masonry or perforated ring units. H3 3.26

- Soakaways serving larger areas shall be lined pits or trench-type soakaways and designed in accordance with BS EN 752-4.

It is an offence to discharge any noxious or polluting material into a watercourse, coastal water or into any drain or sewer connected to a public sewer.

4.4.6.3 Drainage of paved areas

- Surface gradients should direct water draining from a paved area away from buildings. H3 2.2

- Gradients on water resistant surfaces should be designed to permit the water to drain quickly from the surface. H3 2.3

- For very high-risk areas, where ponding would lead to flooding of buildings, the drainage should be designed in accordance with BS EN 752-4. H3 2.5

- Paths, driveways and other narrow areas of paving should be free draining to an absorbent area such as grassland. H3 2.6

- Where water is to be drained on to the adjacent ground, the edge of the paving should be finished above or flush with the surrounding ground to allow the water to run off. H3 2.7

- Where the surrounding ground is not sufficiently permeable to accept the flow, filter drains may be provided. H3 2.8 & 3.33

• Absorbent paving:	H3 2.9
• should be considered for larger paved areas where it is not possible to drain the rainwater to an adjacent pervious area;	H3 2.10 H3 2.11 H3 2.12
• should not be used where excessive amounts of sediment are likely to enter the pavement and block the pores;	
• may also be used as a detention tank prior to flows discharging to a drainage system.	
• Where it is not possible for surfaces to be free draining, gullies or channels (discharging to a drainage system) should be used.	H3 2.14
• Gullies should be provided at low points where water would otherwise cause ponding.	
• Gully gratings should be set approximately 5mm below the level of the surrounding paved area.	H3 2.15
• Provision should be made to prevent silt and grit entering the system.	H3 2.16
• Drainage from large paved areas should be designed in accordance with BS EN 752-4.	H3 2.17

Porous paving should not be used in oil storage areas, or where run-off may be contaminated with pollutants.

4.4.6.4 Gutters and rainwater pipes

• Where the design incorporates valley gutters, parapet gutters, siphonic or drainage systems from flat roofs, and where over-topping of these systems would enable water to enter the building, the design should be carried out in accordance with BS EN 12056.	H3 1.2
• AD-H Table 2 shows the largest effective area which should be drained into the gutter sizes which are most often used.	H3 1.4
• AD-H Table 2 also shows the smallest size of outlet which should be used with the gutter.	

- Where the outlet is not at the end, the gutter should   H3 1.5
  be of the size appropriate to the larger of the areas
  draining into it.
- Where there are two end outlets, they may be up to
  100 times the depth of flow apart.

- Rainwater pipes should discharge into a drain or   H3 1.8
  gully, but may discharge to another gutter or onto
  another surface if it is drained.
- Any rainwater pipe which discharges into a combined
  system should do so through a trap.

- Siphonic roof drainage systems should be designed in   H3 1.11
  accordance with BS EN 12056.

Gutters should be laid with any fall towards the nearest outlet.

- Gutters should be laid so that any overflow in excess   H3 1.7
  of the design capacity (e.g. above-normal rainfall) will
  be discharged clear of the building.

- The size of a rainwater pipe should be at least the   H3 1.10
  size of the outlet from the gutter.

- A down pipe which serves more than one gutter   H3 1.10
  should have an area at least as large as the combined
  areas of the outlets.

- On flat roofs, valley gutters and parapet gutters,   H3 1.7
  additional outlets may be necessary.

- Where a rainwater pipe discharges on to a lower roof   H3 1.9
  or paved area, a pipe shoe should be fitted to divert
  water away from the building.

- The materials used should be of adequate strength   H3 1.16
  and durability, and:
  - all gutter joints should remain watertight under
    working conditions;
  - pipework in siphonic roof drainage systems should
    be able to resist negative pressures in accordance
    with the design;

- gutters and rainwater pipes should be firmly supported;
- different metals should be separated by non-metallic material to prevent electrolytic corrosion.

Note: Gutters and rainwater pipes should be firmly supported without restricting thermal movement.

4.4.6.5 Drain access points

Sufficient and suitable access points should be provided for clearing blockages from drain runs that cannot be reached by any other means.

• Access points constructed to resist the ingress of groundwater (or rainwater) should be provided:	H1 2.49 H1 2.52

- on or near the head of each drain run;
- at a bend;
- at a change of gradient or pipe size;
- at a junction;
- on long runs.

• Access points should have:	H1 2.48

- rodding eyes – capped extensions of the pipes;
- access fittings – small chambers on (or an extension of) the pipes, but not with an open channel;
- inspection chambers – with working space at ground level;
- manholes – deep chambers with working space at drain level.

Note: Inspection chambers and manholes should have removable non-ventilating covers of durable material (e.g. cast iron, steel, precast concrete or uPVC).

• Access points to sewers (serving more than one property) should be in accessible places where they are likely to be required in an emergency (e.g. highways, unfenced front gardens or driveways).	H1 2.51

- Where half round channels are used in inspection chambers and manholes: H1 2.53
  - the branches up to and including 150mm diameter should discharge into the channel in the direction of flow at or above the level of the horizontal diameter;
  - a branch with a diameter larger than 150mm should be set with the soffit level matching that of the main drain;
  - where the angle of the branch is more than 45°, a three-quarter section branch should be used;
  - channels and branches should be benched up at least to the top of the outgoing pipe and at a slope of 1 in 12.
  - the benching should be rounded at the channel with a radius of at least 25mm.

- Inspection chambers and manholes in buildings should have mechanically fixed airtight covers unless the drain itself has watertight access covers.   H1 2.54
- Manholes deeper than 1m should have metal step irons or fixed ladders.

- After laying, gravity drains and private sewers should be tested for water tightness.   H1 2.59

Note: Material alterations to existing drains and sewers are subject to (and covered by) Building Regulations.

All pipework carrying greywater for re-use should be clearly marked with the word 'GREYWATER'.

4.4.6.6 *Junction requirements for walls and floors*

- External cavity walls should have adequate drainage.   E 3.31 & E 3.69

- Pipes and ducts that penetrate a floor separating habitable rooms in different flats should be enclosed for their full height in each flat.   E 3.41, E 3.79, E 3.117 & E 4.46

• The enclosure should:
  • be lined by material with a mass per unit area of at least 15kg/m²;
  • have the duct or pipe wrapped with 25mm unfaced mineral fibre.
	E 3.42
	E4.47
	E 3.118

• Penetrations through a separating floor should have fire protection. • Fire-stopping should be flexible and prevent rigid contact between the pipe and floor.	E 3.43, E 3.82, E 3.119 & E 4.50
• In conversions, piped services passing through a separating floor should be surrounded with sound-absorbent material and enclosed in a duct above and below the floor.	E 4.45

4.4.6.7 *Protection from settlement*

• A drain may run under a building if at least 100mm of granular or other flexible filling is provided round the pipe.	H1 2.23
• Where pipes are built into a structure (e.g. inspection chambers, manholes), suitable measures should be taken to prevent damage or misalignment (see AD-H Diagrams 7a and 7b).	H1 2.24
• All drain trenches should not be excavated lower than the foundations of any building nearby (see AD-H Diagram 8).	H1 2.25

4.4.7 Power to examine and test drains and sewers

• Sewerage undertakers (and/or the Local Authority) may examine and test any drain or private sewer connecting with a public sewer; if they have grounds, the Authority may require the owner of a building to carry out remedial works on a soil pipe, drain or private sewer.	H1 App B.2 H1 App B.4

• Local Authorities have powers to repair or remove blockages on drains or private sewers which are not sufficiently maintained or kept in good repair or are stopped up, provided the cost does not exceed (at the time of writing this book) £250.	H1 App B.5

4.4.7.1 Repair, reconstruction or alterations to underground drains or sewers

Note: Although minor works to drains or sewers are not normally covered under Building Regulations, Local Authorities have other powers to control such works.

Material alterations to existing drains and sewers are subject to Building Regulations.	H1 App B.7

• Notice should be given before repairs or alterations are carried out, and:
   • any person intending to repair, reconstruct or alter a drain must, except in an emergency, give 24 hours' notice of intention to carry out the works to the Local Authority;
   • where the works are carried out in an emergency, they shall **not** cover over the work without giving notice to the Local Authority;
   • the Local Authority must be given free access to inspect the works.
	H1 App B.8

• The Local Authority may use their powers to test the drain, or to require remedial works.	H1 App B.9

4.4.7.2 Sealing or removal of disused drains or sewers

• The Local Authority may require any person demolishing a building to remove or seal any sewer or drain to which the building was connected.	H1 App B.12
• The Local Authority can also require the owner of a building to remove, or otherwise render innocuous, any disused drain or sewer which is prejudicial to health or a nuisance.	H1 App B.13

4.4.8 Fire safety

4.4.8.1 Openings for pipes

It is important that all pipe and joints are properly sealed in order to prevent the spread of fire and smoke to ensure the maintenance of fire resistance.

- Unless they are in a protected shaft, pipe openings that pass through any part of a fire-separating element should be:
  - kept to as few as possible;
  - kept as small as practicable (see AD-B(V1) Table 9.1 and Diagram 9.1);
  - fire-stopped.

  B(V1) 9.4

- Pipes which pass through fire-separating elements (unless the pipe is in a protected shaft), should either:

  B(V1) 9.4

  B(V1) 9.3

  - provide a proprietary sealing system that is capable of maintaining the fire resistance of the wall, floor or cavity barrier;
  - use fire-stopped pipes with a restricted internal diameter (see AD-B(V1) Table 9.1);
  - use sleeving such as a lead, aluminium, aluminium alloy, fibre-cement or uPVC. (See AD-B(V1) Diagram 9.2.)

  B(V1) 9.5

Requirements for buildings other than dwellings – Additional requirements

4.4.9 Fire safety

The basic fire safety requirements for drainage are the same as for dwellings (see above), with the addition of the following requirements.

4.4.9.1 Fire-stopping

• The fire resistance of adjoining compartment walls should be maintained by fire-stopping material such as cement mortar, gypsum-based plaster, glass fibre, crushed rocks or a substance that swells as a result of heat exposure (commonly referred to as intumescent material).	B(V1) 8.22 B(V2) 10.25
• Enclosures for pipes, ducts, conduits or cables for drainage and water supply pipes should be: • kept to as few as possible; • kept as small as practicable; • fire-stopped.	
• Where there is no piped water supply, or the pressure and flow in the water main are insufficient, then an alternative source of supply should be provided.	B(V2) 16.12

4.4.9.2 Protection of openings and fire-stopping

Note: The opening for a pipe should be as small as possible and fire-stopped around the pipe.

• Pipes which pass through a fire-separating element (unless the pipe is in a protected shaft), should either:	B(V2) 10.2
• **p**rovide a protective sealing system which maintains the fire resistance of the wall, floor or cavity barrier; or	B(V2) 10.3
• include fire-stopping around the pipe, keeping the opening as small as possible (see AD-B(V2) Table 10.1); or	B(V2) 10.4
• use a pipe made of high melting-point material (such as lead, aluminum, aluminum alloy, fibre-cement or uPVC). (See AD-B(V2) Diagram 10.1 and Table 10.1 for further details).	B(V2) 10.5

4.4.9.3 Flues and ducts

• The number of openings through a fire-resisting element for pipes, ducts and conduits should be limited, be as small as is practicable and fire-stopped.	B(V2) 10.24

4.4.9.4 Protected shafts and fire-separating elements

• Unless they are in a protected shaft, all pipes passing through a fire-separating element, should either have protective seals or be restricted in diameter or have sleeving made out of a high melting-point metal.	B(V2) 10.2

 Note: If a protective sealing system is not used, then a fire-stop around the pipe, keeping the opening for the pipe as small as possible, should be included.

4.4.9.5 Access to and use of buildings

• The floor of the shower and shower area should be slip-resistant and self-draining.	M2 5.18o

4.4.9.6 Pipes passing through a fire-separating element

Pipes passing through a fire-separating element, should provide one of the following:

* A sealing system that will maintain the fire resistance (of the wall, floor or cavity barrier).
* A fire-stop around the pipe.
* The use of a pipe with a maximum nominal internal diameter of 160mm with a high melting point (as shown in AD-B(V1) Diagram 9.2).

• Openings in cavity barriers (compartment walls or floors) are permitted for the passage of certain types of pipes.	B(V1) 5.24 & B(V2) 9.17
• Openings in compartment walls or floors should be limited to those for pipes (also ventilation ducts and chimneys).	B(V1) 7.20 & B(V2) 10.2
• When openings in the enclosure of a protected shaft are limited, the passage of a pipe is permitted in most circumstances. (For more information, see the provisions of section 10.)	B(V1) 7.29 & B(V2) 8.39

 Note: These conditions apply to both dwellings and buildings other than dwellings.

4.5 Cellars and basements

Unless you have the consent of the Local Authority, you are not allowed to construct a cellar or room in (or as part of) a house, an existing cellar, a shop, inn, hotel or office if the floor level of the cellar or room is lower than the ordinary level of the sub-soil water on, under, or adjacent to the site of the house.

Reminder: The subsections include details for buildings other than dwellings only where they differ from the requirements for dwellings.

Subsection	Description	Page
Requirements for dwellings		
4.5.1	Fire safety	225
4.5.2	Means of escape	226
4.5.3	Resistance to the passage of sound	227
4.5.4	Ventilation	228
4.5.5	Drainage and waste disposal	229
4.5.6	Combustion and fuel storage appliances	230
4.5.7	Protection from falling, collision and impact	230
4.5.8	Security	231
Requirements for buildings other than dwellings – Additional requirements		
4.5.9	Means of escape	231

Reminder: Free downloads of all the diagrams and tables referenced in the subsections are available on-line at the Governments' Planning Portal: www.gov.uk/government/collections/approved-documents.

Requirements for dwellings

4.5.1 Fire safety

4.5.1.1 Fire resistance periods

The building shall be designed and constructed with a provision for the early warning of fire.

4.5.1.2 Venting of heat and smoke from basements

The building should be provided with adequate means for venting heat and smoke from a fire in a basement.

• Each basement space should have one or more smoke outlets, unless the basement is compartmented, in which case, each compartment should have one or more smoke outlets, rather than indirect venting.	B(V1) 16.2
• If basement storeys are fitted with a sprinkler system, a mechanical smoke extraction system may be provided as an alternative to natural venting.	B(V1) 16.11
• Basements containing rooms with doors or windows do not need smoke outlets.	
• Strong rooms do **not** need to be provided with smoke outlets.	B(V1) 16.4

Note: Car parks are not normally expected to be fitted with sprinklers.

4.5.2 Means of escape

4.5.2.1 Escape from basements

There are concessions in respect of fire resistance of elements of structure in basements where at least one side of the basement is open at ground level.

• Basements should have an external door or window suitable for egress from the basement or have a protected stairway leading from the basement to a final exit.	B(V1) 3.9

4.5.2.2 Emergency egress windows and external doors

• Windows and external doors that are intended to enable a person to reach a place free from danger of fire (e.g. a courtyard or back garden) should: • be capable of remaining open without being held. Locks and opening stays may be fitted to escape windows; and	B(V1) 2.10

- the bottom of the openable area should not be more than 1100mm above floor level.
- In a small single-stair building, an escape stair that is the only escape route from an upper storey should not continue down to serve a basement storey. The basement storey should be served by a separate escape stair.

B(V1) 3.71

4.5.2.3 Fire protection of lift installations

- In basements and enclosed car parks, the lift should be within the enclosure of a protected stairway.

B(V1) 3.102

- A lift shaft serving storeys above ground level should *not* serve any basement, if there is only one escape stair serving storeys above ground level.

B(V1) 3.104

4.5.2.4 Firefighting shafts

- In a building with basement storeys that are more than 10m below the fire and rescue service vehicle access level (or if there are two or more basement storeys, each with a minimum area of 900m), the basement should have firefighting shafts.

B(V1) 15.3

- The firefighting shafts do not need to include firefighting lifts.

Note: The firefighting shafts do not need to include firefighting lifts.

4.5.3 Resistance to the passage of sound

- Where any building element functions as a separating element, then the separating element requirements should take precedence.

E Table 2.1

4.5.4 Ventilation

4.5.4.1 Ventilation system for basements

• If a basement is connected to the rest of the dwelling by a large permanent opening such as an open stairway, then the whole dwelling including the basement should be treated as a multi-storey dwelling.	F(V1) 1.39

• If the basement is not connected to the rest of the dwelling by a large permanent opening, then:
  • the part of the dwelling above ground should be considered separately;
  • the basement should be treated as a single-storey dwelling, as if it were above ground.
	F(V1) 1.40

• If the part of the dwelling above ground has no bedrooms, then for the purpose of ventilation requirements:
  • assume that the dwelling has one bedroom; and
  • treat the basement as a single-storey dwelling (with one bedroom) as if it were above ground.	

4.5.4.2 Types of ventilation for basements

• A dwelling that is only a basement should be treated as a single-storey dwelling above ground level, and natural ventilation is inappropriate.	F(V1) 1.41
• If a basement has only one exposed facade then the guidance in the approved documents for ventilation will not be appropriate and expert advice should be sought.	F(V1) 1.39

- If a dwelling with a basement is **not** connected to   F(V1) 1.40
  the rest of the dwelling above ground level by a large
  permanent opening then:
  - the part of the dwelling above ground level should
    be considered separately;
  - the basement should be treated as a single-storey
    dwelling above ground level.
- If the part of the dwelling above ground has no
  bedrooms, then for the purpose of ventilation
  requirements:
  - assume that the dwelling has one bedroom; and   F(V1) 1.41
  - treat the basement as a single-storey dwelling (with
    one bedroom) as if it were above ground.

4.5.5 Drainage and waste disposal

4.5.5.1 Branch discharge pipes

- Branch pipes should discharge into another branch   H 1.7
  pipe or a discharge stack unless appliances discharge
  to a gully.
- Branch pipes should not discharge into open hoppers.

4.5.5.2 Pumping installations

- Pumping installations for use inside buildings should   H 2.37
  be designed in accordance with BS EN 12056-4.

Package pumping installations are available which are suitable for
installation within buildings and floor-mounted units may be particularly suited for installation in basements. These should conform to BS
EN 12050.

4.5.5.3 Surcharging of drains

- Basements containing sanitary appliances, and where   H 2.9
  the risk of flooding by a surcharge from a sewer is
  considered to be high, should have pumped drainage.
- Where the risk of flooding is considered to be low,
  an anti-flooding valve should be installed on the
  basement drainage.

4.5.6 Combustion and fuel storage appliances

4.5.6.1 LPG storage vessels and appliances

- Liquid petroleum gas (LPG) storage vessels and   J 3.5i
  LPG-fired appliances fitted with automatic ignition
  devices or pilot lights must **not** be installed in cellars
  or basements.

4.5.7 Protection from falling, collision and impact

4.5.7.1 General provisions

- Basement areas (or similar sunken areas connected to   K 2(b)
  a building) shall have barriers to protect people in or
  about the building from falling.
- If a building is likely to be used by children under   K 1.39
  five years old, the guarding should stop children from
  easily climbing it, and should not have horizontal
  rails. The construction should prevent a 100mm
  sphere being able to pass through any opening.
- Guarding should be provided at the sides of flights and   K 1.41
  landings where there is a drop of more than 600m.

4.5.7.2 Design of guarding for cellars and basements

- Any wall, parapet, balustrade or similar obstruction   K 1.35
  may be used as guarding.
- Guarding must be capable of resisting, as a
  minimum, the loads given in BS EN 1991-1-1.

4.5.7.3 Handrails for stairs

• If the stairs are 1,000mm or wider, there should be a handrail (whose top shall be 900mm to 1,100mm from the pitch line or floor) on both sides, and the handrail may form the top of a guarding if the heights are matched.	K 1.34

4.5.8 Security

Ground floor, basement and other easily accessible windows (including easily accessible rooflights) should be secure windows.

• Windows should be made to a design that meets the security requirements of British Standards publication PAS 24:2016.	Q 2.2
• Frames should be mechanically fixed to the structure of the building in accordance with the manufacturer's installation instructions.	Q 2.3

Requirements for buildings other than dwellings – Additional requirements

The basic fire safety requirements for cellars and basements are the same as for dwellings (see above), with the addition of the following requirements.

4.5.9 Means of escape

A fire detection system may be required in an unoccupied part of the premises such as a basement.

4.5.9.1 Escape routes

• Basement storeys should be served by separate escape stairs.	B(V2) 3.40 & 5.24
• Final exits should avoid outlets of basement smoke vents.	

Note: In some cases (such as when the building's escape stair does not extend down to the basement), a separate escape stair is required to serve the basement storey.

 If an exit route from a stairway is also the escape route from the ground storey, the width of the exit route may need to be increased.

4.5.9.2 Fire protection of basements

Special measures have to be taken to prevent a basement fire endangering upper storeys, and this can be achieved by ensuring that:

* The basement is served by a separate stairway.
* Escape stairs forming part of the only escape route from upper storeys of a building does not serve any basement storey.
* If there is more than one escape stair from a building's upper storey, only one of the stairs terminates at ground level.

 Note: Other stairs may connect with the basement storey(s) if there is a protected lobby or corridor between the stair(s) and accommodation at each basement level.

4.5.9.3 Fire protection of lift installations

• A lift shaft serving storeys above ground level should ***not*** operate in or to a basement, if the lift shaft is within the enclosure to an escape stair that terminates at ground level.	B(V2) 5.40

 Note: Smoke from a basement should not adversely affect escape routes in upper stories.

4.5.9.4 Firefighting shafts

• Basements more than 10m below the fire and rescue service vehicle access level should have firefighting shafts that contain firefighting lifts.	B(V2) 7

4.5.9.5 *Venting of heat and smoke from basements*

• Heat and smoke from basement fires vented via stairs can cause access problems for firefighting personnel.	B(V2) 18.1
• Each basement space should have one or more smoke outlets. • If a basement is compartmented, each compartment should have one or more smoke outlets rather than indirect venting. • Basement storeys or compartments containing rooms with doors or windows do not require smoke outlets.	B(V2) 18.2

• Smoke outlets from basements that are connected directly to the open air should be provided, unless the basement storey has:
  • a maximum floor area of 200m²; and is
  • a maximum of 3m below the adjacent ground level.
	B(V2) 8

• Strong rooms do not need to be provided with smoke outlets.	B(V2) 8

• Smoke outlets should be:
  • sited at high level in either the ceiling or wall of the space they serve;
  • evenly spaced around the perimeter, to discharge to the open air.
	B(V2) 8

• If basement storeys are fitted with a sprinkler system, a mechanical smoke extraction system may be provided as an alternative to natural venting.	B(V2) 8

See AD-B(V2) Diagram 18.1 for more information concerning fire-resisting construction for smoke outlet shafts.

4.6 Floors

The majority of brick-built buildings are supported on a solid concrete base (called the 'foundation') whilst timber-framed houses are normally built on a concrete foundation with a 'strip' or 'raft' construction to spread the weight.

Reminder: The subsections include details for buildings other than dwellings only where they differ from the requirements for dwellings.

Reminder: Free downloads of all the diagrams and tables referenced in the subsections are available on-line at the Governments' Planning Portal: www.gov.uk/government/collections/approved-documents.

Requirements for dwellings

4.6.1 Types of floors

There are three types of separating floor, as shown in AD-E Diagram 3.1:

1. Concrete base with ceiling and soft floor covering.
2. Concrete base with ceiling and floating floor.
3. Timber frame base with ceiling and platform floor.

4.6.1.1 Floor type 1 – concrete base with ceiling and soft floor covering

• The soft floor covering should not be fixed or glued to the floor.	E 3.27a
• All joints between parts of the floor should be filled.	E 3.27b
• Air paths should be avoided at all points where a pipe or duct penetrates the floor.	E 3.27c
• A separating concrete floor should be built into masonry walls.	E 3.27d

 Penetrations through a separating floor should have fire protection, but the fire-stopping device or equipment should be flexible and prevent rigid contact between the pipe and floor.

• Gaps between the head of a masonry wall and the underside of the concrete floor should be filled with masonry.	E 3.27e
• Flanking transmission from walls connected to the separating floor should be controlled.	E 3.27f
• Floor bases shall not bridge the cavity in a two-cavity masonry wall.	E 3.27a
• Non-resilient floor finishes shall **not** be used.	E 3.27b2
• Soft floor covering should be resilient material with an overall uncompressed thickness of at least 4.5mm.	E 3.28a

4.6.1.1.1 JUNCTION REQUIREMENTS FOR FLOOR TYPE 1

- If the external wall is a cavity wall, the cavity should   E 3.31
  be stopped with a flexible closer. (See AD-E
  Diagram 3.5.)

- The masonry inner leaf of an external cavity   E 3.32
  wall should have a mass per unit area of at least
  120kg/m² excluding finish.

- The floor base (excluding any screed) should be built   E 3.33
  into a cavity masonry external and carried through to
  the cavity face of the inner leaf.

- Where floor type 1.2B is used and the planks are   E 3.34
  parallel to the external wall, the first joint should be
  a minimum of 300mm from the cavity face of the
  inner leaf. See AD-E Diagram 3.5.

- Where the external wall is a cavity wall: E 3.36
  - the outer leaf of the wall may be of any
    construction;
  - the cavity should be stopped with a flexible closer;
  - the wall finish of the inner leaf of the external wall
    should be two layers of plasterboard, each sheet a
    minimum mass per unit area 10kg/m²;
  - all joints should be sealed with tape or caulked
    unenclosed.

- The floor base should be continuous through or   E 3.39
  above an internal masonry wall.

- The mass per unit area of any load-bearing internal   E 3.40
  wall (or any internal wall rigidly connected to
  a separating floor) should be at least 120kg/m²
  excluding finish.

4.6.1.1.2 JUNCTIONS WITH FLOOR PENETRATIONS (EXCLUDING
        GAS PIPES)

Pipes and ducts should be in an enclosure and in all cases:

- The enclosure in a floor should be constructed of   E 3.42
  material with a mass per unit area of at least 15kg/m².

- The enclosure should be either lined or the duct   E 3.42
  (or pipe) within that enclosure wrapped with 25mm
  unfaced mineral fibre.

- Pipes and ducts passing through a separating floor should be fire protected.   E 3.43

- Pipes and ducts penetrating a floor that separates habitable rooms in different flats should be enclosed (see AD-E Diagram 3.6).   E 3.41

4.6.1.1.3 JUNCTIONS WITH A SEPARATING WALL TYPE 1

For floor types 1.1C and 1.2C, two possibilities exist:

- A separating floor type 1.1C base (excluding any screed) should pass through a separating wall type 1. See AD-E Diagram 3.7.   E 3.44

- A separating floor type 1.2B base (excluding any screed) should *not* pass through a separating wall type 1. See AD-E Diagram 3.8.   E 3.45

4.6.1.1.4 JUNCTIONS WITH A SEPARATING WALL TYPE 2

- The mass per unit area of any leaf supports or adjoining the floor should be at least 120kg/m^2 excluding finish.   E 3.46

- The floor base (excluding any screed) should be carried through to the cavity face of the leaf.   E 3.47

- The wall cavity should not be bridged.   E 3.47

- Where floor type 1.2B is used, the first joint should be a minimum of 300mm from the inner face of the adjacent cavity leaf. See AD-E Diagram 3.9.   E 3.48

4.6.1.1.5 JUNCTIONS WITH A SEPARATING WALL TYPE 3

- A separating floor type 1.1C base (excluding any screed) should pass through separating wall types 3.1 and 3.2. (See AD-E Diagram 3.10.)   E 3.49

- A separating floor type 1.2B base (excluding any screed) should *not* be continuous through a separating wall type 3.   E 3.50

- Where a separating wall type 3.2 is used with floor type 1.2B, the first joint should be a minimum of 300mm from the centre line of the masonry core.   E 3.51

- Where a separating wall is type 3.3, the floor base (excluding any screed) should be carried through to the cavity face of the leaf of the core.   E 3.53

- Where a separating wall is type 3.3, the cavity should **not** be bridged.   E 3.53

- Where floor type 1.2B is used with a separating type 3.3, the first joint should be a minimum of 300mm from the inner face of the adjacent cavity leaf of the masonry core.   E 3.54

4.6.1.2 Floor type 2 – concrete base with ceiling and floating floor

- Joints between parts of the floor should be filled to avoid air paths.   E 3.61a

- Air paths should be avoided where a pipe or duct penetrates the floor.   E 3.61b

- A separating concrete floor should be built into the walls if the walls are masonry.   E 3.61c

- All gaps between the head of a masonry wall and the underside of the concrete floor should be filled with masonry.   E 3.61d

- The floor base shall not bridge a cavity in a cavity masonry wall.   E 3.61

- A small gap (tabili. 5mm and filled with a flexible sealant) should be left between the floating layer and wall at all room edges and the skirting.   E 3.63ab

- The floor base (excluding any screed) should be built into a cavity masonry external wall and carried through to the cavity face of the inner leaf.   E3.71
- The cavity should not be bridged.

- If a floor 2.2B is used, the first joint should be a minimum of 300mm from the cavity face of the inner leaf. (See AD-E Diagram 3.16.)   E3.72

- If the external wall is a cavity wall, the wall face of the inner leaf should be two layers of plasterboard with a minimum mass per unit area 10kg/m^2.   E3.74

4.6.1.2.1 JUNCTIONS WITH FLOOR PENETRATIONS (EXCLUDING GAS PIPES)

• Pipes and ducts that penetrate a floor separating habitable rooms in different flats should be enclosed for their full height in each flat. (See AD-E Diagram 3.17.)	E3.79
• The enclosure should be constructed of material that has a mass per unit area of at least 15kg/m². • Either line the enclosure or wrap the duct or pipe within the enclosure with 25mm unfaced mineral wool.	E3.80
• A small gap (sealed with sealant or neoprene) of about 5mm should be left between the enclosure and the floating layer.	E3.81

Note: Gas pipes can be contained in a separate (ventilated) duct or can remain unenclosed.

4.6.1.2.2 JUNCTIONS WITH A SEPARATING WALL TYPE 1

Excluding any screed:

• A separating floor type 2.1C base should pass through a separating wall type 1 as shown in AD-E Diagram 3.18.	E3.84
• A separating floor type 2.2B base should *not* be continuous through a separating wall type 1.	E3.84

4.6.1.2.3 JUNCTIONS WITH A SEPARATING WALL TYPE 2

• The floor base (excluding any screed) should be carried through to the cavity face of the leaf. • The cavity should *not* be bridged.	E3.85

If a floor type 2.2B is used, the first joint should be a minimum of 300mm from the cavity face of the leaf.

4.6.1.2.4 JUNCTIONS WITH A SEPARATING WALL TYPE 3

Excluding any screed:

• A separating floor type 2.1C base should pass through separating wall types 3.1 and 3.2. (See AD-E Diagram 3.19.)	E3.87
• A separating floor type 2.2B base should not be continuous through a separating wall type 3.	E3.88
• If a separating wall type 3.3 is used, the floor base should be carried through to the cavity face of the leaf of the core.	E3.91

Note: At the time of publication, there is no official guidance available on junctions with a separating wall type 4 (i.e. sheer wall) and so it would be advisable to seek specialist advice for this type of situation.

4.6.1.3 Floor type 3 – timber frame base with ceiling and platform floor

Floor joists that are supported on a separating wall should be supported on hangers and not built in.

• There should not be a bridge between the floating layer and the base or surrounding walls.	E3 101
• The floating layer should be a minimum of two layers of board material with a total mass per unit area of 25kg/m².	

4.6.1.3.1 JUNCTIONS WITH AN EXTERNAL CAVITY WALL WITH
MASONRY INNER LEAF

• Where the external wall is a cavity wall: • the outer leaf of the wall may be of any construction; • the cavity should be stopped with a flexible closer;	E3.103
• the masonry inner leaf should be lined with an independent panel;	E3.104
• the ceiling should be taken through to the masonry;	E3.105

- the junction between the ceiling and the   E3.105
  independent panel should be sealed with tape or
  caulked with sealant;

- air paths between floor and wall cavities should be   E3.106
  blocked.

Note: Independent panels are not required if the mass per unit area of the inner leaf is greater than 375kg/m².

4.6.1.3.2 JUNCTIONS WITH AN EXTERNAL CAVITY WALL WITH TIMBER FRAME INNER LEAF

- Where the external wall is a cavity wall, the outer leaf   E3.109
  of the wall may be of any construction and the cavity
  should be stopped with a flexible closer.

- The wall finish of the inner leaf of the external wall   E3.110
  should be two layers of plasterboard (minimum mass
  per unit area 10kg/m²).

4.6.1.3.3 JUNCTIONS WITH INTERNAL FRAMED WALLS

- The spaces between joists should be sealed with full-   E3.114
  depth timber blocking.

- The junction between the ceiling and the internal   E3.115
  framed wall should be sealed with tape or caulked
  with sealant.

4.6.1.3.4 JUNCTIONS WITH FLOOR PENETRATIONS (EXCLUDING GAS PIPES)

- Pipes and ducts that penetrate a floor separating   E3.117
  habitable rooms in different flats should be enclosed for   E3.118 &
  their full height in each flat (see AD-E Diagram 3.21).   119

- Gas pipes may be contained in a separate (ventilated)   E3.120
  duct or can remain unenclosed.

Note: Independent panels are not required if the mass per unit area of the inner leaf is greater than 375kg/m².

4.6.2 Structure

4.6.2.1 Protection from falling

• Where there is a change in floor level between inside and outside of more than 600mm, guarding requirements shown in AD-K are insufficient and guarding of 1.1m should be provided as shown in AD-O Table 3.9.	O 3.6

4.6.2.2 Basic requirements for stability

• Intermediate floors should be constructed so that they provide local support to the walls and act as horizontal support and restrict movement of the wall at right angles to its plane.	A 2A2d
• The imposed loads on roofs, floors and ceilings shall not exceed those shown in AD-A Table 4.	A 2C15

4.6.2.3 Maximum floor area

• As shown in AD-A Diagram 5: • no floor enclosed by structural walls on all sides should exceed 70m²; • no floor *without* a structural wall on one side should exceed 36m².	A 2C14
• The maximum span for any floor supported by a wall is 6m (see AD-A Diagram 10).	A 2C23
• Vertical loading on walls should be distributed where the bearing length for lintels is 150mm or greater. • Where a lintel has a clear span of less than 1,200mm, the bearing length may be reduced to 100mm.	A 2C24a
• Floors should have a distributed load of 2.00Kn/m².	A Table 4

4.6.2.4 Openings and recesses

- No openings should be provided in walls below ground floor, except for small holes for services and ventilation, etc., which should be limited to a maximum area of 0.1m² at not less than 2m centres. — A 2C29

- Openings in compartment walls or floors should be limited to those for pipes, ventilation ducts and chimneys. — B(V1) 7.20

4.6.2.5 Lateral support by floors

- Floors should transfer lateral forces from walls to buttressing walls, piers or chimneys and be secured to the supported wall (see AD-A Diagram 15). — A 2C33a

- Lateral support shall be available for:
  - solid and/or cavity walls longer than 3m that form a junction with the supported wall;
  - internal load-bearing walls at the top of each storey. — A 2C34

- Walls should be strapped to floors above ground level at intervals not exceeding 2m and secured by tension straps conforming to BS EN 845-1 (see AD-A Diagram 15 and Table 9). — A 2C35
- Tension straps should be made of galvanized steel or more resistant materials with a tensile strength no less than 8Kn.

- Points of contact should be in line (see AD-A Diagram 15c). — A 2C35d

- Tension straps need not be provided in houses of not more than two storeys where floors are at or about the same level on each side of a supported wall, and contact between the floors and wall is either continuous or at intervals not exceeding 2m. — A 2C35

4.6.2.5.1 INTERRUPTION OF LATERAL SUPPORT

• If an opening in a floor for a stairway adjoins a supported wall, the maximum permitted length of the opening is 3m.	A 2C37
• Horizontal ties or anchorage of suspended floors to walls shall be provided.	A 5.1b

4.6.2.6 Sizes of certain timber members in floors for dwellings

• Guidance on the sizing of certain members in floors is given in PD 6693-1 and BS 8103-3.	A 2B1

Softwood timber used for roof construction or fixed in the roof space (including ceiling joists within the void spaces of the roof) should be adequately treated to prevent infestation by the house longhorn beetle *(Hylotrupes bajulus L.)*. (See AD-A Paragraph 2B2 and Table 1.)

4.6.2.7 Loft conversions

• If an additional storey is added to an existing two-storey single family dwelling, new floors should have a minimum REI 30 fire resistance.	B(V1) 5.4
• Any floor forming part of the enclosure to the circulation space between the loft conversion and the final exit should also achieve a minimum rating of REI 30.	

4.6.2.8 Conversion to flats

• If an existing building is converted into flats, a review of the existing construction should be carried out to ensure (for example) that retained timber floors meet the relevant provisions for fire resistance.	B(V1) 6.5
• If the converted building has more than four storeys, then it must comply with the *full* requirements of AD-B(V1).	B(V1) 6.7

4.6.3 Site preparation and resistance to moisture

The site should be adequately prepared so as to minimise the hazard of contaminated top-soil or pre-existing foundations, and possible damage to the building from existing vegetation (such as tree roots).

Measures must also be taken to provide suitable sub-soil drainage in order to prevent passage of moisture, or waterborne contaminants, to the building's interior.

Buildings in flood-prone areas **must** be constructed to mitigate the effects of flooding.

- Floors should protect the building and persons in and   C 0.6
  about the building from the effects of:
  - moisture from the ground or from groundwater;
  - precipitation and wind-driven spray;
  - interstitial and surface condensation;
  - spillage of water from or associated with sanitary
    fittings and fixed appliances.

Note: Spills and leaks of water in rooms such as bathrooms and kitchens may damage floor decking or other parts of the structure.

4.6.3.1 Resistance to ground moisture

Mature trees, substances in the ground such as sulphate gasses, or pollutants can all affect ground floor slabs.

- Ground floors should have a maximum thermal   C 4.22a
  transmittance (U-value) of $0.7W/(m^2 \cdot K)$ at any point.

- Floors next to the ground should: C 4.2
  - resist the passage of ground moisture to the upper
    surface of the floor;
  - not be damaged by moisture from the ground;
  - not be damaged by groundwater;
  - resist the passage of ground gases.

- The structural and thermal performance of floors   C 4.3
  next to the ground should not be adversely affected
  by interstitial condensation.

- All floors should be designed so that they do not   C4.4
  promote surface condensation or mould growth.

Note: All floors should be designed so that they do not promote surface condensation or mould growth.

4.6.3.1.1 GROUND SUPPORTED FLOORS

Moisture can rise from the ground to damage floors and the base of walls on any site. To counteract this, a damp-proof course should be continuous with any damp-proof membrane in the walls where there is a risk of moisture from the ground.

- Ground supported floors will meet the requirement   C 4.6
  if the ground is covered with dense concrete laid on a
  hardcore with a damp-proof membrane.
- Unless it is subjected to water pressure, a concrete   C 4.7
  ground supported floor should be built according to
  AD-C Diagram 4.
- A ground floor (or floor exposed from below such as   C 4.21
  above an open parking space or passageway) shall be
  designed as shown in AD-C Diagram 7.

4.6.3.2 Sub-soil drainage

Where the water table can rise to within 0.25m of the lowest floor of the building, or where surface water could enter or adversely affect the building, then either the ground to be covered by the building should be drained by gravity or another means of safeguarding the building should be used.

4.6.3.3 Suspended concrete ground floors exposed to moisture from the ground

- Concrete suspended floors that are next to the   C 4.17
  ground should prevent the passage of moisture to the
  upper surface.

- Concrete floors could either be: C/4.18
  - in situ concrete, at least 100mm thick containing at least 300kg of cement for each m³ of concrete; or
  - precast concrete with or without infilling slabs, and reinforcing steel protected by concrete cover of at least 40mm.

- A suspended concrete floor should incorporate a damp-proof membrane and a ventilated air space. C 4.19

 Note: BS 8313 contains recommendations for the design, construction, installation and maintenance of fixed ducts, suspended flooring, ceiling voids and cavities.

4.6.3.4 Suspended timber ground floors exposed to moisture from the ground

- Any suspended timber floor next to the ground should ensure: C 4.13
  - the ground is covered so as to resist moisture and prevent plant growth;
  - there is a ventilated air space between the ground covering and the timber;
  - there are damp-proof courses between the timber and any material which can carry moisture from the ground.

- Unless it is covered with a floor finish which is highly vapour resistant, a suspended timber floor next to the ground may be built as shown in AD-C Diagram 5. C 4.14

- Ground covering should be unreinforced concrete (at least 100mm thick to mix ST 1 as in BS 8500) laid on a compacted hardcore bed of clean, broken brick or any other inert material. C 4.14a

- A ventilated air space should be provided measuring at least 75mm from the ground covering to the underside of any wall-plates and at least 150mm to the underside of the suspended timber floor (or insulation if provided). C 4.14b C 4.19

- Two opposing external walls should have ventilation openings not less than either 1,500mm²/m run of external wall or 500mm²/m² of floor area, whichever gives the greater opening area.
- Any pipes needed to carry ventilating air should have a diameter of at least 100mm.
- Any softwood boarding should be at least 20mm thick. C 4.15

 See BS 8313 for further details.

4.6.3.5 Protected shaft

- The wall of a flue duct containing flues or appliance ventilation duct(s) should have a fire resistance that is at least half of any compartment wall or compartment floor it passes through or is built into, as shown in AD-B(V1) Diagram 9.5.   B(V1) 9.23

4.6.4 Resistance to the passage of sound

- Internal floors should provide reasonable resistance to sound.   E2
- Floors that have a separating function should achieve the sound insulation values as set out in Tables 0.1a and b.   E 0.1
- New floors within a dwelling (including extensions) should provide a sound insulation value not less than 40 Rw Db.   E 0.9

The relevant parts of the building that should be protected from airborne and impact sound are shown in AD-E Diagrams 0.1 to 0.3.

4.6.5 Junctions between floors and walls

The following points should always be borne in mind:

- Flanking transmission from walls and floors connected to the separating wall should be controlled.
- Where there is a separating floor, there should be a minimum mass per unit area of 120kg/m² excluding finish.

- Where the inner leaf of an external cavity wall is a framed construction and there is a separating floor, the framed inner leaf should be two layers of plasterboard.
- If the floor joists are to be supported on a separating wall, they should be supported on hangers as opposed to being built in.

4.6.5.1 *Junction with wall type 1 – solid masonry*

• Internal concrete floor slabs may only be carried through a type 1 separating wall if the floor base has a mass per unit area of at least 365 kg/m² (see AD-E Diagram 2.11).	E 2.46
• Internal hollow-core concrete plank floors and concrete beams with infilling block floors should **not** be continuous through a type 1 separating wall.	E 2.47
• For internal floors of concrete beams with infilling blocks, avoid beams built into the separating wall unless the blocks in the floor fill the space between the beams where they penetrate the wall.	E 2.48
• The ground floor may be a solid slab, laid on the ground, or a suspended concrete floor (see AD-E Diagram 2.12).	E 2.51
• A suspended concrete floor may only pass under a type 1 separating wall if the floor has a mass of at least 365kg/m².	E 2.52
• Hollow-core concrete plank and concrete beams within filling block floors should not be continuous under a type 1 separating wall.	E 2.53

 Possible sulphate attack from some strata on concrete floor slabs and oversite concrete needs to be considered.

4.6.5.2 *Junction with wall type 2 – cavity masonry*

Cavity walls should **not** be built off a continuous solid concrete slab floor.

• The cavity leaves should be separated below ground-floor level.	E 2.65b
• Cavity walls should not be built off a continuous solid concrete slab floor.	E 2.65c2

- If there is no separating floor with separating wall type 2.3 or 2.4, there is no minimum mass per unit area for internal masonry. E 2.83

- Internal concrete floors should be built into a type 2 separating wall and carried through to the cavity face of the leaf (see AD-E Diagram 2.24). E 2.85

- The ground floor may be a solid slab, laid on the ground, or a suspended concrete floor, but it should **not** be continuous under a type 2 separating wall (see AD-E Diagram 2.24). E 2.88

- A suspended concrete floor should: E 2.89
  - not be continuous under a type 2 separating wall;
  - be carried through to the cavity face of the leaf.

4.6.5.3 *Junction with wall type 3 – masonry between independent panels*

- The panels or the supporting frames should be fixed to the ceiling and floor only. E 2.101c

- If there is no separating floor, the external wall may be finished with plaster or plasterboard of minimum mass per unit area 10kg/m^2. E 2.111

- Spaces between the floor joists for internal timber walls should be sealed with full-depth timber blocking. E 2.120 E 2.124

- For solid masonry cores (wall types 3.1 and 3.2), internal concrete floor slabs may only be carried through a solid masonry core if the floor base has a mass per unit area of at least 365kg/m^2. (See AD-E Diagram 2.34.) E 2.121

- For wall type 3.3 (cavity masonry core), internal concrete floors should be built into a cavity masonry core and carried through to the cavity face of the leaf – provided that the cavity is not bridged. E 2.122

4.6.5.4 Junction with wall type 4 – framed walls with absorbent material

• The wall finish of the inner leaf of the external wall should be two layers of plasterboard if there is a separating floor.	E 2.150b
• Air paths through the wall into the cavity shall be blocked by solid timber or continuous ring beam or joists.	E 2.154
• If the ground floor is a concrete slab laid on the ground, it may be continuous under a type 4 separating wall.	E 2.158
• If the ground floor is a suspended concrete floor, it may only pass under a wall type 4 if the floor has a mass per unit area of at least 365kg/m².	

4.6.5.5 Dwelling houses and flats formed by material change of use

An existing floor in a building that is to undergo a material change of use will have to meet the performance standards set out in the relevant ADs without remedial work.

• Flanking transmission from walls and floors connected to the separating wall should be controlled.	E 4.11
• When extending an existing dwelling with a total useful floor area of over 1,000m², consequential improvements may be required. Guidance in Section 12 should be followed.	L(V1) 10.11

 B7 recommends that photographs should be taken of the ground floor perimeter edge insulation and the ground floor to wall junction.

4.6.6 Floor treatments

4.6.6.1 Floor treatment 1 – independent ceiling with absorbent material

- Independent ceilings with absorbent material should   E 4.27
  have:
  - at least two layers of plasterboard with staggered
    joints;
  - a minimum total mass per unit area 20kg/m²;
  - an absorbent layer of mineral wool laid on the
    ceiling.
- The ceiling should be supported by independent
  joists firmly fixed to the surrounding walls, with or
  without additional support provided by resilient
  hangers attached directly to the existing floor base.
- Where a window head is near to the existing ceiling,   E 4.28
  the new independent ceiling may be raised to form a
  pelmet recess. (See AD-E Diagram 4.4.)

4.6.6.2 Floor treatment 2 – platform floor with absorbent material

- Platform floors with absorbent material (see AD-E   E 4.32
  Diagram 4.6) should be laid between the joists in the
  floor cavity with the floating layer having a minimum
  total mass per unit area of 25kg/m² and each layer
  with a minimum thickness of 8mm.

Note: For the junction between floor treatment 2 and wall treatment 1,
see AD-E Diagram 4.7.

4.6.6.3 Junctions with floor penetrations

- Piped services (excluding gas pipes) and ducts which   E 4.45
  pass through separating floors in conversions should be
  surrounded with sound-absorbent material.

- The enclosure should be constructed using material
  with a mass per unit area of at least 15kg/m² wrapped
  with 25mm unraced mineral wool.   E 4.47

See also AD-E Diagram 4.9.

4.6.7 *Internal walls and floors in new buildings*

There are four main types of approved internal wall constructions currently available:

- *Internal wall types A or B:* Timber or metal   E 5.4
  frames (see AD-E Diagrams 5.1 and 5.2).

- *Internal wall types C or D:* Concrete or aircrete
  blocks (see AD-E Diagrams 5.3 and 5.4).

- *Internal floor types A or B:* Concrete planks   E 5.6
  or concrete beams with infilling blocks (see AD-E
  Diagrams 5.5 and 5.6).

- *Internal floor type C:* Timber or metal joists (see   E 5.7
  AD-E Diagram 5.7).

4.6.7.1 *Pre-completion testing*

- Pre-completion tests should be carried out between   E 1.9
  rooms or spaces that share a common area of
  separating floor.

- Impact sound insulation tests should be carried out   E 1.10
  without a soft covering (e.g. foam-backed vinyl, etc.)
  on the floor.

Occasionally sub-grouping is required in new dwelling houses, flats and rooms for residential purposes with separating floor construction.

- Tests in flats with separating floors (but *not* walls)   E 1.20
  require four individual sound insulation tests (two
  airborne tests and two impact tests) on opposite sides
  of the separating floor.

- Tests in flats with separating floors *and* walls require   E 1.21
  six individual sound insulation tests (four airborne
  tests and two impact tests) on opposite sides of the
  separating floors.

- Some properties (e.g. loft apartments) may be sold before being fitted out, in which case measurements of sound insulation should be made between the available spaces and steps taken to ensure that fitting out will not adversely affect the sound insulation. E 1.28

4.6.8 Combustion appliances and fuel storage systems

Ventilation ducts or vents installed to supply air to combustion appliances should not penetrate floor membranes in a way that will render them ineffective.

- If a factory-made metal chimney penetrates a fire compartment floor, it must not breach the fire-separation requirements by using a factory-made metal chimney or casing the chimney in non-combustible material. J 1.46

4.6.8.1 Hearths for oil-fired appliances

- If the appliance could cause the temperature of the floor below it to exceed 100°C, a more substantial hearth is required such as a hearth of solid non-combustible material at least 125mm thick. J 4.25
J 4.26
J 4.27

4.6.8.2 Location and shielding of connecting flue pipes

- Connecting flue pipes: J 2.14
  - should be used only to connect appliances to their chimneys;
  - should not pass through any floor, except directly into a chimney through either a wall of the chimney or a floor supporting the chimney.

- Constructional hearths should be made of solid, non-combustible material. J 2.24

4.6.9 Ventilation

Note: Extensions of buildings at ground level with a floor area less than 30m² by the addition of a conservatory, porch, covered yard, covered way, or carport open on at least two sides do not have to comply with AD-F (*Ventilation*).

4.6.9.1 Dwelling ventilation

- The minimum whole dwelling ventilation rate should be 0.3 litres per second per m² of internal floor area. — F(V1) 1.24

- Internal doors should be undercut to allow air to flow throughout the dwelling. — F(V1) 1.25

Note: Internal doors should be undercut by 10mm above a floor finish or 20mm above the floor surface where a finish is not fitted.

- Where purge ventilation is delivered through openings in a habitable room, the minimum opening areas should be achieved. — F(V1) 1.29

- Background ventilators should be at least 1700mm above floor level. — F(V1) 1.34

Note: Background ventilators are normally intended to be left open.

4.6.10 Conservation of fuel and power

If work is being undertaken on an existing building which has a total useful floor area of over 1,000m² then, in addition to the principal works needing to comply with the energy-efficiency requirements, consequential improvements (where technically, functionally and economically feasible) will also have to be completed.

- The notional dwelling specification for floors in a new dwelling is $U = 0.13W/(m^2 \cdot K)$. — L(V1) Table 1.1

- The maximum U-value for new floors in new dwellings is $0.18W/(m^2 \cdot K)$. — L(V1) Table 4.1

- Floor insulation should be installed tight to the structure, without air gaps between insulation panels and at edges. — L(V1) 4.15c

- Ground floor wall-to-floor junctions should achieve continuity of insulation: L(V1) 4.17
  - perimeter floor insulation should abut or extend the full depth of the main floor insulation;
  - cavity wall insulation should extend below the damp-proof course and be at one block height below the underside of the floor structure/slab;
  - insulation between timber boards or within sheathing should extend to the floor plate;
  - the wall insulation and the floor perimeter insulation should abut.
- Intermediate floor-to-wall junctions should have continuous insulation in the external wall.

- Two independently controlled heating circuits should be provided in wet heating systems in new dwellings with a floor area of $150m^2$ or greater.   L(V1) 5.14

4.6.10.1 Underfloor heating

- All underfloor heating systems should have controls to adjust the operating temperature.   L(V1) 6.28
- Room thermostats for electric underfloor heating systems should have a manual override.
- Heating systems for screed floors that are greater than 65mm thick should automatically reduce the room temperature at night or when the room is unoccupied.
- Heat loss should be minimised.

- Ground floors and those in contact with the outside of the dwelling should be insulated to limit heat losses to not more than $10W/m^2$.   L(V1) 6.29

- The intermediate floor should have a layer of insulation to reduce downwards heat transmission.   L(V1) 6.31

- Electric cables for direct electric systems should be installed in screeds not exceeding 60mm and for night energy storage systems, at least 65mm.   L(V1) 6.33

- Programmable room thermostats with an override feature should be provided for all direct electric zones of the electric underfloor heating system.   L(V1) 6.35

 Note: All installed equipment in underfloor heating systems should be commissioned in accordance with BS EN 1264-4.

4.6.10.2 *Limiting solar gains*

> • Shading should be provided for buildings in the high-   O 1.6c
> risk locations.

 Note: The maximum glazing area of a building or part of a building is given in AD-O Table 1.1 or Table 1.2.

4.6.10.3 *Removing excess heat*

> • Buildings or parts of buildings with cross-ventilation   O 1.10
> should have: Table 1.3
>   • a total minimum free area that is the greater of
>   6% of the floor area or 70% of the glazing area in
>   high-risk areas;
>   • a bedroom minimum free area that is 13% of the
>   floor area of the room in high-risk areas.
>
> • Buildings or parts of buildings without cross-   O 1.11
> ventilation should have: Table 1.4
>   • a total minimum free area that is the greater of
>   10% of the floor area or 95% of the glazing area
>   in high-risk areas;
>   • a bedroom minimum free area that is 13% of the
>   floor area of the room in high-risk areas.
>
> • Excess heat should be removed from the residential   O 2.10
> building by opening windows, the use of ventilation
> louvres in external walls or a mechanical ventilation
> or cooling system.

4.6.11 *Fire safety*

Top tips for fire safety:

- Install smoke alarms on every level of your home, inside bedrooms and outside sleeping areas.
- Test smoke alarms every month.
- Talk with all family members about a fire escape plan and practice the plan twice a year.
- If a fire occurs in your home, **GET OUT**, **STAY OUT** and **CALL FOR HELP**.

4.6.11.1 Fire detection and fire alarm system

• All dwellings should have a fire detection and alarm system of minimum Grade D2 Category LD3.	B(V1) 1.1

 Smoke alarms should **not** be fixed over any opening between floors (e.g. stairs).

4.6.11.2 Air-circulation systems

• Air-circulation systems for an individual dwelling (or flat) with a floor more than 4.5m above ground level should not have transfer grilles in any wall, door, floor, or ceiling of the stair enclosure.	B(V1) 2.8 B(V1) 2.9 B(V1) 3.23a
• For ducted warm air heating systems, a room thermostat should be sited in the living room at a height between 1,370mm and 1,830mm above the floor.	B(V1) 3.23e

4.6.11.3 Cavity barriers

• Cavity barriers should be provided at every junction between an external cavity wall and every compartment floor (and compartment wall).	B(V1) 5.18

 Note: The compartment wall should be extended to the underside of the floor or roof above.

4.6.11.4 Compartmentation

• Compartment walls should be able to accommodate deflection of the floor when exposed to fire.	B(V1) 7.14
• Where a door is provided from a dwelling into the garage, the door opening must be greater than 100mm above the level of the garage floor (see AD-B(V1) Diagram 5.1).	B(V1) 5.7
• Any floor separating a flat from another part of the building should be treated as a compartment floor with a fire resistance equivalent to that given in AD-B(V1) Appendix B, Table B3.	B(V1) 7.1

Note: Where compartment walls are located within the middle half of a floor between vertical supports, the deflection may be assumed to be 40mm.

• Openings in compartment floors may be made for:	B(V2) 8.31
• fire doorsets;	
• pipes, ventilation ducts, service cables, chimneys, appliance ventilation ducts or ducts encasing one or more flue pipes;	
• refuse chutes of class A1 construction;	
• protected shafts.	
• Adjacent buildings should not be separated by floors.	

Note: Protected shafts are considered to be compartment walls or compartment floors.

4.6.11.5 Means of escape

It is important that people can exit from the dwelling safely.

• All habitable rooms in the upper storey(s) of a dwelling with one stair should be provided with an alternative escape route.	B(V1) 2.4
• A dwelling with two or more storeys more than 4.5m above ground level may either have a protected stairway to an upper storey or have a sprinkler system installed throughout.	B(V1) 2.6

4.6.11.5.1 ESCAPE ROUTES AND STAIRS

• Escape route floor finishes (including the treads of steps and surfaces of ramps and landings) should minimise their slipperiness when wet.	B(V1) 3.39
• Any sloping floor or tier should have a pitch of not more than 35° to the horizontal.	B(V1) 3.40
• Where a storey (or part of a building) is served by a single access stair, then that stair may be external provided that the stair serves a floor not more than 6m above the ground level and the stair meets the provisions in paragraph 3.62.	B(V1) 3.65 B(V1) 3.66 & 3.62

- If the escape stair serves any storey that has a floor   B(V1) 3.82
  level more than 18m above ground or access level,
  then the flights and landings of escape stairs should be
  constructed of materials of class A2-s3, d2 or better.

- All doors on escape routes should be hung so that   B(V1) 3.95
  they are clear of any change of floor level, other than
  a threshold or single step on the line of the doorway.

- Requirement B2 does not include guidance on the   B(V1) B2
  upper surfaces of floors and stairs.

4.6.11.5.2 SIGNAGE AND IDENTIFICATION

This is a new requirement for flats with a top storey that is more than
11m above ground level!

- To assist the fire service in identifying each floor in a   B(V1)15.13
  block of flats with a top storey that is more than 11m
  above ground level, floor identification signs and flat
  indicator signs should be provided.

- The floor identification signs should: B(V1)15.14
  - be located on every landing of a protected stairway
    and every protected corridor/lobby (or open-access
    balcony) into which a firefighting lift opens;
  - be in sans serif typeface with a letter height of at least
    50mm and the height of the numeral that designates
    the floor number should be at least 75mm;
  - be visible from the top step of a firefighting stair
    and, where possible, from inside a firefighting lift
    when the lift car doors open;
  - be mounted between 1.7m and 2m above floor level.

- The floor number designations should meet all of the   B(V1)15.15
  following conditions:
  - the floor closest to the mean ground level should be
    designated as either Floor 0 or Ground Floor;
  - each floor above the ground floor should be
    numbered sequentially beginning with Floor 1;
  - a lower ground floor should be designated as either
    Floor −1 or Lower Ground Floor;
  - each floor below the ground floor should be
    numbered sequentially beginning with Floor −1 or
    Basement 1.

- All floor identification signs should be supplemented by flat indicator signs, which provide information relating to the flats accessed on each storey.   B(V1)15.16

4.6.11.5.3 EMERGENCY ESCAPE WINDOWS AND EXTERNAL DOORS

- The bottom of the openable area of a window or external door providing emergency escape should be a maximum of 1,100mm above the floor.   B(V1) 2.10

4.6.11.5.4 SECURITY

- Only those openings that can be opened securely should be used when determining the free area for ventilation during sleeping hours.   O 3.6

Requirements for buildings other than dwellings – Additional requirements

This section provides the details for the floors in buildings other than dwellings, which in the majority of cases are the same as for dwellings, with the addition of the following requirements.

4.6.12 Fire safety

As a result of the Grenfell Tower disaster in 2017 (a fire which burned for about 60 hours and in which 72 people perished), one of the main changes made by the 2020 amendments to B(V1) and B(V2) was the introduction of a new recommendation for floor identification and flat indication signage within blocks of flats with storeys over 11m (as described in Section 4.6.11.5.2).

A number of other changes were made to provide more details concerning the protection of firefighters and providing users of the building with a means of escape.

4.6.13 Structure

• In small single-storey non-residential buildings and annexes, the floor area of the building or annex should not exceed 36m.	A 2C38a
• Where the floor area of the building or annexe exceeds 10m², the walls should have a mass of not less than 130kg/m².	A 2C38c

Note: There is no surface mass limitation recommended for floor areas of 10m² or less.

• Effective horizontal ties (or effective anchorage of suspended floors) should be provided in the following buildings:	A2 5.1c

  • all buildings not exceeding two storeys to which the public are admitted and which contain floor areas not exceeding 2,000m² at each storey;
  • flats, apartments and other residential buildings not exceeding four storeys;
  • hotels not exceeding four storeys;
  • industrial buildings not exceeding three storeys;
  • offices not exceeding four storeys;
  • retail premises not exceeding three storeys of less than 1,000m² floor area in each storey;
  • single-storey educational buildings.

• The wall of a flue or duct should have a fire resistance (REI) that is at least half of a compartment floor it passes through or is built into.	B(V2) 10.23

Note: Where the floor area of the building or annexe exceeds 10m², the walls should have a mass of not less than 130kg/m²; however, there is no surface mass limitation recommended for floor areas of 10m² or less.

4.6.14 Construction

If one element of structure supports or stabilises another then, as a minimum, the supporting element should have the same fire resistance as the other element.

• Floor areas should be undivided so exits are clearly visible from all parts (except in kitchens, ancillary offices and stores).	B(V2) 4.4
• A suspended ceiling that meets the appropriate provisions may be relied on to add to the fire resistance of the floor.	B(V2) Table B3

Note: Guidance on increasing the fire resistance of existing timber floors is given in BRE Digest 208.

• In flats of more than one storey, any floor that does not contribute to the support of the building should have a fire resistance of 30 minutes.	B(V2) Table B4
• For the floor over a basement (or, for more than one basement, then the floor over the topmost basement) the highest fire resistance figure should be used.	

4.6.14.1 Elements of structure

• Where a roof acts as a floor (e.g. for parking vehicles) or as a means of escape, it is excluded from the definition of 'element of structure' as are: • the lowest floor of the building; • a platform floor; • a loading gallery; • external walls.	B(V2) 7.3a

4.6.14.2 Enclosure of corridors that are not protected corridors

• Partitions should continue to the soffit of the structural floor above, or to a suspended ceiling.	B(V2) 2.25

4.6.14.3 Escape stairs

Every storey with a floor level **more than** 11m above ground level must have an alternative means of escape.

- A single escape stair may serve a building (or part of   B(V2) 3.3
  a building) if the building has no storey with a floor
  level more than 11m above ground level.

Note: If the maximum number of people needing to use escape stairs is unknown, it can be calculated using the floor space factors shown in AD-B(V2) Table D1.

- If the escape stair serves a storey that has a floor level   B(V2) 3.24
  more than 18m above ground or access level – or it is
  external – the flights and landings should be made of
  materials of class A2-s3, d2 or better.

4.6.15 Escape routes

- All doors on escape routes should be hung so that   B(V2) 5.12
  they open with a swing that is clear of any change of
  floor level.
- Escape route floor finishes should minimise their   B(V2) 5.17
  slipperiness when wet.
- Any sloping floor or tier should have a pitch of not   B(V2) 5.18
  more than 35° to the horizontal.

4.6.15.1 Floor space factors

Floor space factors are further described in **AD-B(V2) Table D1**.

- Guidance on appropriate floor space factors for   B(V2) D1
  concourses in sports grounds can be found in
  Concourses published by the Football Licensing
  Authority.
- In buildings other than dwellings, the number of   B(V2) D1
  people is calculated by dividing the area of a room or
  storey(s) (m²) by a floor space factor (m² per person).
  (For further details, see AD-B(V2) Table D1).

4.6.15.2 *Calculating exit capacity*

* If a ground floor storey exit and a stair share a   B(V2) 2.23
  final exit (via a ground floor lobby), then the final
  exit should be wide enough to evacuate people at a
  maximum flow rate equal to, or greater than, from
  the storey exit and stair combined.
* If the number of people entering the lobby from the
  ground storey is more than 60, then the floor distance
  from the foot of the stair (or storey exit) to the final
  exit should be a minimum of 2m.

4.6.15.3 *Cavity barriers*

* Cavity barriers are used to divide any cavity (such as   B(V2) 9.9
  a roof space) in a building. (See AD-B(V2) Table 9.1.)
* The requirement for cavity barriers to be fitted in   B(V2) 6.7
  concealed floor spaces can be reduced by installing
  a fire-resisting ceiling (minimum EI 30) below the
  cavity (see AD-B(V2) Diagram 9.3).

 Note: If the floor cavity is above a fire-resisting ceiling, or below a floor next to the ground, or next to oversite concrete that is not normally accessible by people, then AD-B(V2) Table 9.1 is not applicable.

* The minimum fire resistance requirements for floor   B(V2) 7.1
  structures and gallery structures are set out in AD-
  B(V2) Appendix B, Table B3.
* If there are openings in the floor and materials can
  accumulate in the cavity, then cavity barriers and
  access for cleaning are required.
* Cavity barriers should be provided at the junction   B(V2) 9.3
  between an external cavity wall and every
  compartment floor and compartment wall.
* Cavity barriers should be provided at the junction
  between an internal cavity wall and every
  compartment floor that forms a fire-resisting barrier.
* Compartment walls should extend to the underside   B(V2) 9.4
  of the floor or roof above.

4.6.15.4 Surfaces

- Internal floor surfaces adjacent to the threshold of an   M2 2.7
  accessible entrance should be made of materials that   M2 2.29
  do not impede the movement of wheelchairs, e.g. not
  coir matting.
- Changes in floor materials in accessible entrances
  should not create a potential trip hazard.
- The floor surface in an entrance lobby should help to   M2 2.29f
  remove rainwater from shoes and wheelchairs.
- Any manual controls for powered door systems   M2 2.21g
  should be located between 750mm and 1,000mm
  above floor level.
- The floor surface in an entrance hall or reception   M2 3.6g
  area should be slip-resistant.
- In order to help people with visual impairment to   M2 3.12
  appreciate the size of a space they have entered, or
  to find their way around, there should be a visual
  contrast between the wall and the ceiling, and
  between the wall and the floor.

 If any section has a gradient of 1:20 or steeper, it should be designed as
an internal ramp and in accordance with AD-M2 Diagram 32.

- Where a section of the floor has a gradient steeper   M2 3.14
  than 1:60 but less steep than 1:20, it shall rise no
  more than 500mm without a level rest area at least
  1,500mm long.
- Sloping sections should extend the full width of the
  corridor or, if not, the exposed edge should be clearly
  identified by visual contrast and, where necessary,
  protected by guarding.
- Floor surface finishes with patterns that could be
  mistaken for steps or changes of level should be
  avoided.
- Floor finishes should be slip-resistant.
- Junctions of floor surface materials at the entrance   M2 3.16e
  to the lobby area should not create a potential trip
  hazard.

Note: Special fire-safety measures may be required in a building with an entrance that passes through compartment floors (for further details see Annexes B and C of BS 9999).

4.6.15.5 Compartmentation

Effective compartmentation relies on continuous fire resistance at both the join between elements forming a compartment and also any openings between two compartments.

• Parts of buildings that are occupied mainly for different purposes should be separated from one another by compartment walls and/or compartment floors.	B(V2) 8.3

Compartmentation is not needed if one of the different purposes is a subsidiary to the other.

• Whilst the lowest floor in a building does not need to be a compartment floor.	B(V2) 8.5
• Floors that surround a protected shaft are compartment floors.	B(V2) 8.6
• All floors in residential 'institutional' and residential 'other' buildings should be constructed as compartment floors.	B(V2) 8.8 & 8.10
• Compartment height is measured from finished floor level to the underside of the roof or ceiling.	B(V2) Table 8.1
• All compartment floors should form a complete barrier to fire between the compartments they separate with the appropriate fire resistance (see Appendix B, Tables B3 and B4).	B(V2) 8.15
• Adjoining buildings should only be separated by walls, not floors.	B(V2) 8.18
• At the junction of a compartment floor and an external wall with no fire resistance, the external wall should be restrained at floor level in order to reduce any movement of the wall away from the floor if exposed to fire.	B(V2) 8.23
• Compartment walls should be able to accommodate deflection of the floor when exposed to fire.	B(V2) 8.24

- Openings in compartment floors should be limited.　B(V2) 8.31
- Fire doorsets in a compartment floor should be the same as for the floor in which it is fitted.　B(V2) Table C1
- Maximum nominal internal diameter of pipes passing through a compartment floor are shown in AD-B(V2) Table 10.1.　B(V2) 10.4

4.6.15.5.1 NON-RESIDENTIAL BUILDINGS

- In buildings in a non-residential purpose group, the following should be compartment floors:　B(V2) 8.11
  - if the building's top storey is more than 30m above ground level, every floor;
  - if the building has one or more basements, the floor of the ground storey premises;
  - if the building comprises 'shop and commercial', 'industrial' or 'storage' premises, every floor dividing a building into separate occupancies;
  - if any basement storey is at a depth more than 10m.
- A protected area used for horizontal evacuation from an adjoining area should have a floor area able to accommodate both its own occupants plus those from the largest adjoining area.　B(V2) 2.40

Note: Elements such as floor structures and gallery structures should have, as a minimum, the fire resistance given in AD-B(V2) Appendix B, Table B3.

4.6.16 Ventilation and air transfer

Planned air movement, provided by ventilation systems, is an important part of maintaining good indoor air quality, so when we talk about airtightness, we really mean unintentional air openings, and the uncontrolled exchange of air from outside to inside, and vice versa.

4.6.16.1 Enclosures and storage areas

• Enclosures should be permanently ventilated at the top and bottom and should have a paved impervious floor.	G 1.14
• Communal storage areas should have provision for washing down and draining the floor into a system suitable for receiving a polluted effluent.	G 1.15e
• Gullies should incorporate a trap which maintains a seal even during prolonged periods of disuse.	

4.6.16.2 Small premises

• The floor areas should be generally undivided (except for kitchens, ancillary offices and stores) to ensure that exits are clearly visible from all parts of the floor areas.	M2 3.34
• In a small premises, no storey should have a floor area more than 280m².	B(V2) 4.2
• An open stair may be used as a means of escape in small premises if the stair is a single stair and the floor area of any single storey is a maximum of 90m².	B(V2) 4.8

4.6.16.3 Lifting devices, lifts and stairlifts

• In lifting devices and platforms:	M2 3.28
• the landing call buttons shall be located between 900mm and 1,100mm from the floor of the landing, and at least 500mm from any return wall;	M2 3.43
• the floor of the lifting device should not be of a dark colour.	
• In passenger lifts:	M2 3.34
• car controls should be between 900mm and 1,200mm above the car floor;	
• landing call buttons should be between 900mm and 1,100mm from the landing floor.	
• People using or waiting for a lifting platform should have audible and visual information to tell them that the platform has arrived, and which floor it has reached.	M2 3.37

4.6.17 *Conservation of fuel and power*

4.6.17.1 *Fuel and power*

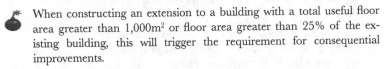

When constructing an extension to a building with a total useful floor area greater than 1,000m^2 or floor area greater than 25% of the existing building, this will trigger the requirement for consequential improvements.

Similarly, if there is a material change of use and/or a change to energy status, the area of openings in the newly created building should not be more than 25% of the total floor area.

4.6.17.2 *Sub-meters and heating zones*

• In buildings with a total useful floor area greater than 1,000m^2, automatic meter reading and data collection facilities should be installed.	L(V2) 5.17
• If the building has a floor area greater than 150m^2, heating should be split into different heating zones and each zone should have separate controls for timing and temperature demands.	L(V2) 6.14

4.6.17.3 *U-values*

• When calculating the building primary energy rate and building emission rate for a building with a swimming pool, the thermal performance of the pool basin should ***not*** be included in the calculation.	L(V2) 2.20

Note: For a building with a swimming pool, the building primary energy rate and building emission rate should be calculated as if the area covered by the pool were replaced with the equivalent area of floor with the same U-value as the pool surround.

- The maximum U-values for new or replacement floors in new and existing buildings and air permeability in new buildings is 0.18W(m²K).

  L(V2) Table 4.1

Note: If meeting the maximum U-values standards in an existing building would reduce by more than 5% the internal floor area of the room bounded by the wall, a lesser provision may be appropriate.

- The limiting U-values for existing floors in existing buildings is:
  - threshold = 0.70;
  - improved = 0.2.

  L(V2) Table 4.2

If meeting this standard would limit head room, or reduce the internal floor area of the room bounded by the wall by more than 5%, then a lesser standard would probably be more appropriate.

4.6.18 Protection from falling, collision and impact

- If the soffit beneath a stair is less than 2m above floor level, the area beneath a stair or ramp should be protected with both guarding and low-level cane detection or by a barrier.

  K 1.8 K 2.7

- A landing may include part of the floor of the building.

  K 1.21a

- In small single-storey non-residential buildings and annexes, the floor area of the building or annex should not exceed 36m².

  A 2C38a

- Where the floor area of the building or annexe exceeds 10m², the walls should have a mass of not less than 130kg/m². There is no limitation recommended for floor areas of 10m² or less.

  A 2C38c

- Effective horizontal ties or anchorage of suspended floors shall be provided (see A 3.5.1c for full details).

  A3 5.1c

4.6.19 *Access to and use of buildings*

4.6.19.1 *Conference and working facilities*

• In audience and spectator facilities, the floor of each wheelchair space should be horizontal.	M2 4.12h
• Wheelchair spaces at the back of a stepped terraced floor are provided in accordance with Figure 7.5.6 or 7.5.7 (the latter is best suited to entertainment buildings, such as cinemas).	M 4.12k
• In lecture/conference facilities, wheelchair users should have access to a podium or stage by means of a ramp or lifting platform.	
• There are exceptions to height requirements for some outlets, e.g. those set into the floor in open-plan offices.	M2 4.25
• Patterned walls in lecture/conference facilities, poor interior lighting or very bright natural backlighting should not have a detrimental effect on the ability of people to receive information from a sign language interpreter or a lip speaker.	M2 4.9

4.6.19.2 *Refreshment facilities*

• In refreshment facilities, bars and counters in all floor areas, even when located at different levels, should be accessible.	M2 4.3
• In refreshment facilities, changes in floor level are acceptable provided the different levels are accessible.	M2 4.15
• Part of the working surface of a bar or serving counter should be permanently accessible to wheelchair users (and at a level of not more than 850mm above the floor) and, where necessary, part of the surface should be at a higher level for people standing.	M2 4.16b
• The worktop of a shared refreshment facility (e.g. for tea making) should be at 850mm above the floor with a clear space beneath which is at least 700mm above the floor (see Figure 7.5.6).	M2 4.16c
• A wheelchair-accessible threshold should be located at the transition between an external seating area and the interior of the facility.	M2 4.16d

4.6.19.3 Sleeping accommodation

• Where a balcony is provided in a wheelchair-accessible bedroom, the door there should be a level threshold and there should be no horizontal transoms between 900mm and 1,200mm above the floor.	4.24
• Wall-mounted socket outlets and switches for permanently wired appliances, etc., should be between 400mm and 1,000mm above the floor.	M2 4.30
• Switches or controls that require precise hand movements should be between 750mm and 1,200mm above the floor.	
• Simple push button controls should not be more than 1,200mm above the floor.	
• Pull cords for emergency alarm systems should be set at 100mm and another set between 800mm and 1,000mm above the floor.	
• Floor, wall and ceiling surface materials should be chosen to help visually impaired people appreciate the boundaries of rooms or spaces and identify access routes.	M2 4.32
• The surface finish of sanitary fittings and grab bars should contrast visually with background wall and floor finishes.	M2 5.4k
• There should be visual contrast between wall and floor finishes.	

4.6.19.4 Car parks

• Any surface finish applied to a floor of a car park should meet the requirements shown in B2 and B4.	B(V2) 11.2

4.7 Walls

A wall consists of piers, columns and parapets. It also includes chimneys if they are attached to the building. It does not include windows, doors and similar openings but, however, does include the joint between their frames and the wall.

Reminder: The subsections include details for buildings other than dwellings only where they differ from the requirements for dwellings.

Reminder: Free downloads of all the diagrams and tables referenced in the subsections are available on-line at the Governments' Planning Portal: www.gov.uk/government/collections/approved-documents.

Requirements for dwellings

4.7.1 Access to and use of dwellings

• Wall-mounted switches, controls and socket outlets shall be accessible to people who have reduced reach.	M4(1), (2) & (3)
• All walls, ducts and boxings to the WC/cloakroom, bathroom and shower room should be strong enough to support grab rails, seats and other adaptations that could impose a load of up to 1.5kN/m².	M4 (2) & (3)
• Additional sanitary facilities beyond those required to comply with this guidance need not have strengthened walls.	

Remember M4(2) and M4(3) (*Access to and use of buildings*) are **voluntary** requirements and only apply in certain circumstances (e.g. where wheelchair access is required).

4.7.2 Structure

As shown in Diagram 1 and Table 2 of A 2C2 and 2C4 of AD-A, there are four main types of separating walls that can be used in order to achieve the required performance standards.

1 **Wall type 1 – solid masonry**
   • The resistance to airborne sound depends mainly on the mass per unit area of the wall.

2 **Wall type 2 – cavity masonry**
   • The resistance to airborne sound depends on the mass per unit area of the leaves and on the degree of isolation achieved. The isolation is affected by connections (such as wall ties and foundations) between the wall leaves and by the cavity width.

3 **Wall type 3 – masonry between independent panels**
   • The resistance to airborne sound depends partly on the type and mass per unit area of the core, and partly on the isolation and mass per unit area of the independent panels.

4   Wall type 4 – framed walls with absorbent material

- The resistance to airborne sound depends on the mass per unit area of the leaves, the isolation of the frames, and the absorption in the cavity between the frames.

4.7.2.1   General requirements

Walls should comply with the relevant requirements of BS EN 1996-2 (and if appropriate) BS EN 1996-1-1.

4.7.3.1   Airtightness and renovation of thermal elements

• With cavity walls, the inner block leaf mortar joint should be fully filled and pointed within the cavity.	L(V1) 4.21e
• Any walls, doors and windows in a conservatory or porch should be insulated and draught-proofed to at least the same extent as in the existing dwelling.	L(V1) 10.13a

Note: Renovation of a thermal element includes providing cavity wall insulation.

4.7.2.3   Basic requirements for stability

• The layout of walls (both internal and external) shall form a robust three-dimensional box structure.	A 2A2b
• Internal and external walls shall be connected either by masonry bonding or by using mechanical connections.	A 2A2c

4.7.2.4   Buttressing walls

• If the buttressing wall (see AD-A Diagram 12) is not itself a supported wall, its thickness (T2) should not be less than: • half the thickness required for an external or separating wall of similar height and length, less 5mm; or	A 2C26a

• 75mm if the wall forms part of a dwelling and does not exceed 6m in total height and 10m in length; and	A 2C26b
• 90mm in all other cases.	A 2C26c
• The length of the buttressing wall should be:	A 2C26
• at least 1/6 of the overall height of the supported wall;	
• bonded or securely tied to the supporting wall and to a buttressing wall, pier or chimney at the other end.	

Note: The size of any opening in the buttressing wall should be restricted, as shown in AD-A Diagram 12.

4.7.3.2 Cavity insulation

• For insulating a cavity wall with UF foam:	D 1.2a
• the inner leaf of a cavity wall should be built of masonry (bricks or blocks);	
• the installation shall be in accordance with BS 5618;	D 1.2e
• the material used shall be in accordance with the relevant recommendations of BS 5617.	D 1.2d

4.7.2.6 End restraint

• The ends of every wall should be bonded or otherwise securely tied throughout their full height to a buttressing wall, pier or chimney.	A 2C25
• Long walls may be provided with intermediate buttressing walls, piers or chimneys dividing the wall into distinct lengths within each storey.	

Note: Each distinct length is considered to be a supported wall for the purposes of the Building Regulations.

• Intermediate buttressing walls, piers or chimneys should provide lateral restraint to the full height of the supported wall.	A 2C25

4.7.2.7 Lateral support by roofs and floors

- A wall in each storey of a building should: A 2C32
  - extend to the full height of that storey;
  - have horizontal lateral supports.

Note: See also AD-A Table 9.

4.7.2.8 Loading on walls

- The maximum span for any floor supported by a wall   A 2C23
  is 6m, where the span is measured centre to centre of
  bearing (see AD-A Diagram 10).
- Vertical loading on walls should be distributed.   A 2C24a
- The difference in the level between one side of the   A 2C24b
  wall and the other side should be less than four times
  the thickness of the wall (see AD-A Diagram 11).
- The combined dead and imposed load should not   A 2C24c
  exceed 70kN/m at the base of the wall.
- Walls should not be subjected to lateral load other   A 2C24d
  than from wind (and that covered by differences in
  ground level).

4.7.2.9 Masonry units – construction and workmanship

Walls should be properly bonded and solidly put together with mortar
and constructed of masonry units conforming to the following standards:

- Clay bricks or blocks to BS EN 771-1.   A 2C20
- Calcium silicate bricks or blocks to BS EN 771-2.
- Concrete bricks or blocks to BS EN 771-3 or BS EN
  771-4.
- Manufactured stone to BS EN 771-5.
- Square dressed natural stone to the appropriate
  requirements described in BS EN 771-6.

4.7.2.9.1 MORTAR

Mortar should be one of the following:

- Mortar designation (iii) according to BS EN 1996-1-1. A 2C22
- Strength class M4 according to BS EN 998-2.
- 1:1:5 to 6 CEM I, lime, and fine aggregate.

4.7.2.10 Openings, recesses, overhangs and chases

- The number, size and position of openings and A 2C28
  recesses should not impair the stability of a wall.

Openings below ground floor (e.g. for services, ventilation, etc.) should be limited to a maximum area of 0.1 m² at not less than 2m centres (see AD-A Diagram 14 and Table 8).

- The amount of any projection should not impair the A 2C31
  stability of the wall.
- Vertical chases should not be deeper than 1/3 of the A 2C30a
  wall thickness.
- Horizontal chases should not be deeper than 1/6 of A 2C30b
  the thickness of the leaf of the wall.
- Chases should not be positioned so as to impair the A 2C30c
  stability of the wall.

4.7.2.11 Piers and chimneys providing restraint

The overall thickness should **not** be less than twice the required thickness of the supported wall (see AD-A Diagram 13).

- Chimneys should be at least twice the thickness, A 2C27b
  measured at right angles to the wall.
- Piers should have a minimum width of 190mm. A 2C27a
- Piers should measure at least three times the thickness
  of the supported wall.

4.7.3.3 Wall ties in separating and external cavity masonry walls

Wall ties for use in masonry cavity walls are either tie type A or B.

> • ***Tie type A***   E 2.19
> The leaves of a cavity masonry wall construction
> should be connected by either butterfly ties
> conforming to BS 1243:1978 or wall ties with an
> appropriate measured dynamic stiffness for the cavity
> width.

BS 5628-3 limits this tie type and spacing to cavity widths of 50mm to 75mm with a minimum masonry leaf thickness of 90mm.

> • ***Tie type B*** (for use only when tie type A does not   E 2.20
> satisfy the requirements)
> The leaves of a cavity masonry wall should be
> connected by double-triangle ties as described in
> BS 1243or wall ties with an appropriate measured
> dynamic stiffness for the cavity width.

In external cavity masonry walls, tie type B decreases the airborne sound insulation due to flanking transmission via the external wall leaf, as compared to tie type A.

4.7.3.1.2 CORRIDOR WALLS

> • Separating walls should be used between corridors   E 2.25
> and rooms in flats, in order to control flanking
> transmission and to provide the required sound
> insulation.

It is highly likely that the amount of sound insulation gained by using a separating wall will be reduced by the presence of a door.

> • Noisy parts of the building should preferably have a   E 2.27
> lobby, double door or high-performance doorset to
> contain the noise.

4.7.3.1.3 REFUSE CHUTES

• A wall separating a habitable room (or kitchen) from a refuse chute should have a mass per unit area (including any finishes) of at least 1,320kg/m².	E 2.28

4.7.3 Types of walls

4.7.3.1 Wall type 1 – solid masonry

When using a solid masonry wall, the resistance to airborne sound depends mainly on the mass per unit area of the wall. As shown in Table 4.3 below, there are three different categories of solid masonry walls.

4.7.3.1.1 WALL TYPE 1 – GENERAL REQUIREMENTS

• Fill and seal all masonry joints with mortar.	E 2.32a
• Lay bricks frog up to achieve the required mass per unit area and avoid air paths.	E 2.32b
• Use bricks/blocks that extend to the full thickness of the wall.	E 2.32c
• Ensure that an external cavity wall is stopped with a flexible closer at the junction with a separating wall.	E 2.32d
• Control flanking transmission from walls and floors connected to the separating wall.	E 2.32e
• Stagger the position of sockets on opposite sides of the separating wall.	E 2.32f

• Ensure flue blocks:
  • will not adversely affect the sound insulation;
  • use a suitable finish.
	E 2.32g

4.7.3.1.2 WALL TYPE 1 – JUNCTIONS WITH AN EXTERNAL CAVITY
      WALL WITH MASONRY INNER LEAF

Where the external wall is a cavity wall:

• The outer leaf of the wall may be of any construction.	E 2.36a
• The cavity should be stopped with a flexible close (see AD-E Diagram 2.5).	E 2.36b

Table 4.3 Wall type 1 – categories

Wall type 1 **Category 1.1** **Solid masonry** Dense aggregate concrete block, plaster on both room faces	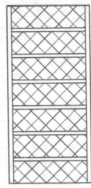	Minimum mass per unit area (including plaster) 415kg/m² 13mm Plaster on both room faces Blocks laid flat to the full thickness of the wall For example: Size 215mm laid flat Density 1840kg/m³ Coursing 110mm Lightweight Plaster 13mm

Wall type 1
Category 1.2
  Dense aggregate
  concrete
Dense aggregate concrete, cast in situ, plaster on both room faces
	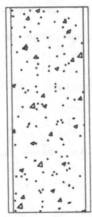	Minimum mass per unit area (including plaster) 415kg/m²
		Plaster on both room faces
		For example:
		Concrete 190mm
		Density 2200 kg/m³
		Lightweight Plaster 13mm

Wall type 1 **Category 1.3 Brick** Brick, plaster on both room faces		Minimum mass per unit area (including plaster) 375kg/m² 13mm plaster on both room faces Bricks to be laid frog up, coursed with headers For example: Size 215mm laid flat Density 1610kg/m³ Coursing 75mm Lightweight Plaster 13mm

The separating wall should be joined (bonded or tied) to the inner leaf of the external cavity wall by one of the methods shown in AD-E Diagrams 2.6 or 2.7.

• The masonry inner leaf should have a mass per unit area of at least 120kg/m² excluding finish unless there are openings in the external wall (see AD-E Diagram 2.8) that are: • not less than 1m high; • on both sides of the separating wall at every storey; • not more than 700mm from the face of the separating wall on both sides.	E 2.38

4.7.3.1.3 WALL TYPE 1 – JUNCTIONS WITH AN EXTERNAL CAVITY WALL WITH TIMBER FRAME INNER LEAF

• Where the external wall is a cavity wall: • the outer leaf of the wall may be of any construction;	E 2.40a
• the cavity should be stopped with a flexible closer. (See AD-E Diagram 2.9.)	E 2.40b
• Where the inner leaf of an external cavity wall is of framed construction, the framed inner leaf should: • abut the separating wall;	E 2.41a1
• be tied to the separating wall with ties at no more than 300mm centres vertically.	E 2.41b1
• The wall finish of the framed inner leaf of the external wall should be: • one layer of plasterboard; or	E 2.41a2
• two layers of plasterboard where there is a separating floor;	E 2.41b2
• each sheet of plasterboard should be of minimum mass per unit area 10kg/m²;	E 2.41c
• all joints should be sealed with tape or caulked with sealant.	E 2.41d

4.7.3.1.4 WALL TYPE 1 – JUNCTIONS WITH INTERNAL TIMBER FLOORS

- If the floor joists are to be supported on a type 1 separating wall, then they should be supported on hangers (as opposed to being built in).   E 2.45

4.7.3.1.5 WALL TYPE 1 – JUNCTIONS WITH INTERNAL CONCRETE FLOORS

- An internal concrete floor slab may only be carried through a type 1 separating wall if the floor base has a mass per unit area of at least $365kg/m^2$.   E 2.46
- Internal hollow-core concrete plank floors and concrete beams with infilling block floors should not be continuous through a type 1 separating wall.   E 2.47

4.7.3.1.6 WALL TYPE 1 – JUNCTIONS WITH CONCRETE GROUND FLOORS

- The ground floor may be a solid slab laid on the ground, or a suspended concrete floor.   E 2.51
- A concrete slab floor on the ground may be continuous under a type 1 separating wall.
- A suspended concrete floor may only pass under a type 1 separating wall if the floor has a mass of at least $365kg/m^2$.   E 2.52
- Hollow-core concrete plank and concrete beams within filling block floors should not be continuous under a type 1 separating wall.   E 2.53

4.7.3.1.7 WALL TYPE 1 – JUNCTIONS WITH CEILING AND ROOF

- Where a type 1 separating wall is used, it should be continuous to the underside of the roof.   E 2.55
- The junction between the separating wall and the roof should be filled with a flexible closer which is also suitable as a fire-stop. (See AD-E Diagram 2.13.)   E 2.56

- Where the roof or loft space is not a habitable room (and there is a ceiling with a minimum mass per unit area of 10kg/m^2 with sealed joints), the mass per unit area of the separating wall above the ceiling may be reduced to 150kg/m^2.
  E 2.57

- If lightweight aggregate blocks of density less than 1,200kg/m^3 are used above ceiling level, one side should be sealed with cement paint or plaster skim.
  E 2.58

- Where there is an external cavity wall, the cavity should be closed at eaves level with a suitable flexible material (e.g. mineral wool). (See AD-E Diagram 2.14.)
  E 2.59

4.7.3.2 Wall type 2 – cavity masonry

4.7.3.2.1 WALL TYPE 2 – GENERAL REQUIREMENTS

- Fill and seal all masonry joints with mortar.
  E 2.65a

- Keep the cavity leaves separate below ground-floor level.
  E 2.65b

- Ensure that any external cavity wall is stopped with a flexible closer at the junction with the separating wall.
  E 2.65c

- Control flanking transmission from walls and floors connected to the separating wall.
  E 2.65d

- Stagger the position of sockets on opposite sides of the separating wall.
  E 2.65e

- Ensure that flue blocks do not adversely affect the sound insulation and that a suitable finish is used over the flue blocks.
  E 2.65f

- The cavity separating wall should **not** be converted to a type 1 (solid masonry) separating wall by inserting mortar or concrete into the cavity between the two leaves.
  E 2.65a2

- A change should not be made to a solid wall construction in the roof space.
  E 2.65b2

- Cavity walls should **not** be built off a continuous solid concrete slab floor.
  E 2.65c2

• Deep sockets and chases should **not** be used in a separating wall.	E 2.65d2
• Deep sockets and chases in a separating wall should **not** be placed back-to-back.	E 2.65d2
• Wall ties used to connect the leaves of a cavity masonry wall should be tie type A.	E 2.66

4.7.3.2.2 WALL TYPE 2 – JUNCTIONS WITH AN EXTERNAL CAVITY WALL WITH MASONRY INNER LEAF

• Where the external wall is a cavity wall:	
• the outer leaf of the wall may be of any construction;	E 2.73a
• the cavity should be stopped with a flexible closer.	E 2.73b
• The separating wall should be joined to the inner leaf of the external cavity wall.	E 2.74
• The masonry inner leaf should have a mass per unit area of at least 120kg/m² excluding finish.	E 2.75
• There is no minimum mass requirement where separating wall type 2.1, 2.3 or 2.4 is used **unless** there is also a separating floor.	E 2.76

4.7.3.2.3 WALL TYPE 2 – JUNCTIONS WITH AN EXTERNAL CAVITY WALL WITH TIMBER FRAME INNER LEAF

• Where the external wall is a cavity wall:	E 2.77a
• the outer leaf of the wall may be of any construction;	
• the cavity should be stopped with a flexible closer.	E 2.77b

Where the inner leaf of an external cavity wall is of framed construction, the framed inner leaf should abut the separating wall and be tied to it with ties at no more than 300mm centres vertically and the wall finish of the inner leaf of the external wall should be:

• One layer of plasterboard (two where there is a separating floor).	E 2.78a2

- Each sheet of plasterboard to be of minimum mass per unit area 10kg/m².   E 2.78c2

- All joints should be sealed with tape or caulked with sealant.   E 2.78d2

4.7.3.2.4 WALL TYPE 2 – JUNCTIONS WITH INTERNAL TIMBER FLOORS

If the floor joists are to be supported on the separating wall, they should be supported on hangers (as opposed to being built in).

4.7.3.2.5 WALL TYPE 2 – JUNCTIONS WITH INTERNAL CONCRETE FLOORS

Internal concrete floors should generally be built into a type 2 separating wall and carried through to the cavity face of the leaf.

The cavity should **not** be bridged.

4.7.3.2.6 WALL TYPE 2 – JUNCTIONS WITH CONCRETE GROUND FLOORS

- The ground floor may be a solid slab laid on the ground, or a suspended concrete floor.   E 2.88
- A concrete slab floor on the ground should not be continuous under a type 2 separating wall.

- A suspended concrete floor should not be continuous under a type 2 separating wall.   E 2.89
- A suspended concrete floor should be carried through to the cavity face of the leaf.
- The cavity should **not** be bridged.

4.7.3.2.7 WALL TYPE 2 – JUNCTIONS WITH CEILING AND ROOF SPACE

- A type 2 separating wall should be continuous to the underside of the roof.   E 2.91

- The junction between the separating wall and the roof should be filled with a flexible closer that is also suitable to act as a fire-stop.   E 2.92

- If lightweight aggregate blocks (with a density less than 1,200kg/m^3) are used above ceiling level, then one side should be sealed with cement paint or plaster skim.

  E 2.94

- The cavity of an external cavity wall should be closed at eaves level with a suitable flexible material (e.g. mineral wool).

  E 2.95

- A rigid connection between the inner and external wall leaves should be avoided.

4.7.3.2.8 GUIDANCE FOR OTHER WALL TYPE 2 JUNCTIONS

- Junctions with internal masonry walls that abut a type 2 separating wall should have a mass per unit area of at least 120kg/m^2 excluding finish.

  E 2.81

When there is a separating floor, the internal masonry walls should also have a mass per unit area of at least 120kg/m^2 excluding finish.

4.7.3.3 Wall type 3 – masonry between independent panels

4.7.3.3.1 WALL TYPE 3 – GENERAL REQUIREMENTS

- Fill and seal all masonry joints with mortar.

  E 2.101a

- Control flanking transmission from walls and floors connected to the separating wall.

  E 2.101b

- The panels and any frame should not be in contact with the core wall.

  E 2.99

- The panels and/or supporting frames should be fixed to the ceiling and floor only.

  E 101c

- All joints should be taped and sealed.

  E 2.101d

- Flue blocks shall not adversely affect the sound insulation.

  E 2.101e

- A suitable finish should be used over the flue blocks (see BS 1289-1).

  E 2.101e

- Freestanding panels and/or the frame should not be fixed, tied or connected to the masonry core.

  E 2.101

- Wall ties in cavity masonry cores, used to connect the leaves of a cavity masonry core together, should be tie type A. E 2.102

- The minimum mass per unit area of independent panels (excluding any supporting framework) should be 20kg/m². E 2.104

- Panels should be either at least two layers of plasterboard with staggered joints or a composite panel consisting of two sheets of plasterboard separated by a cellular core. E 2.104

- Panels that are not supported on a frame should be at least 35mm from the masonry core. E 2.104

- Panels that are supported on a frame should have a gap of at least 10mm between the frame and the masonry core. E 2.104

4.7.3.3.2 WALL TYPE 3 – JUNCTIONS WITH AN EXTERNAL CAVITY WALL WITH MASONRY INNER LEAF

- If the external wall is a cavity wall: E 2.108
  - the outer leaf of the wall may be of any construction;
  - the cavity should be stopped with a flexible closer.

- If the inner leaf of an external cavity wall is masonry: E 2.109
  - the inner leaf of the external wall should be bonded or tied to the masonry core;
  - the inner leaf of the external wall should be lined with independent panels.

- If there is a separating floor, the masonry inner leaf (of the external wall) should have a minimum mass per unit area of at least 120kg/m² excluding finish. E 2.110 / E 2.111

- If there is no separating floor, the external wall may be finished with plaster or plasterboard of minimum mass per unit area 10kg/m².

4.7.3.3.3 WALL TYPE 3 – JUNCTIONS WITH INTERNAL FRAMED WALLS

- Load-bearing (framed) internal walls should be fixed to the masonry core through a continuous pad of mineral wool.   E 2.115
- Non-load-bearing internal walls should be butted to the independent panels.   E 2.116
- All joints between internal walls and panels should be sealed with tape or caulked with sealant.   E 2.117

4.7.3.3.4 WALL TYPE 3 – JUNCTIONS WITH INTERNAL MASONRY WALLS

- Internal walls that abut a type 3 separating wall should not be of masonry construction.   E 2.118

4.7.3.3.5 WALL TYPE 3 – JUNCTIONS WITH TIMBER GROUND FLOORS

- Floor joists supported on a separating wall should be supported on hangers, as opposed to being built in.   E 2.123
- The spaces between floor joists should be sealed with full-depth timber blocking.   E 2.124

4.7.3.3.6 WALL TYPE 3 – JUNCTIONS WITH INTERNAL TIMBER FLOORS

- If the floor joists are to be supported on the separating wall, they should be supported on hangers, as opposed to being built in.   E 2.119
- Spaces between the floor joists should be sealed with full-depth timber blocking.   E 2.120

4.7.3.3.7 WALL TYPE 3 – JUNCTIONS WITH CONCRETE GROUND
FLOORS

• The ground floor may be a solid slab laid on the ground, or a suspended concrete floor.	E 2.126
• For wall types 3.1 and 3.2 (solid masonry cores):	
• a concrete slab floor on the ground may be continuous under the solid masonry core of the separating wall;	E 2.127
• hollow-core concrete planks (and concrete beams with infilling block floors) should **not** be continuous under the solid masonry core of the separating wall.	E 2.129

A suspended concrete floor may only pass under the solid masonry core
if the floor has a mass per unit area of at least 365kg/m^2.

• For wall type 3.3 (cavity masonry core):	
• a concrete–masonry core of a slab floor on the ground should **not** be continuous under the cavity masonry core of the separating wall;	E 2.130
• a suspended concrete floor should **not** be continuous under the cavity type 3.3 separating wall;	E 2.131
• a suspended concrete floor **should** be carried through to the cavity face of the leaf, but the cavity should not be bridged.	E 2.132

4.7.3.3.8 WALL TYPE 3 – JUNCTIONS WITH INTERNAL CONCRETE
FLOORS

• For wall types 3.1 and 3.2 (i.e. those with solid masonry cores), internal concrete floor slabs may only be carried through a solid masonry core if the floor base has a mass per unit area of at least 365kg/m^2.	E 2.121
• For wall type 3.3 (cavity masonry core):	E 2.122
• internal concrete floors should generally be built into a cavity masonry core and carried through to the cavity face of the leaf;	
• the cavity should not be bridged.	

4.7.3.3.9 WALL TYPE 3 – JUNCTIONS WITH INTERNAL MASONRY FLOORS

Internal walls that abut a type 2 separating wall should not be of masonry construction.

As no official guidance is currently available on junctions with an external cavity wall with timber frame inner leaf and external solid masonry wall junctions, it is best to seek specialist advice.

4.7.3.3.10 WALL TYPE 3 – JUNCTIONS WITH CEILING AND ROOF SPACE

• The masonry core should be continuous to the underside of the roof.	E 2.133
• The junction between the separating wall and the roof should be filled with a flexible closer that is also suitable as a fire-stop.	E 2.134
• The junction between the ceiling and independent panels should be sealed with tape or caulked with sealant.	E 2.135
• If there is an external cavity wall, the cavity should be closed at eaves level with a suitable flexible material (e.g. mineral wool). (See AD-E Diagram 2.36.)	E 2.136
• Rigid connections between the inner and external wall leaves should be avoided where possible.	E 2.136
• If a rigid material is used, then it should only be rigidly bonded to one leaf.	E 2.136
• For wall types 3.1 and 3.2 (solid masonry core):	
• if the roof or loft space is not a habitable room, independent panels may be omitted in the roof space;	E 2.137
• if lightweight aggregate blocks with a density less than $1,200\text{kg/m}^3$ are used above ceiling level, one side should be sealed with cement paint or plaster skim;	E 2.138
• for wall type 3.3 (cavity masonry core), if the roof or loft space is not a habitable room, the independent panels may be omitted in the roof space, but the cavity masonry core should be maintained to the underside of the roof.	E 2.139

4.7.3.4 Wall type 4 – framed walls with absorbent material

4.7.3.4.1 GENERAL REQUIREMENTS

• If a fire-stop is required in the cavity between frames, then it should either be flexible or only be fixed to one frame.	E 2.146
• Stagger the position of sockets on opposite sides of the separating wall, and use a similar thickness of cladding behind the socket box.	E 2.146b1
• Do not locate sockets back-to-back.	E 2.146b2
• Ensure that each layer of plasterboard is independently fixed to the stud frame.	E 2.146c
• Ensure there is a minimum edge-to-edge stagger of 150mm. • Plasterboard should not be chased.	E 2.146b2
• The lining shall be two or more layers of plasterboard. • If a masonry core is used for structural purposes, the core should only be connected to one frame. • The minimum distance between inside lining faces shall be 200mm. • Plywood sheathing may be used in the cavity if required for structural reasons.	E 2.147

4.7.3.4.2 WALL TYPE 4 – JUNCTIONS WITH AN EXTERNAL CAVITY WALL WITH TIMBER FRAME INNER LEAF

• If the external wall is a cavity wall: • the outer leaf of the wall may be of any construction; • the cavity should be stopped between the ends of the separating wall and the outer leaf with a flexible closer. • The wall finish of the inner leaf of the external wall should:	E 2.149

- be one layer of plasterboard (or two layers of plasterboard if there is a separating floor);
- each sheet of plasterboard should be of minimum mass per unit area 10kg/m^2;
- all joints should be sealed with tape or caulked with sealant.

E 2.150a
E 2.150b

4.7.3.4.3 WALL TYPE 4 – JUNCTIONS WITH INTERNAL TIMBER FLOORS

- Air paths through the wall into the cavity shall be blocked using solid timber blockings, continuous ring beams or joists.

E 2.154

4.7.3.4.4 WALL TYPE 4 – JUNCTIONS WITH TIMBER GROUND FLOORS

- Air paths through the wall into the cavity shall be blocked using solid timber blockings, continuous ring beams or joists.

E 2.156

4.7.3.4.5 WALL TYPE 4 – JUNCTIONS WITH CONCRETE GROUND FLOORS

- If the ground floor is a concrete slab laid on the ground, it may be continuous under a type 4 separating wall.
- If the ground floor is a suspended concrete floor, it may only pass under a wall type 4 if the floor has a mass per unit area of at least 365kg/m^2.

E 2.158

4.7.3.4.6 WALL TYPE 4 – JUNCTIONS WITH CEILING AND ROOF SPACE

- The wall should preferably be continuous to the underside of the roof.

E 2.160

- The junction between the separating wall and the roof should be filled with a flexible closer.

E 2.161

- The junction between the ceiling and the wall linings   E 2.162
  should be sealed with tape or caulked with sealant.

- If the roof or loft space is not a habitable room (and
  there is a ceiling with a minimum mass per unit area
  of 10kg/m² with sealed joints), then:
  - either the linings on each frame may be reduced to   E 2.162a
    two layers of plasterboard; or
  - the cavity may be closed at ceiling level without   E 2.162b
    connecting the two frames rigidly together.

- External wall cavities should be closed at eaves level   E 2.163
  with a suitable material.

4.7.4 Ventilation

4.7.4.1 Installing energy-efficiency measures

- Building work aimed at improving energy efficiency   F(V1) 3.6
  should not reduce infiltration and cause the dwelling
  to become under-ventilated.

- When making changes to thermal elements, AD-F   F1 3.8
  Table 3.1 should be used to calculate the number of
  major and minor energy-efficiency measures involved
  in a project.

AD-F Appendix D provides a useful checklist for determining the ventilation provision in an existing dwelling which can be used (before energy-efficiency measures are implemented) to help establish compliance with the minimum standards of requirement.

4.7.4.2 Humidity reduction

By keeping the home well ventilated, high levels of humidity are controlled and those who suffer with allergies and asthma will see a reduction in symptoms.

- In a properly heated dwelling subject to typical moisture generation, there should be no visible mould on the inner surfaces of the external walls.   F(V1) App B B2
- Mould can grow whether the dwelling is occupied or unoccupied, and the average relative humidity in a room should always be less than that shown in AD-F Table B3.

4.7.4.3 Background ventilators for continuous mechanical extract ventilation

- Where there is no external wall in a habitable room, it can be ventilated (provided that there is a permanent opening) through either: F(V1) 1.65
  - another habitable room; or
  - a conservatory;
  which can provide purge ventilation and/or background ventilation.

4.7.4.4 Natural ventilation with background ventilators and intermittent extract fans

- If a wet room has no external walls, the intermittent extract fan should extract at least four air changes per hour (4 ach) in order to meet purge ventilation standards.   F(V1) 1.48
- All rooms with external walls should have background ventilators.   F(V1) 1.52

4.7.5 Site preparation and resistance to moisture

4.7.5.1 External cavity walls

The outer leaf of cavity external walls shall be separated from the inner leaf by a drained air space.

- A cavity external wall could include: C 5.13
  - outer leaf masonry (bricks, blocks, stone or manufactured stone);
  - a cavity of at least 50mm wide, bridged only by wall ties.
- Where a cavity is to be partially filled, the residual cavity should not be less than 50mm wide (see AD-C Diagram 11).

- A full or partial fill insulating material may be placed   C 5.15
  in the cavity between the outer leaf and an inner leaf of masonry.

4.7.5.2 Tension straps

- Tension straps (conforming to BS EN 845-1) should   A 2C35
  be used to strap walls to floors above ground level, at intervals not exceeding 2m and as shown in AD-A Diagram 15.

- Tension straps should have corrosion resistance (e.g. galvanized steel or austenitic stainless steel).
- The declared tensile strength of tension straps should not be less than 8kN.

- Tension straps need not be provided in the   A 2C35a
  longitudinal direction of joists in houses of not more than two storeys if the joists:
  - are no more than 1.2m centres;
  - have at least 90mm bearing on the supported walls;
  - are carried on the supported wall by joist hangers   A 2C35b
    (in accordance with BS EN 845-1 – see AD-A Diagram 15);
  - are incorporated at not more than 2m centres;   A 2C35c
  - are used when a concrete floor has at least 90mm bearing on the supported wall (see AD-A Diagram 15d).

Note: Where floors are at the same level on each side of a supported wall, and contact between the floors and wall is continuous, then the points of contact should be in line on plan (see AD-A Diagram 15e).

4.7.5.2.1 GABLE WALL TENSION STRAPS

- Gable walls should be strapped to roofs by tension   A 2C35d
  straps, as shown in AD-A Diagram 16(a) and (b).
- Vertical strapping at least 1m in length should be
  provided at eaves level at intervals not exceeding 2m,
  as shown in AD-A Diagram 16 (c) and (d).

Vertical strapping may be omitted if: the roof has a pitch of 15° or more; is tiled or slated; is capable of resisting wind gusts; and the main timber members spanning on to the supported wall are at not more than 1.2m centres.

4.7.5.3 Thickness of walls in certain small buildings

This section applies to small residential buildings not more than three storeys high (types of wall are shown in AD-A Table 2).

- Walls should comply with the requirements of BS   A 2C3c
  EN 1996-2.

The recommended thicknesses of different types of walls are:

- Solid walls constructed of coursed brickwork or   A 2C6
  blockwork: at least as thick as 1/16 of the storey
  height.
- Walls constructed in uncoursed stone, flints, clunches   A 2C7
  or bricks, or other burnt or vitrified material: not less
  than 1.33 times the thickness of the storey height.
- All cavity walls in coursed brickwork or blockwork: A 2C8
  eaves at least 90mm thick and cavities at least 50mm
  wide.
- For external walls, compartment walls and separating
  walls in cavity construction: the combined thickness
  of the two leaves plus 10mm should be at least as
  thick as 1/16 of the storey height.
- Walls that provide vertical support to other walls: A 2C9
  should not be less than the thickness of any part of
  the wall to which it gives vertical support.

- All internal load-bearing walls (except compartment   A 2C10
  and/or separating walls): should have a thickness not
  less than 5mm.

- The minimum thickness and maximum height of   A 2C11
  parapet walls should be as shown in AD-A Diagram 4.

 Walls constructed of bricks or blocks shall be in accordance with BS EN 1991-1-4.

4.7.5.3.1 SMALL BUILDING DIMENSIONS

- The height of a wall or a storey should be measured   A 2C18
  in accordance with the rules set out in AD-A
  Diagram 8.

- The maximum height of the building should be less   A 2C4i
  than 15m (see AD-A Diagram 7).

- The height of the building should not exceed twice   A 2C4ii
  the least width of the building (see AD-A Diagram 8).

Where the projection exceeds the width, the height of the wing should not be greater than twice the minimum width of the wing.

4.7.5.4 Wall cladding

The guidance given below relates to **all** forms of cladding, including curtain walling and glass façades.

- Cladding is securely fixed to and supported by   A 3.2
  the structure of the building and capable of safely
  sustaining wind loads and any differential movement
  of the supporting structure.

- The cladding shall be securely fixed to, and   A 3.2b
  supported by, the structure of the building using both
  vertical support and horizontal restraint.

If the cladding is required to support other fixtures (e.g. handrails or fittings such as antennas and signboards), loads and forces arising from such fixtures and fittings must be taken into account.

4.7.5.5 Wall ties

There are two types of wall tie that can be used in masonry cavity walls:

1 Type A (butterfly ties), which are the norm.
2 Type B (double-triangle ties), which are used only in external masonry cavity walls.

Stainless steel cavity wall ties are specified for **all** houses regardless of their location.

• Wall ties should have a horizontal spacing of 900mm and a vertical spacing of 450mm.	A 2C8
• Wall ties should also be provided (spaced not more than 300mm apart vertically) within a distance of 225mm from the vertical edges of all openings, movement joints and roof verges.	A 2C8
• Wall ties should comply with BS EN 845-1 and AD-A Table 5.	A 2C19

4.7.6 Fire safety

The latest editions of AD-B1 and AD-B2 are aimed at reducing the spread of fire over external walls and roofs to limit the potential of a tragedy such as Grenfell Tower happening again by:

- **Resisting fire spread over external walls.**
  It is essential that the external envelope of a building does not enable undue fire spread from one part of a building to another part.
- **Resisting fire spread from one building to another.**
  The external envelope of a building should not provide a medium for undue fire spread to adjacent buildings or be readily ignited by fires in adjacent buildings.
- **Using fire-stopping and sealing systems.**
  Only proprietary fire-stopping and sealing systems that have been shown (by test) to maintain the fire resistance of the wall or other element are used.

Proprietary fire-stopping and sealing systems are available and may be used.

4.7.6.1 Cavities and concealed spaces

Cavities in a building's construction enable fire spread and should be limited. Therefore, buildings shall be designed and constructed so that the unseen spread of fire and smoke within concealed spaces in their structure and fabric is prevented, and flame spread over wall or ceiling surfaces shall be controlled by providing appropriate lining materials.

4.7.6.1.1 CAVITY BARRIERS

• Cavity barriers should be provided at:
  • the edges of cavities and around openings;
  • between an internal or external cavity wall, compartment floor and any wall and door assembly forming a fire-resisting barrier.
	B(V1) 5.18
	B(V1) 8.3

• Cavity barriers can be made in a stud wall, partition, or around openings, using a mixture of steel, timber, polythene-sleeved (or compressed) mineral wool, calcium silicate, cement, or gypsum-based boards.	B(V1) 5.21
• If the wall is an external cavity wall (see AD-C Diagram 8), the cavity should be taken down at least 225mm below the level of the lowest damp-proof course, or a damp-proof tray should be provided.	C5.5
• A protected stairway enclosure requires cavity barriers.	B(V1) 2.5
• If there is a cavity above a fire-resisting protected escape route, then this should be enclosed on the lower side by a fire-resisting ceiling and be extended throughout the building.	B(V1) 8.5

4.7.6.2 Compartmentation

To prevent the spread of fire within a building, whenever possible:

• Walls separating semi-detached houses, or houses in terraces, should be constructed as a compartment wall and the houses should be considered as separate buildings.	B(V1) 5.5

- Compartment walls that are common to two or more   B(V1) 8.4
  buildings should:
  - extend to the underside of the floor or roof above
    if it is not possible to fit a cavity barrier;
  - be constructed as a compartment wall;
  - run the full height of the building in a continuous
    vertical plane.

The lowest floor in a building does not need to be constructed as a compartment floor.

4.7.6.2.1 JUNCTION OF A COMPARTMENT WALL WITH ANOTHER COMPARTMENT WALL

- Compartment walls should be provided to: B(V1) 7.1
  - separate a flat from another part of the building;
  - enclose a refuse storage chamber;
  - act as a wall common to two or more buildings;
  - separate parts of a building occupied mainly for   B(V1) 7.3
    different purposes.
- Compartment walls should meet the underside of the   B(V1) 5.11
  roof covering or deck with fire-stopping material.   B(V1) 7.15

4.7.6.2.2 JUNCTION OF A COMPARTMENT WALL WITH A ROOF

If a fire penetrates a roof near a compartment wall, there is a risk that it will spread over the roof to the adjoining compartment.

- To reduce the risk of fire spreading over the roof   B(V1) 5.12
  from one compartment to another, a 1,500mm wide
  zone of the roof, either side of the wall, should be
  covered (see AD-B(V1) Diagram 5.2a) by double-
  skinned material that is 300mm wide and centred
  over the wall.

4.7.6.2.3 OPENINGS IN COMPARTMENT WALLS SEPARATING BUILDINGS OR OCCUPANCIES

• Openings in compartment walls or compartment floors should be limited to:	B(V1) 7.19
• doors that have the appropriate fire resistance;	
• the passage of pipes, ventilation ducts, service cables, chimneys, appliance ventilation ducts or ducts encasing one or more flue pipes;	
• fire doorsets;	B(V1) 7.20
• refuse chutes;	
• protected shafts.	

Note: For buildings containing flats, see section 6.6.7.

4.7.6.2.4 CONSTRUCTION OF COMPARTMENT WALLS

Adjoining buildings should **only** be separated by walls, not floors.

• All compartment walls should form a complete barrier to fire (see Tables B3 and B4 in Appendix B).	B(V1) 7.5
• If trussed rafters bridge the wall, they should be designed so that failure of any part of the truss due to a fire in one compartment will not cause failure of any part of the truss in another compartment.	B(V1) 7.6
• Timber beams, joists, purlins and rafters may be built into or carried through a masonry or concrete compartment wall if the openings for them are kept as small as practicable and then fire-stopped.	B(V1) 8.16
• Junctions between a compartment floor and an external wall that has no fire resistance (such as a curtain wall) should be restrained at floor level.	B(V1) 7.13
• Adjoining buildings should only be separated by walls, not floors, which should run the full height of the building, in a continuous vertical plane to the underside of the roof (see AD-B(V1) Diagram 5.2).	B(V1) 7.8
• Where services could provide a source of ignition, the risk of fire developing and spreading into adjacent compartments should be controlled.	B(V1) 7.7

Consideration should also be given to the effect of services that may be built into the construction that could adversely affect its fire resistance.

For instance, where downlighters, loudspeakers and other electrical accessories are installed, additional protection may be required to maintain the integrity of a wall or floor.

4.7.6.3 Construction of an external wall

The external walls of the building shall adequately resist the spread of fire over the walls and from one building to another, having regard to the height, use and position of the building. In addition, the external walls of the building should have sufficient fire resistance to prevent fire spread across the boundary.

4.7.7 Fire resistance and fire-stopping

> • To prevent the spread of fire within a building, the   B(V1) B2
>   internal linings shall:
>   • be capable of resisting the spread of flame over
>   their surfaces; and
>   • have, if ignited, either a reasonable rate of heat
>   release or fire growth.

All openings for pipes, ducts, conduits or cables to pass through any part of a fire-separating element should be as few in number as possible, small in size, and fire-stopped. Joints between fire-separating elements should be fire-stopped. Where a compartment wall meets another compartment wall, the junction should maintain the fire resistance of the compartmentation.

 Note: Materials used for fire-stopping should be reinforced with (or supported by) materials of limited combustibility (e.g. mineral wool, cement mortar, gypsum plaster).

4.7.7.1 Garage fire doors

Fire doors **only** need to be provided with self-closing devices if they are between a dwelling and an integral garage.

4.7.7.2 Load-bearing elements of structure

If the structure is essential for the stability of an external wall which needs to have fire resistance, it is excluded from the definition of element of structure.

4.7.7.3 Single-storey buildings

Although most elements of structure in a single-storey building may **not** need fire resistance, fire resistance will be needed if the element:

- is part of (or supports) an external wall;
- is part of (or supports) a compartment wall, including a wall common to two or more buildings, or a wall between a dwelling and an attached or integral garage;
- supports a gallery.

4.7.7.4 Smoke alarms

There should be a smoke alarm in the circulation space within 7.5m of the door to **every** habitable room. Wall-mounted units may also be used, provided they are positioned above the level of the doorway opening into the space.

4.7.8 Effects of moisture

• A wall shall be erected to prevent undue moisture from the ground reaching the inside of the building and (if it is an outside wall) adequately resisting the penetration of rain and snow to the inside of the building.	C 3
• A damp-proof membrane over a ground-supported floor may be either above or below the concrete, and continuous with the damp-proof course in walls.	C 4.7c C 5.5a
• A suspended timber ground floor should have a ventilated air space at least 75mm from the ground underneath any wall-plates and at least 150mm to underneath the suspended timber floor (or insulation if provided).	C 4.19b
• Two opposing external walls should have ventilation openings placed so that the ventilating air will have a free path between opposite sides and to all parts.	

Openings should not be less than either 1,500mm²/m² run of external wall or 500mm²/m² of floor area, whichever is the greater.

- A suspended concrete ground floor should have a
  ventilated air space measuring at least 150mm from
  the ground covering to the underside of the floor (or
  insulation if provided).   C 4.19b
- Two opposing external walls should have ventilation
  openings placed so that the ventilating air will have a
  free path between opposite sides and to all parts.
- The openings should be not less than either
  1,500mm^2/m^2 run of external wall or 500mm^2/m^2
  of floor area, whichever is the greater.

4.7.8.1 Cracking of external walls

Cladding can be designed to protect a building from precipitation (often
driven by the wind) either by holding it at the face of the building or by
stopping it from penetrating beyond the back of the cladding.

The possibility of severe rain penetrating through cracks in external
masonry walls should be taken into account when designing a building.

4.7.8.2 Framed external walls

- Cladding shall be separated from the insulation or
  sheathing by a vented and drained cavity with a
  membrane that is vapour open, but resists the passage   C 5.17
  of liquid water, on the inside of the cavity (see AD-C
  Diagram 11).

4.7.8.3 Joints between walls and doors/window frames

- Joints between walls and doors and window frames
  should not be damaged by precipitation or permit
  precipitation to reach other parts of the building that   C 5.29
  would be damaged by it.
- Damp-proof courses should be provided to direct   C 5.30
  moisture towards the outside.
- Direct plastering of the internal reveal of any
  window frame should only be used with a backing of   C 5.31
  expanded metal lathing or similar.

- In areas of the country that are exposed to very   C 5.32
  severe driving rain:
  - the frame should be set back behind the outer leaf
    of masonry, as shown in AD-C Diagram 13;
  - an insulated finned cavity closer may be used.

4.7.8.4 Impervious cladding systems for walls

- Cladding systems for walls should: C 5.19
  - not be damaged by precipitation;
  - resist the penetration of precipitation to the inside
    of the building.
- Cladding that is designed to protect a building from   C 5.21
  precipitation shall be:
  - impervious;
  - moisture-resisting;
  - jointless or have sealed joints (i.e. to allow for
    structural and thermal movement).
- If the cladding has overlapping dry joints, it shall be   C 5.21
  backed by a material which will direct precipitation
  that enters the cladding towards the outer face.

Each sheet, tile and section of cladding should be securely fixed (according to guidance contained in BS 8000-6:2013).

- Where cladding is supported by timber components   C 5.27
  (or is on the façade of a timber-framed building), the
  space between the cladding and the building should
  be ventilated to ensure rapid drying of any water that
  penetrates the cladding.

4.7.8.5 Interstitial condensation (external walls)

Specialist advice should be sought when designing swimming pools and other buildings where interstitial condensation in the walls (caused by high internal temperatures and humidities) can cause the generation of high levels of moisture on external walls.

- External walls shall be designed and constructed in   C 5.34
  accordance with Clause 8.3 of BS 5250.

4.7.8.6 Resistance to moisture in walls

• Internal and external walls shall have a damp-proof course of bituminous material, polyethylene, engineering bricks or slates in cement mortar to prevent the passage of moisture.	C 5.5
• If the wall is an external wall, the damp-proof course should be at least 150mm above the level of the adjoining ground (see AD-C Diagram 8), unless the building has been designed to protect the wall.	C 5.5b
• If the wall is an external cavity wall (see AD-C Diagram 9a), then either:	
• the cavity should be taken down at least 225mm below the level of the lowest damp-proof course; or	C 5.5c
• a damp-proof tray should be provided to prevent precipitation passing into the inner leaf (see AD-C Diagram 9b), with weep holes every 900mm.	C 5.5c

All walls should:

• Resist the passage of moisture from the ground to the inside of the building.	C 5.2a
• Not be damaged by moisture from the ground.	C 5.2b
• Not carry moisture from the ground to any part that would be damaged by moisture.	C 5.2b

External walls should:

• Resist rain penetrating components of the structure that could be damaged by moisture.	C 5.2c
• Resist rain penetrating to the inside of the building.	C 5.2d
• Be designed and constructed so that their structural and thermal performance is not affected by interstitial condensation.	C 5.2e
• Not promote surface condensation or mould growth.	C 5.2f

4.7.8.7 Solid external walls

Solid walls should be capable of holding moisture arising from rain and snow until it can be released in a dry period without penetrating to the inside of the building or causing damage to the building.

- Solid external walls exposed to **very severe** conditions should be protected by external impervious cladding.   C 5.9

- Solid external walls exposed to **severe** conditions may be built with: C 5.9a
  - brickwork (or stonework) at least 328mm thick;
  - dense aggregate concrete blockwork at least 250mm thick;
  - lightweight aggregate (aerated autoclaved concrete blockwork) at least 215mm thick;
  - the exposed face of the bricks or blocks should be rendered with a rendering mix of 1 part of cement, 1 part of lime and 6 parts of well-graded sharp sand (nominal mix 1:1:6) unless the blocks are of dense concrete aggregate, in which case the mix may be 1:½:4.   C 5.9b

- Adequate protection should be provided at the top of walls, etc. (see AD-C Diagram 10).   C 5.9c

- Unless the protection and joints are a complete barrier to moisture, a damp-proof course should also be provided.   C 5.9c

- A solid external wall may be insulated on the inside or on the outside.   C 5.10

4.7.8.8 Surface condensation and mould growth (external walls)

External walls shall be designed and constructed so that:

- The thermal transmittance (U-value) does not exceed $0.7W/(m^2 \cdot K)$ at any point.   C 5.36a

4.7.9 Electrical safety

- Electrical installation work (see AD-P Diagram 2)   P 2.5
  may be in special locations such as:
  - adjacent to the position of the shower head where
    it is attached to a wall or ceiling at a point higher
    than 2.25m from that level;
  - where there is no bath tub or shower tray, from the
    centre point of the shower head where it is attached
    to the wall or ceiling to a distance of 1.2m.
- Electrical installation work is notifiable work and
  must be notified to a Building Control Body.

4.7.10 Protection from falling, collision and impact

4.7.10.1 Critical locations

- Critical locations include large uninterrupted areas   K 7.1
  of transparent glazing which form, or are part of, the
  internal or external walls.

4.7.10.2 Guarding

- You can use any wall, parapet, balustrade or similar   K 2.36
  obstruction as guarding providing it is, as a minimum:
  - 900mm for stairs, landings, ramps and edges of
    internal floors;
  - 1,100mm for external balconies and edges of roofs.

4.7.11 Security

- Lightweight framed walls should incorporate a   Q 1.6
  resilient layer to reduce the risk of anyone breaking
  through the wall and accessing the locking system.
- The resilient layer should be timber sheathing at least
  9mm thick, expanded metal or a similar resilient
  material.
- The resilient layer should be to the full height of the
  door and 600mm either side of the doorset.

This requirement is only applicable to new dwellings, although it may be referred to for new buildings other than dwellings.

4.7.12 *Resistance to the passage of sound*

• All **new** walls constructed within a dwelling (flat or room used for residential purposes) shall meet the laboratory sound insulation values set out in AD-E Table 6.6.10.	E 0.9
• Walls for rooms for residential purposes (plus dwellings and flats that have a separating function) should achieve the sound insulation values, as set out in AD-E Table 1A.	E 0.1

Any building work involving a historic building is a virtual minefield and you would be well advised to seek professional advice well before embarking on any construction or reconstruction work. (Also refer to BS 7913.)

4.7.12.1 *Testing*

• Tests should be carried out between rooms or spaces that share a common area formed by a separating wall or separating floor.	E 1.9
• Impact sound insulation tests should be carried out without a soft covering (e.g. carpet, foam-backed vinyl, etc.) on the floor as follows:	E 1.10
• in dwelling houses, one set of tests made up of two individual sound (i.e. airborne) insulation tests;	E 1.19
• in flats with separating floors but not walls, one set of tests comprising four individual sound insulation tests (two airborne tests and two impact tests);	E 1.20
• in flats with separating floors and walls, one set of tests comprising six individual sound insulation tests (four airborne tests and two impact tests).	
• It is preferable that each set of tests contains individual tests in bedrooms and living rooms.	E 1.23

4.7.13 Physical infrastructure for high-speed electronic communications networks

• AD-R Diagram 1 shows a possible arrangement for the physical infrastructure for a single-occupancy building: • the access point is on an outside wall and is connected by a through-wall duct to a network termination point.	R 1.7
• AD-R Diagram 2 shows a possible arrangement for the physical infrastructure for a multi-dwelling building: • multi-dwelling buildings should have a common access point and dedicated vertical and horizontal service routes so that service providers can connect from the access point to the network termination point in each dwelling.	R 1.7

AD-R (*High-speed electronic communications networks*) took effect on 1 January 2017 for use in England. It does not apply to work subject to a Building Notice, Full Plans Application or Initial Notice submitted before 1 January 2017.

4.7.14 Drainage and waste disposal

• Where a drain or sewer running below a building is less than 2m deep, the foundation should be extended locally so that the drain or sewer passes through the wall.	H4 1.10
• Where a drain or sewer runs through a wall or a foundation, suitable measures should be taken to prevent damage or misalignment.	H4 1.11
• Where eaves drop systems are used, they should be designed to protect the fabric of the building from ingress of water, caused by water splashing on the external walls.	H4 1.13
• Surface gradients should direct water draining from a paved area away from buildings.	H3 2.2

4.7.15 Combustion appliances and fuel storage systems

4.7.15.1 Carbon monoxide alarm siting

- When a carbon monoxide alarm is fitted on a wall, it should be as high up as possible (above any doors and windows).   J 2.36

4.7.15.2 Permanently open air vents

- Ventilation openings should **not** be made in fire-resisting walls other than external walls.   J 1.10
- Where a building is to be altered for different use (e.g. converted into flats), the fire resistance of walls of existing masonry chimneys may need to be improved.   J 1.31

4.7.15.3 Thermal expansion

- Where a factory-made metal chimney passes through a wall, sleeves should be provided to prevent damage to the flue or building through thermal expansion.   J 1.43

4.7.15.4 Walls adjacent to hearths

- Walls that are not part of a fireplace recess or a prefabricated appliance chamber, but are adjacent to hearths or appliances, also need to protect the building from catching fire. A way of achieving this requirement is shown in AD-J Diagram 30.   J 2.32

4.7.16 Conservation of fuel and power

In accordance with the 2014 revision of AD-L (*Conservation of fuel and power*) and Regulation 7, reasonable provision shall be made for the conservation of fuel and power in buildings by limiting heat gains and losses through thermal elements and other parts of the building fabric.

4.7.16.1 U-values

• If a wall that separates extensions contains either a conservatory or porch, they may be exempt from energy-efficiency requirements if the separating wall has been retained or, if removed, been replaced with a similar wall with the same specification.	L(V1) 0.14b

• The notional U-value dwelling specification is:
  • for new walls in existing dwellings: a maximum 0.18W/(m²·K) U-value;
  • for external walls in a new dwelling: also 0.18W/(m²·K); and
  • for party walls in a new dwelling: 0W/(m²·K).
	L(V1) Table 1.1

Note: Contrary to previous assumptions, party cavity walls may not be zero heat-loss walls as air flow in the cavity provides a heat-loss mechanism.

• Limiting U-value for existing walls (with cavity insulation) in existing dwellings is 0.70W/(m²·K) – with an improved threshold of 0.55W/(m²·K).	L(V1) Table 4.3
• The threshold U-value for new walls (with internal or external insulation) in existing dwellings is 0.70W/(m²·K) – with an improved threshold of 0.30W/(m²·K).	

4.7.16.2 Continuity of insulation

• Window or door units should be attached to the insulation layer of the external wall.	L(V1) 4.15
• Wall insulation should be fitted without any air gaps and tight to the structure, cavity closers, lintels and cavity trays.	
• Roof insulation should be installed tight to the structure, without air gaps, and should extend to the wall insulation.	

- Wall insulation should be installed to the top of the wall-plate in places such as the eves, and roof insulation should be continuous with the wall insulation.
- Roofs insulated at rafter level (or at the eaves) should extend to the top of the external wall.
- Buildings should be constructed so that thermal bridging (including at a party wall) is limited by: L(V1) 4.16–4.17
  - lightweight blockwork in the inner leaf of a cavity wall; or
  - intermediate floor-to-wall junctions.
- Ground floor wall-to-floor junctions should achieve continuity of insulation: L(V1) 4.17
  - cavity wall insulation should extend below the damp-proof course and be at least the equivalent of one full block height below the underside of the floor structure/slab and beyond the depth of the floor insulation;
  - insulation between timber boards or within sheathing should extend to the floor plate;
  - the wall insulation and the floor perimeter insulation should abut.
- Intermediate floor-to-wall junctions should have continuous insulation in the external wall.

It is recommended that photographs should be taken for each dwelling on a development as a record of the construction, and should show all the wall-to-floor junctions.

Requirements for buildings other than dwellings – Additional requirements

The basic requirements for walls in buildings other than dwellings are the same as for dwellings, with the addition of the following requirements.

4.7.17 Access to and use of buildings

4.7.17.1 Conference facilities and sleeping accommodation

• Walls in lecture/conference facilities should not have a detrimental effect on the ability of people to receive information from a sign language interpreter or a lip speaker.	M2 4.9
• Wall-mounted socket outlets, telephone points and TV sockets should be between 400mm and 1,000mm above the floor.	M2 4.30
• Switches for permanently wired appliances should be between 400mm and 1,200mm above the floor, unless needed at a higher level for particular appliances.	
• All switches and controls that require precise hand movements should be between 750mm and 1,200mm above the floor.	
• Simple push button controls should not be more than 1,200mm above the floor.	
• Wall surface materials and finishes should help visually impaired people appreciate the boundaries of rooms or spaces, identify access routes and receive information.	M2 4.32

4.7.17.2 Doors and screens

• A space alongside the leading edge of a door should be provided to·enable a wheelchair user to reach and grip the door handle, then open the door without releasing hold on the handle and without the footrest colliding with the return wall.	M2 2.15
• Manually operated non-powered entrance doors should have an opening force at the leading edge of the door of not more than 30N at the leading edge from 0° (the door in the closed position) to 30° (open), and not more than 22.5N at the leading edge from 30° to 60° of the opening cycle.	M2 2.17
• Manually operated non-powered entrance doors should have an unobstructed space of at least 300mm on the pull side of the door between the leading edge of the door and any return wall.	

- Internal doors should: M2 3.10
  - have an unobstructed space of at least 300mm
    on the pull side of the door unless the door has
    power-controlled opening or provides access to a
    standard hotel bedroom;
  - ensure the door frames contrast visually with the
    surrounding wall;
  - ensure glass or fully glazed doors are clearly
    differentiated from any adjacent glazed wall or
    partition by the provision of a high-contrast strip
    at the top and on both sides.

To help people with visual impairment to appreciate the size of a space
they have entered, or to find their way around, there should be a visual
contrast between the wall and the ceiling, and between the wall and
the floor.

4.7.17.3 Handrails

- A ramped access should have a surface width   M2 1.26
  between walls, upstands, etc., of at least 1.5m.
- A stepped access should have flights whose surface   M2 1.33
  width between enclosing walls, strings or upstands is
  not less than 1.2m.
- Handrails should be spaced away from the wall and   M2 1.35
  rigidly supported in a way that avoids impeding
  finger grip.
- There should be a clearance of between 50mm and   M2 1.37j
  75mm between the handrail and any adjacent wall
  surface.

4.7.17.4 Lifting devices, lifts and stairlifts

- Landing call buttons should be located between   M2 3.28
  900mm and 1,100mm from the floor of the landing
  and at least 500mm from any return wall.
- Handrails should be provided on at least one wall
  of the lifting device, with its top surface at 900mm
  (nominal) above the floor and located so that it does
  not obstruct the controls or the mirror.

• The use of visually and acoustically reflective wall surfaces should be avoided.	M2 3.32
• Passenger lift car controls should be located between 900mm and 1,200mm (preferably 1,100mm) from the car floor and at least 400mm from any return wall. • Lift landing and car doors should be distinguishable visually from the adjoining walls.	M2 3.34
• The use of visually and acoustically reflective wall surfaces should be minimised within the lifting platform to prevent discomfort for people with visual and hearing impairment.	M2 3.42
• Lifting platform controls should be located between 800mm and 1,100mm from the floor of the lifting platform and at least 400mm from any return wall. • Doors should be distinguishable visually from the adjoining walls.	M2 3.43

4.7.17.5 Unisex wheelchair-accessible toilets

• The surface finish of sanitary fittings and grab bars should contrast visually with background wall finishes. • There should be visual contrast between wall and floor finishes.	M2 5.4k
• The transfer space in unisex wheelchair-accessible toilets should be kept clear to the back wall.	M2 5.8
• Where the horizontal support rail on the wall adjacent to the WC is set with the minimum spacing from the wall, an additional drop-down rail should be provided on the wall side, at a distance of 320mm from the centre line of the WC. • Where the horizontal support rail on the wall adjacent to the WC is set so that its centre line is 400mm from the centre line of the WC, there is no additional drop-down rail.	M2 5.10
• In WC compartments within separate-sex toilet washrooms, there should be a 450mm diameter manoeuvring space between the swing of the door, the WC pan and the side wall of the compartment.	M2 5.14

- There should be a vertical grab bar on the rear wall
  and space for a shelf and fold-down changing table.
- Wheelchair-accessible changing and shower facilities   M2 5.18
  should be provided with wall-mounted drop-down
  support rails and wall-mounted slip-resistant tip-up
  seats (not spring loaded).

4.7.18 Structure

- Walls should comply with the requirements of BS   A 2C3a
  EN 1996-2, BS EN 1996-1-1 and BSI Published
  Document PD 6697.

- Walls of a small single-storey non-residential building   A 2C38
  or annexe should be constructed in brickwork or
  blockwork, and if the floor area exceeds 10m², the
  walls should have a mass of not less than 130kg/m².

- Walls should be properly bonded and solidly put   A 2C20
  together with mortar and constructed of masonry
  units conforming to:
  - clay bricks or blocks to BS EN 771-1;
  - calcium silicate bricks or blocks to BS EN 771-2;
  - concrete bricks or blocks to BS EN 771-3 or BS
    EN 771-4;
  - manufactured stone to BS EN 771-5;
  - square dressed natural stone to the appropriate
    requirements described in BS EN 771-6.

- Walls should have a minimum thickness of 90mm.   A 2C38
- Walls with one or two major openings should in
  addition have piers, as shown in AD-A Diagrams 18b
  and c.
- Walls which do not contain a major opening but
  exceed 2.5m in length or height should be bonded or
  tied to piers for their full height at not more than 3m
  centres (see AD-A Diagram 18).

Only one or two major openings, not more than 2.1m in height, are permitted in one wall of a building or annexe.

- External and internal load-bearing and parapet walls   A 2C2
  must extend to the full storey height.

- Walls should be tied horizontally at no more than   A 2C38
  2m centres to the roof structure at eaves level, base
  of gables and along roof slopes, as shown in AD-A
  Diagram 19.
- Where ties are used to connect piers to walls, they
  should be flat, 20mm × 3mm in cross-section, in
  stainless steel, placed in pairs and be spaced at not
  more than 300mm centre vertically.

- Wall ties should comply with BS EN 845-1.   A 2C19

- Where straps cannot pass through a wall, they should   A 2C20
  be adequately secured to the masonry using suitable
  fixings.

4.7.18.1 Wall cavity barriers

The external wall of a building should not provide a medium for fire spread, since this is likely to be a risk to health and safety. Combustible materials and cavities in external walls and attachments to them can present such a risk, particularly in tall buildings. The guidance below is designed to reduce the risk of vertical fire spread, as well as the risk of ignition from flames coming from adjacent buildings.

- Cavity barriers should be provided: B(V2) 9.3
  - at the junction between an external cavity wall and
    a compartment wall;
  - at the junction between an external cavity wall and
    every compartment wall and compartment floor;
  - around openings (such as windows, doors and
    exit/entry points for services).

AD-B(V2) Table 9.1 sets out the maximum dimensions for undivided cavities.

Cavity barriers do not need to be provided between double-skinned corrugated or profiled insulated roof sheeting, provided the sheeting is a material of limited combustibility.

4.7.18.1.1 CONSTRUCTION AND FIXINGS FOR CAVITY BARRIERS

Every cavity barrier should be constructed to provide at least 30 minutes' fire resistance. A cavity barrier should, wherever possible, be tightly fitted to a rigid construction and mechanically fixed in a position that is unlikely to be made ineffective by the movement of the building. Cavity barriers in a stud wall or partition, or provided around openings, may be formed of steel, timber, polythene-sleeved mineral wool, mineral wool slab, or cement-based or gypsum-based boards.

4.7.19 Conservation of fuel and power

• Any wall that separates a conservatory or porch from the building, that has been retained or, if removed, replaced with a similar wall, is exempt from the requirements.	L(V2) 0.18

Note: The exemption for conservatory and porch extensions only applies where the existing walls, doors and windows which separate the conservatory from the building are retained or, if removed, are replaced by walls, windows and doors which meet the energy-efficiency requirements.

• Thermal bridging, including at the party wall, should be limited.	L(V2) 4.9b
• Limiting U-values for new or replacement walls in new and existing buildings and air permeability in new buildings = 0.26Wmk. • Limiting U-values for curtain walling in new and existing buildings and air permeability in new buildings = 1.6Wmk.	L(V2) Table 4.1
• Limiting U-values for existing walls with cavity insulation in existing buildings: threshold = 0.70 and improved = 0.55. • Limiting U-values for existing walls with external or internal insulation in existing buildings: threshold = 0.70 and improved = 0.30.	L(V2) Table 4.2
• Any wall in an extension should be insulated and draught-proofed to at least the same extent as in the existing building.	L(V2) 10.13a

- Vehicle access doors, display windows and similar glazing and smoke vents can be as large an area of wall or roof as required for the purpose.　L(V2) 10.09

- Renovating a thermal element includes providing cavity wall insulation (the diagrams in Chapter 6.6 apply to buildings other than dwellings).　L(V2)11.2e

4.7.19.1 Overheating

- Excess heat should be removed from the residential building by opening windows, the use of ventilation louvres in external walls, or a mechanical ventilation or cooling system.　O 2.10

 Note: This requirement may need a larger amount of purge ventilation than that listed in AD-F(V2) to remove excess heat from an indoor environment.

4.7.20 Fire safety

- The potential for fire to spread over external walls and roofs should be restricted.　B(V2) B4

- Smoke alarms/detectors should be mounted at least 300mm from walls.　B(V2) 1.14

Wall-mounted units may be used provided that the units are above the level of doorways opening into the space.

- Escape stairs should be protected by a smoke control system.　B(V2) 3.15a

- Where an air-circulation system circulates air only within an individual flat with an internal protected stairway or entrance hall, transfer grilles should not be fitted in any wall enclosing a protected stairway or entrance hall.　B(V2) 10.9

- The wall between each flat and the corridor should be a compartment wall.　B(V2) 2.24

- Storeys may be divided into two refuges by a compartment wall.   B(V2) Diagram 3.1

- The choice of materials for walls and ceilings can significantly affect the spread of a fire and its rate of growth, even though they are not likely to be the materials first ignited.

- Load-bearing walls and compartment walls are treated as elements of structure.   B(V2) B3.iii

External walls, such as curtain walls or other forms of cladding which transmit only self-weight and wind loads and do not transmit floor load, are not regarded as load bearing.

4.7.20.1 Care homes – horizontal evacuation

- Progressive Horizontal Evacuation (PHE) in a care home needs to be sub-divided into protected areas separated by compartment walls and compartment floors (see AD-B(V2) Diagram 2.11).   B(V2) 2.35

Protected areas provide a place of relative safety, from which further evacuation can be made if necessary.

- Each storey, used for the care of residents, should be divided into at least three protected areas by compartment walls.   B(V2) 2.36

- Every protected area should have a minimum of two exits to adjoining protected areas.   B(V2) 2.37

4.7.20.2 Compartmentation

A building should be sub-divided into compartments separated from one another by walls and/or floors of fire-resisting construction.

The following walls need to be constructed as compartment walls in buildings:

- Walls common to two or more buildings.   B(V2) 8.2
- Walls dividing buildings into separated parts.   to 8.4
- Constructions enclosing places of special fire hazard.

- Walls bordering a protected shaft.   B(V2) 8.6

- Parts of a building that are occupied mainly for   B(V2) 8.11
  different purposes.

In addition to:

- Walls enclosing a refuse storage.
- Walls separating a flat from any other part of the building; and from
  parts used for a non-residential purpose (e.g. offices and shops, etc.).

4.7.20.2.1 FIRE RESISTANCE

Fire resistance should be continuous at the join between compartmentation, and any openings between two compartments should not reduce the fire resistance.

- All compartment walls should: B(V2) 8.15
  - form a complete barrier to fire between the
    compartments they separate;
  - have the appropriate fire resistance, as given in
    AD-B(V2) Appendix B, Tables B3 and B4.
- Elements of structure, such as structural frames,   B(V2) 7.1
  beams, columns, load-bearing walls (internal and
  external), floor structures and gallery structures,
  should have at least the fire resistance given in
  AD-B(V2) Appendix B, Table B3.

Note: Storerooms in shops should be separated from retail areas by fire-resisting construction.

4.7.20.2.2 CONSTRUCTION OF COMPARTMENT WALLS AND FLOORS

Any part of an external wall which has less fire resistance than that shown in AD-B Appendix A is considered to be an unprotected area.

- Compartment walls that are common to two or more   B(V2) 8.18
  buildings should run the full height of the building in
  a continuous vertical plane.

Adjoining buildings should only be separated by walls, not floors.

- Compartment walls in a top storey beneath a roof should be continued through the roof. B(V2) 8.21

- Where a compartment wall meets another compartment wall or an external wall, the junction should maintain the fire resistance of the compartmentation. B(V2) 8.22

- At the junction of a compartment floor with an external wall that has no fire resistance (such as a curtain wall), the external wall should be restrained at floor level to reduce the movement of the wall away from the floor when exposed to fire. B(V2) 8.23

- Compartment walls should be able to accommodate predicted deflection of the floor above. B(V2) 8.24

- Compartment walls should be taken up to meet the underside of the roof covering or deck, with fire-stopping where necessary. B(V2) 8.25
- The compartment wall should be continued across any eaves cavity.

To reduce the risk of fire spread over a roof or zone of the roof, material of limited combustibility (as set out in AD-B(V2) Diagram 8.2a) should be laid 1,500mm on either side of the wall.

- Any openings in a compartment wall which is common to two or more buildings should be limited to: B(V2) 8.30
  - a door that is part of a means of escape in case of fire;
  - the passage of a pipe.

- Openings in other compartment walls should be limited to those for fire doorsets, pipes, ventilation ducts, service cables, chimneys, refuse chutes of class A1 construction, atria or protected shafts. B(V2) 8.31

- Openings in compartment walls should be limited at the junction between an internal (and external) cavity wall and every compartment floor and compartment wall. B(V2) 8.34 / B(V2) 9.3
- Any compartment wall should be carried up through a ceiling or roof cavity to maintain the standard of fire resistance.

- Pipes which pass through a fire-separating element should have a proprietary sealing system which has been shown by test to maintain the fire resistance of the wall.   B(V2) 10.2
- Where sealing is not used, the nominal diameter of a pipe should conform to AD-B Table 14.

- If a flue or duct passes through or is built into a compartment wall, the wall of the flue should have a fire resistance of at least half that of the wall (see AD-B(V1) Diagram 10.4).   B(V2) 10.23

4.7.20.3 External escape routes

- Where an external escape route is beside an external wall of the building, if part of that wall is within 1,800mm of the escape route, then it should be of fire-resisting construction up to a height of 1,100mm above the paving level of the route.   B(V2) 2.30

- An escape route over a flat roof is compliant if: B(V2) 2.32
  - it is part of the same building from which escape is being made;
  - the route leads to a storey exit or an external escape route;
  - it is fire resisting (minimum REI 30);
  - the route is clearly defined and guarded by walls and/or protective barriers.

4.7.20.4 External walls

- External walls of the building should have sufficient fire resistance to prevent fire spread across the boundary.   B(V2) B4

- Any unprotected areas of an external wall on, or within 1,000mm of, the relevant boundary should meet the conditions in AD-B(V1) Diagram 13.5, and the rest of the wall should be fire resisting from both sides.   B(V2) 13.7

- Where a portal-framed building is near a relevant   B(V2)
  boundary, the external wall near the boundary may   13.15
  need fire resistance to restrict the spread of fire
  between buildings.

- Internal and external load-bearing walls should   B(V2) 12.2
  maintain their load-bearing function in the event of
  fire.

- The external surfaces of the building should have   B(V2) 12.5
  the appropriate fire resistance given AD-B(V1) Table
  12.1.

- Where a mixed-use building includes Assembly   B(V2) 12.6
  and Recreation Purpose Group(s) accommodation,
  the external surfaces of walls should meet the
  requirements of AD-B Diagram 40c.

4.7.20.4.1 EXTERNAL WALLS ADJACENT TO PROTECTED
          STAIRWAYS

Precautions should be taken against the spread of fire in the following
situations:

- Depending on how the external walls are configured,   B(V2) 3.29
  a fire in one part of a building could subject the
  external wall of a protected stairway to heat. (See
  AD-B(V1) Diagram 3.3.)

- If a protected stairway projects beyond the adjoining   B(V2) 3.30
  external wall of the building, then the minimum
  distance between an unprotected area of the building
  enclosure and an unprotected area of the stair
  enclosure should be 1,800mm.

 Note: Natural smoke outlets should be at a high level in the ceiling or
wall of the space they serve.

4.7.20.5 Boundaries

The fire resistance of a building's wall depends on its distance from the
relevant boundary (see AD-B(V1) Diagram 13.1), and it is this distance
that determines whether a fire on an adjoining site would have any effect
on the building's wall.

Note: To assist your calculation, refer to AD-B(V1) Diagram 13.2 (whose centre line shows where further development is unlikely) and AD-B(V1) Diagram 13.3 (which indicates the notional protection boundary between two buildings).

• Parts of an external wall with less fire resistance than the amount given in AD-B(V1) Appendix B, Table B4, are called 'unprotected areas'.	B(V2) 13.5
• If a fire-resisting external wall has a surface material that is worse than class B-s3, d2 and is more than 1mm thick, then that part of the wall is classified as an unprotected area equating to half its area (see AD-B(V1) Diagram 13.4), and should meet the conditions shown in AD-B(V1) Diagram 13.5 (with both sides of the remainder of the wall to be fire resistant).	B(V2) 13.6 &13.7

In other words: external surface materials facing the boundary should be class B-s3, d2 or better.

4.7.20.6 *Inner rooms*

• The enclosures (walls or partitions) of an inner room should be stopped at least 500mm below the ceiling.	B(V2) 2.11
• A vision panel not less than 0.1m² should be located in the door or wall of an inner room.	

4.7.20.7 *Protected shafts*

• A protected shaft should from a complete barrier to fire between compartments connected to that particular shaft, and it should have the amount of fire resistance stated in AD-B(V2) Appendix B, Table B3. (For more details, see BS 5588-5.)	B(V2) 8.33

4.7.20.8 *Wall linings*

To restrict the spread of fire within the building, it is important that the internal linings shall:

- resist the spread of flame over their surfaces;
- have (if ignited) a rate of heat release (or fire growth) that, in the circumstances, is acceptable.

- Surface linings of walls should meet the requirements   B(V2) 6.1
  of AD-B(V2) Table 6.1.
- Wallcoverings should conform to BS EN 15102.

Note: Any flexible membrane covering a structure (other than an air-supported structure) should comply with the recommendations given in Appendix A to BS 7157.

4.7.20.9 Cavity barriers

The external wall of a building should not provide a medium for fire spread, since this is likely to be a risk to health and safety. Combustible materials and cavities in external walls and attachments to them can present such a risk, particularly in tall buildings. The guidance below is designed to reduce the risk of vertical fire spread, as well as the risk of ignition from flames coming from adjacent buildings.

- Cavity barriers should be provided: B(V2) 6.1
  - at the junction between an external cavity wall and
    a compartment wall;
  - at the junction between an external cavity wall and
    every compartment wall and compartment floor;
  - around openings (such as windows, doors and
    exit/entry points for services).

AD-B(V2) Table 9.1 sets out the maximum dimensions for undivided cavities.

Cavity barriers do not need to be provided between double-skinned corrugated or profiled insulated roof sheeting, provided the sheeting is a material of limited combustibility.

4.7.20.9.1 CONSTRUCTION AND FIXINGS FOR CAVITY BARRIERS

Every cavity barrier should be constructed to provide at least 30 minutes' fire resistance. A cavity barrier should, wherever possible, be tightly fitted to a rigid construction and mechanically fixed in a position that is unlikely to be made ineffective by the movement of the building. Cavity barriers in a stud wall or partition, or provided around openings, may be formed of steel, timber, polythene-sleeved mineral wool, mineral wool slab, or cement-based or gypsum-based boards.

4.7.21 External fire resistance

A wall common to two or more buildings should be capable of resisting the spread of fire between buildings.

B(V2) B3(2)

Note: See AD-B(V2) in Appendix B, Table B3 which provides details of the minimum fire resistance that load-bearing walls (internal and external) should have.

* Fire resistance is required for elements that are part of, or support for:
  * an external wall;
  * a compartment wall; and
  * a wall that is common to two or more buildings.

B(V2) B26

Note: External walls, such as curtain walls or other forms of cladding, which transmit only self-weight and wind loads and do not transmit floor load, are excluded from the definition of 'element of structure'.

* The risk of vertical fire spread, as well as the risk of ignition from flames coming from adjacent buildings, largely depends on combustible materials and cavities in external walls (and attachments to them), particularly in tall buildings.
* To guard this risk:
  * the external wall of a building should ***not*** provide a medium for fire spread (if it is likely to be a risk to health and safety);
  * buildings of any height or use should carefully consider the choice of materials used for the external walls, or attachments to the walls; and
  * the external surfaces of external walls should comply with AD-B(V2) Table 12.1.

B(V2) 12.1–12.5

These provisions apply to each wall individually:

- In a building with a storey greater than 18m in height (see B(V2) Appendix D, Diagram D6), any insulation product or filler material used in the construction of an external wall – unless it is a masonry cavity wall – should be class A2-s3, d2 or better. — B(V2) 12.6

- Membranes used as part of the external wall construction above ground level should achieve a minimum of class B-s3, d0. — B(V2) 12.16

- Thermal breaks (used as part of the external wall construction to restrict thermal bridging) should not span two compartments, and their size should be limited to the minimum required to restrict thermal bridging.

Note: Although the regulations apply to materials which become part of an external wall, consideration should also be given to other attachments to the wall which could impact on the risk of fire spread over the wall.

- The fire resistance of a wall depends on its distance from the boundary that the wall faces (this is referred to as the 'relevant boundary'). — B(V2) 13.4

- Where a portal-framed building is near a relevant boundary, the external wall near the boundary may need fire resistance to restrict the spread of fire between buildings. — B(V2) 13.15

- In cases where the external wall of the building cannot be wholly unprotected, the rafter members of the frame, as well as the column members, may need to be fire protected.

Note: Best practice guidance for green walls (also called 'living walls') can be found in *Fire Performance of Green Roofs and Walls*, published by the Department for Communities and Local Government.

4.7.21.1 Unprotected areas and fire resistance

Parts of an external wall with less fire resistance than the appropriate amount given in AD-B(V2) Appendix B, Table B4, are called 'unprotected areas'.	B(V2) 13.5
• Where a fire-resisting external wall has a surface material that is worse than class B-s3, d2, and is more than 1mm thick, that part of the wall should be classified as an unprotected area equating to half its area.	B(V2) 13.6
• Unprotected areas should meet the conditions shown in AD-B(V2) Diagram 13.4, and the rest of the wall should be fire resisting from both sides.	B(V2) 13.7
• External surface materials facing the boundary should be class B-s3, d2 or better.	
• Unprotected areas should not exceed the result given by one of the methods in Section 13.17 and the rest of the wall (if any) should be fire resisting – ***but only*** from the inside of the building.	B(V2) 13.8

4.7.21.2 Exclusions to fire resistance

• When assessing unprotected areas, the external walls of protected stairways may be excluded.	B(V2) 13.9
• In an otherwise protected wall, small unprotected areas may be ignored where they meet the conditions in AD-B(V2) Diagram 13.5.	B(V2) 13.10
• Parts of walls of compartmented buildings that are more than 30m above mean ground level may be ignored.	B(V2) 13.11

4.8 Ceilings

The ceilings of all buildings are extremely important and it is crucial that they are constructed according to the regulations. There are a few warning signs which can indicate a little more than basic ceiling mainten-ance will be required. These signs can potentially lead to ceiling collapse, so always be on the lookout for crack marks, cracking sounds, bubbling and sagging.

Reminder: The subsections include details for buildings other than dwellings only where they differ from the requirements for dwellings.

Subsection	Description	Page
Requirements for dwellings		
4.8.1	Access to and use of dwellings	333
4.8.2	Structure	334
4.8.3	Fire safety	338
4.8.4	Resistance to the passage of sound	339
4.8.5	Ventilation	342
4.8.6	Conservation of fuel and power	342
4.8.7	Electrical safety	343
Requirements for buildings other than dwellings – Additional requirements		
4.8.8	Access to and use of buildings	344
4.8.9	Fire safety	344
4.8.10	Ventilation	347
4.8.11	Conservation of fuel and power	348

Reminder: Free downloads of all the diagrams and tables referenced in the subsections are available on-line at the Governments' Planning Portal: www.gov.uk/government/collections/approved-documents.

Requirements for dwellings

4.8.1 Access to and use of dwellings

4.8.1.1 Category 3 – wheelchair-user dwellings

- The ceiling structure of the following rooms should be strong enough to allow for the fitting of an overhead hoist capable of carrying a load of 200kg to: M 3.35c
  - every bedroom;
  - all bathrooms; and
  - all WC/cloakrooms.

Note: Requirements M3.35 and 3.36 are optional requirements for wheelchair-user dwellings.

4.8.2 Structure

4.8.2.1 Ceiling treatments

Each floor type requires one of the ceiling treatments described below.

Using a better performing ceiling than that described in this guidance should improve the sound insulation of the floor, provided there is no significant flanking transmission.

4.8.2.1.1 CEILING TREATMENT A

This is an independent ceiling with absorbent material.

- Ceiling treatment A should meet the following   E 3.19
  specification:
  - at least two layers of plasterboard with staggered
    joints;
  - a minimum total mass per unit area of
    plasterboard of 20kg/m²;
  - an absorbent layer of mineral wool (minimum
    thickness 100mm and minimum density 10kg/m³),
    laid in the cavity formed above the ceiling.
- The ceiling should be supported by one of the
  following methods:
  - *floor types 1 and 2:* using independent joists
    that are only fixed to the surrounding walls with
    a clearance of at least 100mm between the top
    of the plasterboard forming the ceiling and the
    underside of the base floor;
  - *floor type 3:* using independent joists, fixed to
    the surrounding walls with additional support
    provided by resilient hangers attached directly to
    the floor: a clearance of at least 100mm should be
    left between the top of the ceiling joists and the
    underside of the base floor.

Ceiling treatment A is essential with a timber-framed   E 3.100
base and platform floor (i.e. floor type 3.1A).

Remember to seal the perimeter of the independent ceiling with tape (or similar sealant), so as to avoid a rigid or direct connection between the independent ceiling and the floor base!

4.8.2.1.2 CEILING TREATMENT B

This is plasterboard on proprietary resilient bars with absorbent material.

• Ceiling treatment B should meet the following specification:
  • a single layer of fixed, 10kg/m² plasterboard;
  • fixed using proprietary resilient metal bars with absorbent material (on concrete floor,s these bars should be fixed to timber battens);
  • an absorbent layer of mineral wool (minimum density 10kg/m³) filling the ceiling void.
	E 3.21

• Ceiling treatment 1.2B should be used in conjunction with floor type 1.2B: concrete planks (solid or hollow) – see AD-E Diagram 3.4.	E 3.30
• Ceiling treatment B should be used in conjunction with floor type 2.2B: concrete planks (solid or hollow) – see AD-E Diagrams 3.14 and 3.15.	E 3.68

4.8.2.1.3 CEILING TREATMENT C

This is plasterboard on timber battens or proprietary resilient channels with absorbent material.

• Ceiling treatment C should meet the following specification:
  • a single layer of plasterboard with a minimum mass per unit area of 10kg/m²;
  • fixed using timber battens or proprietary resilient channels.
	E 3.21

• Ceiling treatment C or better should be used in conjunction with floor type 1.1C: solid concrete slab (cast in situ, with or without permanent shuttering) – see AD-E Diagrams 3.3, 3.12 and 3.13.	E3.29 E3.67

Electric cables give off heat when in use and special precautions may be required when they are covered by thermally insulating materials.

4.8.2.2 Ceiling joists

Softwood timber used for roof construction including ceiling joists should be adequately treated to prevent infestation by the house longhorn beetle *(Hylotrupes bajulus L.).*

- Preservative treatment for softwood timber is   A 2B2
  particularly recommended in the following areas:
  - the Borough of Bracknell Forest, in the parishes of
    Sandhurst and Crowthorne;
  - the Borough of Elmbridge;
  - the District of Hart, in the parishes of Hawley and
    Yateley;
  - the District of Runnymede;
  - the Borough of Spelthorne;
  - the Borough of Surrey Heath;
  - the Borough of Rushmoor, in the area of the
    former district of Farnborough; and
  - the Borough of Woking.

Notes:

1 · Guidance on suitable preservative treatments is given in The Wood
  Protection Association's manual *Industrial Wood Preservation: Specification
  and Practice* (2012).

2   Guidance on the sizing of certain members in ceilings is
  given in the TRADA *Span Tables for Solid Timber Members
  in Floors, Ceilings and Roofs (Excluding Trussed Rafter Roofs) for
  Dwellings.*

3   See also BS EN 1995–1-1, BSI Published Document PD 6693–1
  and BS 8103–3.

4.8.2.3 Imposed loads on ceilings

- The imposed loads for ceilings are as shown in AD-A   A 2C15
  Table 4.

4.8.2.4 Linings

• The surface linings of ceilings should meet the classifications in AD-B(V1) Table 4.1.	B(V1) 4.1
• The fixings of panels used to form a suspended ceiling should pass through the panel and support it from the lower face.	B(V1) 4.10

 Any part of a ceiling which slopes at an angle greater than 70° should be considered as a wall.

4.8.2.5 Stairs

• Where there is no cupboard under the stairs, it will be necessary to construct an independent ceiling below the stairs.	E 4.37

4.8.2.6 Work on existing buildings

• **Stair treatment 1** (stair covering and an independent ceiling with absorbent material): to be used where a timber stair performs a separating function.	E 4.9
Note: The resistance to airborne sound depends mainly on the mass of the stair, and the mass and isolation of any independent ceiling.	
• **Floor treatment 1** (see AD-E Diagram 4.3) (independent ceiling with absorbent material): the resistance to airborne and impact sound depends on the combined mass, isolation and airtightness of the existing floor and the independent ceiling.	E 4.26
• Independent ceilings with absorbent material should have: • at least two layers of plasterboard with staggered joints (with minimum total mass per unit area of 20kg/m³); • an absorbent layer of mineral wool laid on the ceiling (with minimum thickness of 100mm and density of 10kg/m³).	E 4.27

- The ceiling should be supported by:
  - independent joists fixed only to the surrounding walls; or
  - independent joists fixed with additional support (provided by resilient hangers attached directly to the existing floor base) to the surrounding wall.
- If a window head is near to the existing ceiling, the new independent ceiling may be raised to form a pelmet recess (see AD-E Diagram 4.4).   E 4.28

Note: If there is an existing lath and plaster ceiling, it should be retained, provided it satisfies Building Regulation AD-B (Fire safety).

 Do not create a rigid or direct connection between the independent ceiling and the floor base.

4.8.3 Fire safety

Note: A ceiling will not include trap doors and their frames, frames of windows or rooflights and frames in which glazing is fitted, architraves, cover moulds, picture rails, exposed beams and similar narrow members.

4.8.3.1 Cavity barriers

Cavities in a building's construction enable fire spread and should be limited.

- Cavity barriers should be fixed so that if a suspended ceiling collapses, it will not cause the barrier to fall.   B(V1) 5.23

4.8.3.2 Escape routes

4.8.3.2.1 COMMON ESCAPE ROUTES

- Smoke vents should extend a minimum of 2.5m above the ceiling of the highest storey.   B(V1) 3.51

4.8.3.2.2 PROTECTED ESCAPE ROUTES AND STAIRWAYS

• A protected stairway enclosure requires cavity barriers or a fire-resisting ceiling (see AD-B(V1) Diagram 2.3).	B(V1) 2.5
• Transfer grilles should **not** be fitted in a ceiling of a protected stair.	B(V1) 2.9
• Any protected escape route with a cavity above the fire-resisting construction should:	B(V1) 8.5
• be enclosed on the lower side by a fire-resisting ceiling (minimum EI 30) which extends throughout the building, compartment or separated part (see AD-B(V1) Diagram 8.3);	
• have no openings, except the occasional one for a fire doorset, pipe, cable or conduit (unless fitted with a fire damper);	B(V1) 5.24
• extend throughout the building or compartment; and	
• not be easily demountable.	

4.8.3.3 Lighting diffusers

• Diffusers may be part of a luminaire or used below sources of light. (See AD-B(V1) Diagram 4.1.)	B(V1) 4.15
• Diffusers may be incorporated in ceilings to rooms and circulation spaces (but not to protected stairways), provided that the surface linings of walls and ceilings meet the classifications shown in AD-B(V1) Table 4.2 and Diagram 4.2.	B(V1) 4.16
• A ceiling constructed from TP(a) flexible panels should have a maximum area of 5m² and be supported on all sides.	B(V1) 4.17

4.8.4 Resistance to the passage of sound

4.8.4.1 Ceiling and roof junctions

Any junction between the ceiling and walls should be taken through the masonry and sealed with tape or caulked with sealant.

4.8.4.2 Wall type 1 – solid masonry

- If lightweight aggregate blocks with a density less   E 2.58
  than 1,200kg/m³ is used above ceiling level, then one
  side should
  be sealed with cement, paint or plaster skim.
- If the roof or loft space is not a habitable room (and   E 2.57
  there is a ceiling with a minimum mass per unit area
  of 10kg/m² with sealed joints), the mass per unit
  area of the separating wall above the ceiling may be
  reduced to 150kg/m² (see AD-E Diagram 2.13).

4.8.4.3 Wall type 2 – cavity masonry

Note: Where a type 2 separating wall is used, it should be continuous to the underside of the roof.

- Where the roof or loft space is not a habitable room   E 2.93
  and there is a ceiling with a minimum mass per unit
  area of 10kg/m² with sealed joints, then the mass per
  unit area of the separating wall above the ceiling may
  be reduced to 150kg/m² (but it **must** still be a cavity
  wall).
- If lightweight aggregate blocks (with a density less   E 2.94
  than 1,200kg/m³) are used above ceiling level, one side
  should be sealed with cement, paint or plaster skim.

Remember to fix the supporting frame of panels to the ceiling and floor only.

4.8.4.4 Wall type 3 – masonry between independent panels

- The junction between the ceiling and independent   E 2.135
  panels should be sealed with tape or caulked with
  sealant.

- For wall types 3.1 and 3.2 (solid masonry core):   E 2.137
  - if the roof or loft space is not a habitable room (but the ceiling has sealed joints and a minimum mass of 10kg/m² per unit area), independent panels may be omitted in the roof space and the mass per unit area of the separating wall above the ceiling may be a minimum of 150kg/m²;
  - if lightweight aggregate blocks with a density less   E 2.138 than 1,200kg/m³ are used above ceiling level, one side should be sealed with cement, paint or plaster skim.
- For wall type 3.3 (cavity masonry core):   E 2.139
  - if the roof or loft space is not a habitable room (and the ceiling has a minimum mass per unit area of 10kg/m² with sealed joints), independent panels may be omitted, but the cavity masonry core should be maintained to the underside of the roof.

Note: Where it is necessary to connect two leaves together, do **not** use ties of greater cross-section than 40mm × 3mm fixed to the studwork at, or just below, ceiling height.

4.8.4.5 Wall type 4 – framed walls with absorbent material

- The junction between the ceiling and the wall linings   E 2.162 should be sealed either with tape or caulked with sealant.
- If the roof or loft space is not a habitable room (and there is a ceiling with a minimum mass per unit area of 10kg/m² with sealed joints), then the cavity may be closed at ceiling level without connecting the two frames rigidly together.

4.8.4.6 Interaction with floors

- In ***floor type 1***, the resistance to airborne sound   E 3.23 partly depends on the mass per unit area of the ceiling.

- In *floor type 2*, the resistance to airborne and impact sound partly depends on the isolation of the floor and ceiling.   E 3.56

- In *floor type 3*, the resistance to airborne and impact sound partly depends on the structural floor base and the isolation of the platform floor and the ceiling.   E 3.94

4.8.5 Ventilation

- For multi-storey car parks, each storey should be ventilated by permanent openings at each parking level.   F(V1) F1

Note: The following regulations do not apply to a building or space within a building:

- into which people do not normally go;
- which is used solely for storage;
- which is a garage used solely in connection with a single dwelling.

- Extract ventilation terminals and fans, not including cooker extract hoods, should be installed as high as is practicable in the room and a maximum of 400mm below the ceiling. See also AD-F B(V1) Tables 1.1 and 1.2.   F(B(V1)) 1.20

When making changes to thermal elements, use AD-F(V1) Table 3.1 to calculate the number of major and minor energy-efficiency measures that need to be addressed. This should then be compared to AD-F(V1) Diagram 3 to decide whether further ventilation is required.

4.8.6 Conservation of fuel and power

In considering heat losses via party walls, it is important to remember that, wherever the blockwork of a masonry party wall penetrates insulation at ceiling level, a thermal bridge is likely to exist – even when the party wall U-value is zero. Any bridging at the party wall should be evaluated and then taken into account, along with other thermal bridges.

Any solution to a party wall bypass should not contravene other parts of the regulations, in particular AD-E (*Sound*).

- To limit thermal bridging in new dwellings: L(V1) 4.17
  - loft insulation at the eaves should extend beyond the wall insulation (taking into consideration the pitch of the roof);
  - at gables and party walls, the insulation should extend to the wall at gables;

If the space between the wall and a joist is less than 100mm, perimeter insulation may also be required.

- Loft hatches should be designed and installed to L(V1) 4.21i
  ensure optimum airtightness.

- Whenever possible, cost-effective U-value L(V1) C4
  improvements should be made when undertaking Table C1
  renovation works to thermal elements. For example:
  - replacing any existing insulation at ceiling level which is less than 250mm, in poor condition, or likely to be disturbed or removed during planned work;
  - inserting new 250mm mineral fibre or cellulose fibre as a quilt laid between and across ceiling joists.

Note: Roof insulation should be installed when the eaves are still accessible.

4.8.7 Electrical safety

Notifiable work (i.e. to the Building Control Body) in rooms containing baths and showers that effect ceilings and walls is shown in Diagram 2.

Requirements for buildings other than dwellings – Additional requirements

The basic requirements for ceilings in buildings other than dwellings are the same as for dwellings, with the addition of the following requirements.

Note: The Workplace (Health, Safety and Welfare) Regulations 1992 also contain some requirements which affect building design.

4.8.8 Access to and use of buildings

- In order to help people with visual impairment to appreciate the size of a space they have entered, or to find their way around, there should be a visual contrast between the wall and the ceiling.   M2 3.12
M2 4.32
- Attention to surface finishes should be coupled with good natural and artificial lighting design.

4.8.8.1 Car parks

Each storey should be ventilated by permanent openings at each parking level.   B(V2) 11.4

Note: These permanent openings in car parks can be at ceiling level.

4.8.9 Fire safety

4.8.9.1 Cavities

A cavity (see Diagram 9.1) is considered to be any concealed space, and as these provide a ready route for smoke and fire, they need careful consideration.

- Cavity barriers should be used to divide any cavity, including roof spaces (see AD-B(V2) Table 9.1 for details concerning the maximum dimensions of these cavities).   B(V2) 9.9

Note: See Table 9.1 for the maximum dimension of cavities in buildings other than dwellings.

- If a single room has a ceiling cavity exceeding the dimensions shown in AD-B(V2) Table 9.1, cavity barriers need only be provided on the line of the enclosing walls/partitions of that room.   B(V2) 9.11

- Cavity barriers should be unaffected by: B(V2) 9.16
  - any movement of the building owing to subsidence, shrinkage or temperature change;
  - failure of material or construction to which cavity barriers abut (such as a suspended ceiling that continues over a fire-resisting wall);
  - failure of cavity barrier fixings.

4.8.9.2 Fire resistance

- Table B3 (para 18) gives the specific requirements for the minimum requirements of the fire resistance of ceilings (i.e. 30 mins). B(V2) b.24

4.8.9.3 Escape routes

- If a corridor is used for a means of escape but is not a protected corridor. B(V2) 2.25
- Any partitions should continue to the soffit of the structural floor above, or to a suspended ceiling.
- If a cavity exists above the enclosures of the corridor used as a means of escape but not a protected corridor: B(V2) 2.27
  - the cavity on the lower side of the suspended ceiling should be fire resistant and one that extends throughout the building, compartment or separated part.

4.8.9.4 Inner rooms

As an inner room at risk of fire might mean a fire occurring in an access room (see Diagram 2.3):

- Inner room enclosures (walls or partitions) should stop a minimum of 500mm below the ceiling. B(V2) 2.11

4.8.9.5 Internal linings

- The surface linings of ceilings should meet the classifications shown in AD-B(V2) Table 6.1.

  B(V2) 6.1

- For the purpose of this requirement, a ceiling includes **all** of the following:
  - any part of a wall at 70° or less to the horizontal;
  - glazed surfaces;
  - the underside of a gallery; and
  - the underside of a roof exposed to the room below.

  B(V2) 6.5

- On the other hand, it does **not** include any of the following:
  - architraves, cover moulds, picture rails, exposed beams and similar narrow members;
  - trap doors and their frames; or
  - the frames of windows, glazing or rooflights.

  B(V2) 6.6

4.8.9.6 Lighting diffusers

Note: The following requirements apply to lighting diffusers forming part of a ceiling. They do **not** apply to diffusers of light fittings attached to the soffit of a ceiling or suspended beneath a ceiling. (See AD-B(V2) Diagram 6.1.)

- Diffusers may be part of a luminaire or used below sources of light.

  B(V2) 6.16

- Diffusers constructed of thermoplastic material may be incorporated in ceilings to rooms and circulation spaces, but **not** to protected stairways if:
  - the upper surfaces of the thermoplastic panels, wall and ceiling surfaces exposed in the space above the suspended ceiling have been constructed in accordance with AD-B(V2) Table 6.1;
  - the diffusers have been classified as either TP(a) rigid (no restrictions on their extent) or TP(b) (limited in their extent).

  B(V2) 6.17a

4.8.9.7 Resisting fire spread

• The need for cavity barriers in a concealed floor or roof space can be reduced by installing a fire-resisting ceiling (minimum EI 30) below the cavity.	B(V2) 6.7
• Where panels are used to form a suspended ceiling, the fixing should pass through the panel and support it from the lower face (see AD-B(V2) Diagram 6.3).	B(V2) 6.11
• Cavities above the fire-resisting construction should be enclosed on the lower side by a fire-resisting ceiling (minimum EI 30) that extends throughout the building, compartment or separated part.	B(V2) 9.5

4.8.9.7.1 SUSPENDED OR STRETCHED-SKIN CEILINGS

• A ceiling constructed from flexible panels should have a maximum area of 5m² and be supported on all sides.	B(V2) 6.18

4.8.10 Ventilation

• For naturally ventilated offices that do not use mechanical supply and extract ventilation:
  • extract ventilators should be sited on the wall as high as practicable; and
  • passive stack ventilation terminals should be located in the ceiling.
	F(V2) 1.30

4.8.10.1 Venting of heat and smoke from basements

• Natural smoke outlets serving a basement should be sited at high level in either the ceiling or wall of the space they serve, and evenly distributed around the perimeter.	B(V2) 18.5

4.8.11 Conservation of fuel and power

• Reasonable provision shall be made for the conservation of fuel and power by: • limiting heat gains and losses; and • providing fixed building services.	L(V2) Schedule 1
• If a thermal element is being renovated, or new thermal elements being installed in an existing building, then it should achieve the U-values shown in AD-L(V2) Table 4.2.	L(V2) Table 4.2

4.9 Roofs

The roof of a brick-built house is normally an aitched (sloping) roof comprising rafters fixed to a ridge board, braced by purlins, struts and ties, and secured to wall-plates bedded on top of the walls. They are then usually clad with slates or tiles to keep the rain out.

Timber-framed houses usually have trussed roofs – prefabricated triangulated frames that combine the rafters and ceiling joists – which are lifted into place and supported by rails. The trusses are joined together with horizontal and diagonal ties. A ridge board is not fitted, nor are purlins required. Roofing-felt battens and tiling are applied in the usual way.

Softwood timber used for roofs or fixed in the roof space (including any ceiling joists within the void spaces of the roof) should be adequately treated to prevent infestation by the house longhorn beetle *(Hylotrupes bajulus L.)*

Guidance on the sizing of timber floors and roofs for traditional house construction (known as the Timber Tables) are published by TRADA.

Reminder: The subsections include details for buildings other than dwellings only where they differ from the requirements for dwellings.

Subsection	Description	Page
Requirements for dwellings		
4.9.1	Structure	349
4.9.2	Fire safety	352
4.9.3	Resistance to the passage of sound	359
4.9.4	Ventilation	361

Subsection	Description	Page
4.9.5	Rainfall protection	363
4.9.6	Combustion appliances	364
4.9.7	Conservation of fuel and power	365
4.9.8	Electrical safety	367
4.9.9	Security	368
Requirements for buildings other than dwellings – Additional requirements		
4.9.10	Structure	368
4.9.11	Fire safety	371
4.9.12	Conservation of fuel and power	374

Reminder: Free downloads of all the diagrams and tables referenced in the subsections are available on-line at the Governments' Planning Portal: www.gov.uk/government/collections/approved-documents.

Requirements for dwellings

4.9.1 Structure

Note: Unless they serve the function of a floor (such as a roof terrace), roofs are not treated as elements of structure.

• The maximum height of a residential building, from ground level to the highest point of any roof, should ***not*** be greater than 15m.	A 2C4i
• Roofs shall be constructed so that they:	
• provide local support to the walls;	A 2A2d
• act as horizontal diaphragms capable of transferring the wind forces to the buttressing elements of a building.	
• Special consideration should be given if large panels of glass cladding are used in roofs that are not divided into smaller areas by load-bearing frames.	A 3.9

Note: See BS 8297 and BS 8298 for further details.

4.9.1.1 Timber

A traditional cut timber roof (such as one that uses rafters, purlins and ceiling joists) generally has sufficient built-in resistance to instability and wind forces (which can be variable).

Softwood timber used for roofs or fixed in the roof space (including any ceiling joists within the void spaces of the roof), should be adequately treated to prevent infestation by the house longhorn beetle *(Hylotrupes bajulus L.)*.

4.9.1.2 Imposed loads

Imposed loads on roofs are not permanent and can be variable.

• The building shall be constructed so that the applied loading and the actual wind load are both transmitted to the ground, safely, and without causing any changes to the structural integrity of the roof or any supporting elements.	A 1
• For design purposes, the following imposed loads (see AD-A Table 4) should be considered: • 1.00kN/m² for spans not exceeding 12m; • 1.5kN/m² for spans not exceeding 6m.	A 2C15

4.9.1.3 Lateral support

Lateral support should be provided to help prevent sideways movement.

• Roofs should act to transfer lateral forces from walls to buttressing walls, piers or chimneys.	A 2C33a
• Roofs should be secured to the supporting wall.	A 2C33b

Note: The requirements for lateral restraint of roofs are shown in AD-A Diagram 15 and Table 9.

- Vertical strapping (at least 1m in length) is required in   A 2C36
  a gable wall unless the roof complies with all of the
  following:
  - it has a pitch of 15° or more;
  - it is tiled or slated;
  - it is of a type known by local experience to be
    resistant to wind gusts;
  - it has main timber members spanning on to the
    supported wall at not more than 1.2m centres.

Note: Walls shall be tied to the roof structure vertically and have a horizontal lateral restraint at roof level.

4.9.1.4 Openings

Where a roof opening for a stairway adjoins a supported wall and interrupts the continuity of lateral support:

- The maximum permitted length of the opening is 3m.   A 2C37a
- Connections (other than by anchor) should be   A 2C37b
  throughout the length of each portion of the wall
  (and on each side of the opening).
- Connections via mild steel anchors should be spaced   A 2C37c
  closer than 2m on each side of the opening.
- There should be no other interruption of lateral   A 2C37d
  support.

Where a masonry chimney is inadequately supported by ties or securely restrained in any way:

- The chimney's height, if measured from the highest   A 2D1
  point of intersection with the roof surface, gutter,
  etc. (see AD-A Diagram 20), should not exceed 4.5W,
  provided the density of the masonry is greater than
  1,500kg/m^3.

4.9.1.5 Pitch

- Vertical parts of a pitched roof, such as dormer   B(V1)
  windows, should be included only if the slope of the   11.14
  roof exceeds 70°.

It is a matter of personal choice and judgement whether a continuous run of dormer windows that occupies most of a steeply pitched roof should be treated as a wall rather than a roof.

4.9.1.6 Roof covering

Recovering of roofs is one way of extending the useful life of buildings. However, it should be remembered that roof structures may be required to carry underdrawing or insulation at a later date.

To underdraw a roof or ceiling is to cover the underside of exposed beams and joists, either with boards or with lath and plaster.

- All materials used to cover roofs shall be capable of   A 4.1
  safely withstanding the concentrated imposed loads
  upon roofs specified in BS EN 1991–1-1.
- Special consideration should be given if large panels   A 3.9
  of glass cladding are used in roofs that are not
  divided into smaller areas by load-bearing frames.

The most frequently used materials for pitched roofs are ceramic roof tiles, natural slates (either fibre reinforced, clay or concrete), reed or bitumen felt.

Alternatively, a pitched roof could consist of a self-supporting sheet made out of galvanized steel, aluminium, fibre-reinforced cement, or pre-painted (coil-coated) steel or aluminium with a PVC or PVF2 coating.

4.9.2 Fire safety

The roof of the building should adequately resist the spread of fire over the roof and from one building to another.

A structure that **only** supports a roof is not an 'element of structure', unless the roof performs the function of a floor (such as a roof terrace), a means of escape, or is essential for the stability of an external wall that needs to be fire resistant.

In some cases, structural members **within** a roof may be essential for the structural stability system of the building. In these cases, the structural members in the roof do not just support a roof and must therefore demonstrate the relevant fire resistance for the building.

4.9.2.1 Balconies and flat roofs

• Where a flat roof forms part of a means of escape, it should:
   • be part of the same building from which escape is being made;
   • lead to a storey exit or external escape route;
   • be of fire-resisting construction (minimum REI 30).
	B(V1) 2.13
	B(V1) 3.10

• A balcony or flat roof intended to form part of an escape route should be provided with guarding, etc., in accordance with AD-K.	B(V1) 2.14 B(V1) 3.11
• Any alternative exit from a flat may lead to a final exit, via a common stair if necessary, through a door to an escape route over a flat roof.	B(V1) 3.22

4.9.2.2 Cavity barriers

Concealed spaces or cavities in the construction of roofs will provide an easy route for smoke and flame to spread, which, because it is concealed, will present a greater danger than would a more obvious weakness in the fabric of the building. A cavity barrier is **any** construction which is provided to restrict the movement of smoke or flame within a concealed space.

• It is not appropriate to complete a line of compartment walls by fitting cavity barriers above them. Instead, the compartment wall should be extended to the underside of the floor or roof above.	B(V1) 5.19

- In roof spaces, where cavity barriers are fixed to roof   B(V1)
  members, there is no expectation of fire resistance   5.23c
  from roof members which are provided as a means
  of support.

The need for cavity barriers in some concealed roof spaces can be reduced by using a fire-resisting ceiling.

4.9.2.3 Compartment walls

To prevent the spread of fire within a building, whenever possible the
building should be sub-divided into compartments separated from one
another by walls and/or floors of fire-resisting construction.

- A compartment wall should be taken up to meet the   B(V1) 5.11
  underside of the roof covering or deck, with fire-
  stopping at the wall/roof junction.

- Compartment walls common to two or more   B(V1) 5.10
  buildings should run the full height of the building in
  a continuous vertical plane and be continued through
  any roof space to the underside of the roof.

- Compartment walls should extend to the underside   B(V1) 8.4
  of the floor or roof above.

- Compartment walls in a top storey beneath a roof   B(V1) 5.8
  should be continued through the roof space.

4.9.2.3.1 CONSTRUCTION OF COMPARTMENT WALLS

- Compartment walls common to two or more   B(V1) 7.8b
  buildings should be continued through any roof
  space to the underside of the roof (see AD-B(V1)
  Diagram 5.2).

- Double-skinned insulated roof sheeting with a   B(V1) 7.17
  thermoplastic core should incorporate a band of
  material (rated class A2-s3, d2 or better) with a
  minimum width of 300mm, centred over the wall.

4.9.2.3.2 DOUBLE-SKINNED CORRUGATED OR PROFILED ROOF SHEETING

- A double-skinned corrugated or profiled insulated roof sheeting does not require a cavity barrier between the sheeting if:
  - the sheeting is rated class A2-s3, d2 or better;
  - both insulating layer surfaces are rated class C-s3, d2 or better;
  - the sheeting makes contact with both the inner and outer skins of cladding.

B(V1) 8.7

4.9.2.3.3 JUNCTIONS WITH COMPARTMENT WALLS

- If a fire penetrates a roof near a compartment wall, there is a risk that it will spread over the roof to the adjoining compartment. To reduce this risk either:
  - the wall should be extended up through the roof for a height of at least 375mm above the top surface of the adjoining roof covering (see AD-B(V1) Diagram 5.2c); or
  - a 1,500mm wide zone of the roof, either side of the wall, should have a covering classified as BROOF(t4), on a substrate or deck of a material rated class A2-s3, d2 or better. (See AD-B(V1) Diagram 5.2a.)

B(V1) 5.12

4.9.2.4 *Escape routes*

- A flat roof forming part of a means of escape should comply with the following provisions:
  - the roof should be part of the same building from which escape is being made;
  - the route across the roof should lead to a storey exit or external escape route;
  - the part of the roof forming the escape route and its supporting structure (together with any opening within 3m of the escape route) should provide 30 minutes fire resistance.

B(V1) 3.30
B(V1) 3.69

Note: If the fire-resisting construction of a protected　　B(V1) 8.5
escape route is either of the following:
  • not carried to full storey height;
  • not carried to the underside of the roof covering;
then the cavity above or below the fire-resisting
construction should be fitted with cavity barriers on the
line of the enclosure (see AD-B(V1) Diagram 8.3).

The escape route should be clearly defined and guarded to protect
against falling.

4.9.2.5 *Performance of roofs*

If one element of structure supports or stabilises another, as a minimum,
the supporting element should have the same fire resistance as the other
element.

• Any part of a roof should achieve the minimum performance as detailed in Section 12.	B(V1) 10.15
• Roof covering describes one or more layers of material, but not the actual roof structure as a whole.	B(V1) 12.1
• Provisions for the fire properties of roofs are detailed in: 　• Requirement AD-B(V1) – roofs that form a means of escape; 　• Requirement AD-B2(V1) – internal surfaces of rooflights; 　• Requirement AD-B3(V1) – roofs that are used as a floor; 　• Section 11 – the circumstances in which a roof is subject to the provisions for space separation.	B(V1) 12.2
• Table 12.1 sets out separation distances (i.e. the minimum distance from the roof, or part of the roof, to the relevant boundary) by the type of roof covering and the size and use of the building.	B(V1) 12.3
• Rooflights performance is specified in a similar way to that of roof coverings.	B(V1) 12.4
• Plastic rooflights may also be used provided their lower surface has a minimum class D-s3, d2 rating (see AD-B(V1) Table 12.2 and Diagram 12.1).	B(V1) 12.5

- The limitations for using thermoplastic materials with a TP(a) rigid or TP(b) classification are shown in AD-B(V1) Table 12.3.   B(V1) 12.6

- Other than for the purposes of AD-B(V1) Diagram 5.2, polycarbonate or uPVC rooflights achieving a minimum rating of class C-s3, d2 can be regarded as having a BROOF(t4) classification.   B(V1) 12.7

- When used in rooflights, unwired glass (a minimum of 4mm thick) can be regarded as having a BROOF(t4) classification.   B(V1) 12.8

- If the performance of thatch or wood shingles cannot be established, they should be regarded as EROOF(t4).   B(V1) 12.9

- Consideration can be given to thatched roofs being closer to the relevant boundary than shown in AD-B(V1) Table 12.1 if all of the following precautions are incorporated in the design:
  - the rafters are overdrawn with construction having not less than 30 minutes fire resistance;
  - the guidance given in AD-J is followed;
  - the smoke alarm installation (see Section 1) extends to the roof spaces.

- AD-B(V1) uses the European classification system for roof covering set out in BS EN 13501-5; however, there may be some products lawfully on the market using the classification system set out in previous editions. Where this is the case, AD-B(V1) Table B2 can be used.   B(V1) B18

  A plant room (or space dedicated for mechanical equipment and associated electrical equipment) on the roof may need greater fire resistance than the elements of structure that support it.

4.9.2.6 *Precipitation*

Roofs have the following functions:

- Roofs should resist the penetration of precipitation to the inside of the building.   C 6.2a
- Roofs should not be damaged by precipitation.   C 6.2b

- Roofs should not carry precipitation to any part of the building that would be damaged by it.
- Roofs should be designed and constructed so that their structural and thermal performance is not adversely affected by interstitial condensation.

C 6.2c

4.9.2.7 Resisting fire spread from one building to another

- The external envelope of a building should not provide a medium for undue fire spread to adjacent buildings or be readily ignited by fires in adjacent buildings. Flame spread over the roof and/or fire penetration from external sources through the roof should be restricted.

B(V1) B4

- Portal frames are often used in single-storey industrial and commercial buildings where there may be no need for fire resistance of the structure (requirement B3). However, where a portal-framed building is near a relevant boundary, the external wall (particularly if it is attached to a flat roof) may need fire resistance to restrict the spread of fire between buildings.

B(V1) 11.15

4.9.2.8 Resistance to damage from interstitial condensation

 Roofs shall be designed and constructed in accordance with Clause 8.4 of BS 5250 and BS EN ISO 13788.

- Cold deck roofs shall be ventilated.
- Any parts of a roof that have a pitch of 70° or more shall be insulated as though they were walls.

C 6.11

- Gaps and penetrations for pipes and electrical wiring should be filled and sealed.
- A draught seal should be provided to loft hatches.

C 6.12

4.9.2.8.1 RESISTANCE TO SURFACE CONDENSATION AND MOULD GROWTH

Roofs shall be designed and constructed so that the thermal transmittance (U-value) does not exceed $0.35W/(m^2 \cdot K)$ at any point.

- Ensure that junctions between elements and   C 6.14b
  the details of openings (such as windows) are in
  accordance with the recommendations in the report
  on robust construction details (see BRE IP17/01[145]).

4.9.3 *Resistance to the passage of sound*

In order for the construction to be fully effective regarding its resistance
to the passage of sound, care should be taken to correctly detail the junctions
between the separating wall and the roof.

If there is an external cavity wall, the cavity should be closed at eaves
level with a suitable flexible material (such as mineral wool).

- If lightweight aggregate blocks (with a density less   E 2.58
  than 1,200 kg/m^3) are used above ceiling level, then
  one side should be sealed with cement, paint or
  plaster skim.

4.9.3.1 *Ceiling and roof junctions*

Where a type 1 separating wall is used, it should be continuous to the
underside of the roof.

- The junction between the separating wall and the   E 2.56
  roof should be filled with a flexible closer which is
  also suitable as a fire-stop.

4.9.7.4 *Resistance to moisture from the outside*

Roofs should be designed to protect the building from precipitation
either by holding the precipitation at the face of the roof or by stopping
it from penetrating beyond the back of the roofing system.

- Roofs should be: C 6.6
  - weather resistant (such as natural stone or slate,
    cement-based products, fired clay and wood);
  - moisture resistant (such as bituminous and plastic
    products);
  - ideally jointless (so as to disallow structural and
    thermal movement).

• Roofs that have overlapping dry joints should be weather resistant and backed by a material (such as roofing felt).	C 6.7
• Materials that can deteriorate rapidly without special care (such as paint, any coating, surfacing or rendering) should be avoided on a roof.	C 6.5

 Note: Each sheet, tile and section of roof should be fixed in accordance with the guidance contained in BS 8000-6.

4.9.7.5 Site preparation and resistance to moisture

Roofs are exposed to moisture in three ways:

1 precipitation from the outside;
2 interstitial condensation;
3 condensation or mould growth on the internal surface.

 Note: There are particular issues when working in historic buildings to ensure that moisture ingress to the roof structure is limited and the roof can breathe. Where it is not possible to provide dedicated ventilation to pitched roofs, it is important to seal existing service penetrations in the ceiling and to provide draught-proofing to any loft hatches. Any new loft insulation should be kept sufficiently clear of the eaves so that any adventitious ventilation is not reduced.

4.9.3.4 Wall type 1 – solid masonry

• Where the roof or loft space is not a habitable room, the mass per unit area of the separating wall above the ceiling may be reduced to 150kg/m². (See AD-E Diagram 2.13.)	E 2.57

4.9.3.5 Wall type 2 – cavity masonry

Where a type 2 separating wall is used, it should be continuous to the underside of the roof.

• The junction between the separating wall and the roof should be filled with a flexible closer that is also suitable as a fire-stop. (See AD-E Diagram 2.25.)	E 2.92

If the roof or loft space is not a habitable room (and there is a ceiling with a minimum mass per unit area of $10kg/m^2$ with sealed joints), the mass per unit area of the separating wall above the ceiling may be reduced to $150kg/m^2$ – **but** it should still be a cavity wall.

4.9.3.6 Wall type 3 – masonry between independent panels

If a type 3 separating wall is used, the masonry core should be continuous to the underside of the roof.

• The junction between the separating wall and the roof should be filled with a flexible closer that is also suitable as a fire-stop. (See AD-E Diagram 2.35.)	E 2.134

4.9.3.7 Wall type 4 – framed walls with absorbent material

• When a type 4 separating wall is used, the wall should preferably be continuous to the underside of the roof.	E 2.160
• The junction between the separating wall and the roof should be filled with a flexible closer.	E 2.161
• The junction between the ceiling and the wall linings should be sealed with tape or caulked with sealant.	E 2.162
• External wall cavities should be closed at eaves level with suitable material.	E 2.163

There need only be one frame in the roof space provided there is a lining of two layers of plasterboard (each sheet of minimum mass per unit area $10kg/m^2$) on both sides of the frame.

4.9.4 Ventilation

To promote cross-ventilation:

- Roofs with a pitch of 15° or more should have ventilation openings at least 10mm wide at eaves level.
- Roofs with a pitch of less than 15° should have ventilation openings at least 25mm in two opposite sides.

4.9.4.1 Energy efficiency – ventilating roof voids

When carrying out energy-efficiency measures on an existing dwelling's loft and roof spaces, initially you should calculate the number of major and minor energy-efficiency measures involved. This calculation should include the following:

- Have any additional energy-efficiency measures been fitted since the 'original' dwelling was constructed?
- Are further energy-efficiency measures planned for the future?

As shown in AD-F Table 3, some of the most important energy-efficiency measures concern:

• Renewal (or potential renewal) of loft insulation.	F(V1)
• Effective edge sealing at junctions and penetration.	Table 3.1
• Changing a cold loft (insulation at ceiling level) to a warm loft (insulation at roof level).	
• Sealing around structural or service penetrations through walls, floors or ceilings/roofs.	

Note: Placing outlet terminals at the ridge of the roof is the preferred option, as it is not prone to wind gusts and/or certain wind directions.

• Continuity of insulation across roofs should be as tight as possible to the structure, without air gaps, and should extend to the wall insulation.	L(V1) 4.15–4.17

Note: For roofs insulated at ceiling level, the long-term protection of the insulation layer should be considered:
- insulation at rafter level of the leaves of roofs should extend to the top of the external wall;
- voids between insulation at the top of the external wall and the cavity wall (or timber frame)should be fully filled with insulation;
- where the roof is insulated at ceiling level, loft hatches should ensure optimum airtightness.   L(V1) 4.21 i

The renewal of a roof covering on a pitched roof should have a U-value of $0.16 \text{W}/(\text{m}^2{\cdot}\text{K})$.

4.9.5 Rainfall protection

Where the design incorporates valley gutters, parapet gutters, or siphonic or drainage systems from flat roofs, and where over-topping of these systems would have particularly high consequences such as water entering the building, wetting of insulation, or cause other dampness, the design should be carried out in accordance with **BS EN 12056**.

• The flow into a gutter depends on the area of surface being drained and whether the surface is flat or pitched and (if it is pitched) on the angle of pitch: • for roofs with a 300 pitch, multiply the plan area of the portion by 1.29; • for roofs with a 450 pitch, multiply the plan area of the portion by 1.50; • for roofs with a 600 pitch, multiply the plan area of the portion by 1.879; • for roofs with a 700 pitch or more, multiply the elevational area by 0.5.	H1.3
• Where rainwater from a roof with an effective area greater than 25m² discharges through a single downpipe onto a lower roof, a distributor pipe should be fitted to the shoe. • Where a rainwater pipe discharges onto a lower roof or paved area, a pipe shoe should be fitted.	H 1.9
• On flat roofs, valley gutters, and parapet gutters, additional outlets may be necessary.	H 1.7

4.9.5.1 Termination of discharge pipe

• The discharge pipe from the tundish should terminate in a safe place where there is no risk to persons in the vicinity of the discharge.	G 3.61
• Acceptable discharge arrangements are: • discharges to a trapped gully with the end of the pipe being below fixed grating and above the water seal;	G 3.62

- downward discharges at low level (e.g. up to
  100mm above external surfaces such as car parks,
  hardstandings, grassed areas, etc.) are acceptable
  providing that a wire cage or similar guard is
  positioned to prevent contact, while maintaining
  visibility;
- discharges at high level (e.g. into a metal hopper
  and metal downpipe) with the end of the discharge
  pipe clearly visible, or onto a roof capable of
  withstanding high temperature discharges of water
  and 3m from any plastic guttering system.

As the discharge would consist of high temperature water and steam,
asphalt, roofing felt and non-metallic rainwater goods may be damaged
by such discharges.

4.9.6 *Combustion appliances*

- There should be a permanent means of safe access to   J 1.60
  appliances for maintenance.
- Roof space installations of gas-fired appliances
  should comply with the requirements of BS 6798.
- The outlet from a flue should be above the roof of   J 2.1
  the building in a position where the products of
  combustion can discharge freely and will not present
  a fire hazard, whatever the wind conditions.

Note: Positions for flue outlets for solid-fuel appliances (easily ignited
roof coverings) are shown in Diagram 18.

4.9.6.1 *Tests and examinations*

- A visual inspection should be carried out of the   J E3b
  accessible parts to identify deterioration in the
  structure, connections or linings which could affect
  the flue's gas-tightness and safe performance with the
  proposed combustion appliance.

Note: In particular, the interior of the flue and the exterior of the
chimney, including in the roof space, should be examined.

4.9.7 *Conservation of fuel and power*

The main changes made by the 2016 amendments include the withdrawal of Regulations 29 to 33 of the Building Regulations 2010 and their replacement with Regulation 7A of the Energy Performance of Buildings (England and Wales) Regulations 2012, as well as changes in the wording of Regulations 24, 25, 26, 26A, 27 and 27A of the Building Regulations 2010.

4.9.7.1 *Extensions*

Extensions to dwellings should either use newly constructed thermal elements or make use of existing (or new) roof windows and rooflights, provided that they meet the requirements for the conservation of fuel and power.

4.9.7.2 *Historic and traditional buildings*

Building inspectors will require the use of sympathetic treatment when restoring the historic character of a building that has been subject to previous inappropriate alteration (e.g. when replacing a roof or a rooflight).

When undertaking work on any type of historic or traditional type of building, the aim should always be to improve energy efficiency as far as is reasonably practicable, without changing the character of the host building or increasing the risk of long-term deterioration of the building's fabric or fittings.

If you are unsure of exactly what renovations, work or extensions you are permitted to complete on a historic or traditional building, then it is best to seek specialist advice, particularly if you are:

- Restoring the actual historic character of a building when it has been subjected to previous inappropriate alteration (such as replacement windows, doors and rooflights).
- Rebuilding a former historic building (for example, following a fire or filling a gap site in a terrace).
- Making changes to enable the fabric of a historic building to 'breathe', in order to control moisture.

4.9.7.3 *Non-notifiable work*

If you want to install new thermal insulation in a roof space or loft space and this is the only work being carried out and the work is not carried out so as to comply with any requirement in the Building Regulations, then this will class as non-notifiable work.

4.9.7.4 Party walls

• In a room-in-roof design, the insulation layer may follow the sloping roof sections to a horizontal ceiling, then continue at ceiling level.	L(V1) 4.17
• At the eaves of roofs insulated at rafter level, insulation should extend to the top of the external wall.	
• At gables and party walls, if the space between the wall and joist is less than 100mm, perimeter insulation may be required.	

Note: AD-L(V1) Table 2.1 provides details to assist in determining a party wall's U-value for the type of construction adopted.

4.9.7.5 Roof windows and rooflights

Roof windows and rooflights are all controlled fittings.

• When installing roof windows or rooflights, the controlled fitting should be well fitted and reasonably draught proof.	L(V1) 4.23
• For rooflights, U-values should be calculated based on a horizontal position.	L(V1) 4.5
• New or replaced roof windows or rooflights should meet the standards in AD-L(V1) 4.17, Table 4.2 ('Limiting U-values for new fabric elements in existing dwellings').	

4.9.7.6 Siting of pedestrian guarding

• The edges of any part of a roof (including rooflights and other openings) should have some form of safety guarding.	K 3.1
• The height of the guarding on an external balcony and an edge of roof should be 1,100mm. (See AD-K Diagram 3.10.)	

4.9.7.7 Siting of vehicle barriers

- Barriers (3.75mm high) should be provided on any building roof edge to which vehicles have access.   K 4.1
- The height of the guarding on any building roof edge should be at least 3.75mm. (See AD-K Diagram 4.1.)

4.9.7.8 Thermal bridges

The building fabric should be constructed so that there are no reasonably avoidable thermal bridges in the insulation layers caused by gaps within the various elements, at the joints between elements, and at the edges of elements.

A suitable approach to showing that this requirement has been achieved would be to adopt Accredited Construction, details of which are on: www.gov.uk.

4.9.7.9 Upgrading thermal elements

Extensions to dwellings should either use newly constructed thermal elements or make use of existing (or new) roof windows and rooflights, provided that they meet the requirements for the conservation of fuel and power.

Reductions in thermal performance can occur where is no connection between the air barrier and the insulation layer, and the cavity between them is subject to air movement. To avoid this problem:

- The insulation layer should share a common border with the air barrier at all points in the building envelope.
- The space between the insulation layer and air barrier should be filled with solid material such as in a masonry wall.

4.9.8 Electrical safety

Part P applies to electrical installations that are outside the dwelling – for example photovoltaic panels on roofs – as well as to electrical installations within a roof space.

4.9.9 Security

Any easily accessible windows (including rooflights) fitted in new dwellings should be secure windows fitted within 2m vertically of a flat or sloping roof (with a pitch of less than 30°) that is within 3.5m of ground level.

Requirements for buildings other than dwellings – Additional requirements

The basic requirements for roofs in buildings other than dwellings are the same as for dwellings, with the addition of the following requirements.

 Note: The potential for fire to spread over roofs should be restricted.

4.9.10 Structure

- The underside of a roof exposed to the room below will be categorised as a ceiling.
- The height of the top storey excludes roof-top plant areas and any top storeys consisting exclusively of plant rooms.
- 'Roof covering' describes one or more layers of material, but not the roof structure as a whole.

4.9.10.1 Small single-storey non-residential buildings and annexes

- Access to the roof should only be for the purposes of   A 2C38
  maintenance and repair.
- Roofs should be braced at rafter level, horizontally at eaves level and at the base of any gable by roof decking, rigid sarking or diagonal timber bracing in accordance with BS EN 1995-1-1.
- Walls should be tied to the roof structure vertically and horizontally, and with horizontal lateral restraint at roof level.
- The roof structure of an annexe should be secured to the structure of the main building at both rafter and eaves level.
- Where straps cannot pass through a wall, they should be secured to the masonry using suitable fixings.

- Isolated columns should be tied to the roof structure.
- Walls should be tied horizontally at no more than 2m centres to the roof structure at eaves level, base of gables and along roof slopes.

- Vertical strapping may be omitted if the roof:
  - has a pitch of 15° or more;
  - is tiled or slated;
  - is of a type known by local experience to be resistant to wind gusts;
  - has main timber members spanning onto the supported wall at not more than 1.2m centres.

4.9.10.2 *Cavity barriers*

- For buildings other than dwellings, because most enclosures are not carried to the underside of the roof covering at the top storey, the potential for smoke to bypass the enclosure should be restricted by fitting cavity barriers, dividing the storey, or enclosing the cavity on the lower side by a fire-resisting ceiling.	B(V2) 2.27
- The need for cavity barriers in concealed roof spaces can be reduced by installing a fire-resisting ceiling (see AD-B(V2) Diagram 9.3).	B(V2) 6.7

- If a cavity exists above or below a partition between bedrooms in 'residential (institutional)' and 'residential (other)' buildings because the enclosure is not a fire-resisting partition, then:
  - barrier openings should be kept to a minimum;
  - all penetrations should be sealed with fire-stopping material.
	B(V2) 9.18

- Cavity barriers should be used to divide any cavity (including roof spaces) (see AD-B(V2) Table 9.1).	B(V2) 9.9
- Any cavity between a roof and ceiling should be no greater than 20m in any direction (see AD-B(V2) Table 9.1).	

 It must be remembered that roof members which are provided to support cavity barriers in roof spaces do not normally possess any fire resistance.

4.9.10.2.1 DOUBLE-SKINNED CORRUGATED OR PROFILED ROOF SHEETING

• Cavity barriers are not required between double-skinned corrugated or profiled insulated roof sheeting if;
  • the sheeting is rated class A2-s3, d2 or better;
  • the surfaces of the insulating layer are rated class C-s3, d2 or better; and
  • both surfaces of the insulating layer make contact with the inner and outer skins of cladding.
	B(V2) 9.8

4.9.10.2.2 MAXIMUM DIMENSIONS OF CAVITIES

• The maximum dimensions of undivided cavities are shown in AD-B(V2) Table 9.1.	B(V2) 9.3
• A compartment wall should extend to the underside of the floor or roof above.	B(V2) 9.4

4.9.10.3 Compartmentation

• Compartment walls common to two or more buildings should be continued through any roof space to the underside of the roof.	B(V2) 8.18
• Compartment walls in a top storey beneath a roof should be continued through the roof space.	B(V2) 8.21
• The compartment height is measured from finished floor level to the underside of the roof or ceiling.	B(V2) Table 8.1
• A compartment wall should meet the underside of the roof covering or deck, with fire-stopping to maintain the continuity of fire resistance, and be continued across any eaves.	B(V2) 8.25
• To reduce the risk of fire spreading over the roof from one compartment to another, a 1,500mm wide zone of the roof, either side of the wall, should have a covering classified as BROOF(t4) (see AD-B(V2) Diagram 8.2a).	B(V2) 8.26

- Class B-s3, d2 used as a substrate to the roof covering   B(V2) 8.27
  and any timber tiling battens, fully bedded in mortar
  or other suitable material for the width of the wall,
  may extend over the compartment wall in buildings
  that are no more than 15m high (see AD-B(V2)
  Diagram 8.2b).

- Double-skinned, insulated roof sheeting, a minimum   B(V2) 8.28
  of 300mm in width, centred over the wall, should
  incorporate a band of material rated class A2-s3, d2
  or better.

- A compartment wall may extend through the roof in   B(V2) 8.29
  the following circumstances:
  - where the height difference between the two roofs
    is less than 375mm;
  - 200mm above the top surface of the adjoining
    roof covering if the height difference between
    the two roofs is 375mm or more and the roof
    coverings either side of the wall are of a material
    classified as **BROOF**(t4).

Note: Rooflights should be at least 1,500mm from the compartment
wall (see AD-B(V2) Diagrams 8.2a and b plus Table 14.3).

4.9.11 Fire safety

4.9.11.1 Escape stairs

- Access to an external escape stair may be via a flat   B(V2) 3.33
  roof, provided the flat roof does not serve a public
  building and is fire resistant.

- Where an escape route over a flat roof is provided,   B(V2) 2.32
  the roof should:
  - be part of the same building;
  - lead to a storey exit or external escape route;
  - be fire resistant;
  - be clearly defined; and
  - be guarded by walls and/or barriers to protect
    from falling.

- A flat roof that is in the open air, protected (or remote) from any fire risk, and has its own means of escape is considered a safe refuge (see AD-B(V2) Diagrams 3.1 and 3.2). B(V2) 3.5

- If the fire-resisting construction of a protected escape route is not carried to the underside of the roof covering on the top storey, then the cavity above or below the fire-resisting construction should be either: B(V2) 9.5
  - fitted with cavity barriers on the line of the enclosure; or
  - enclosed on the lower side by a fire-resisting construction (see AD-B(V2) Diagram 9.3).

4.9.11.2 Mechanical ventilation and air-conditioning systems

- Terminals of exhaust points should be sited away from final exits, cladding, and roofing materials of class B-s3, d2 or worse, as well as openings into the building. B(V2) 10.6

4.9.11.3 Resisting fire spread

Roofs (including dormer windows) with a pitch more than 70° to the horizontal should be assessed in accordance with the following:

- Building roofs shall adequately resist the spread of fire from one building to another (see AD-B(V2) Table B4). B(V2) B26

- A structure that supports only a roof is excluded from the definition of 'element of structure' unless the roof performs the function of a floor, a means of escape, or is essential for the stability of an external wall that needs to be fire resisting. B(V2) 7.3

- In some cases, the structural members within a roof may be essential for the structural stability system of the whole building, in which case they do not just support a roof, but they also play a part in the actual fire resistance for the whole building. See further guidance in paragraph 7.2a of AD-B(V2).

- The compartment height is measured from finished floor level to the underside of the roof or ceiling (see AD-B(V2) Diagram 8.1). B(V2) 8.11

Note: It is a matter of judgement whether a continuous run of dormer windows that occupies most of a steeply pitched roof should be treated as a wall rather than a roof.

4.9.11.3.1 RESISTING FIRE SPREAD OVER ROOF COVERINGS

• Separation distance is the minimum distance from the roof, or part of the roof, to the relevant boundary. AD-B(V2) Table 14.1 sets out separation distances by the type of roof covering and the size and use of the building.	B(V2) 14.3 Table 14.1
• If the performance of thatch or wood shingles cannot be established, they should be regarded as having an EROOF(t4) classification as shown in AD-B(V2) Table 14.1.	B(V2) 14.9

4.9.11.4 Rooflights

The performance of rooflights is specified in a similar way to the performance of roof coverings and should meet the following classifications:

• Plastic rooflights should be a minimum class D-s3, d2 rating (as shown in AD-B(V2) Table 14.2 and Diagram 14.1).	B(V2) 6.8
• If plastic rooflights do not meet the rating above, they should meet the relevant classification shown in AD-B(V2) Table 6.1.	
• In rooms and circulation spaces other than protected stairways, rooflights may be constructed of thermoplastic material whose: • lower surface is classified as TP(a) rigid or TP(b); • size and location follow the limits in AD-B(V2) Table 6.2.	B(V2) 6.15
• Rooflights deigned for rooms in industrial and other non-residential purpose group buildings should be evenly distributed (at least 1,800mm apart) with a total area not exceeding 20% of the room area (see AD-B(V2) Diagram 14.31 supported by Table 14.1).	B(V2) 14.5

• Polycarbonate or uPVC rooflights with a minimum rating of class C-s3, d2 can be regarded as having a BROOF(t4) classification.	B(V2) 14.7

4.9.11.5 Rooftop plant rooms

• If a plant room on the roof needs greater fire resistance than the structural elements that support it, then the structure of the roof will have to be further fire proofed.	B(V2) B26(iii)

4.9.11.6 Special applications

• Any flexible membrane covering a structure, other than an air-supported structure, should comply with Appendix A of BS 7157.	B(V2) 6.9
• Guidance on the use of PTFE-based materials for tension-membrane roofs and structures is given in BRE report BR 274.	B(V2) 6.10

4.9.12 Conservation of fuel and power

• Limiting the effects of solar gains for new residential buildings in summer will depend on the glazing system used (see AD-L(V2) Table 4.3 for full details).	L(V2) 4.16 Table 4.3
• If a window or a rooflight is enlarged (or a new one created), their area should not exceed 20% or the total area of the roof.	L(V2) 10.5
• If new or replacement roof windows or rooflights are fitted, then: • units should be draught proofed and meet the minimum standards shown in AD-L(V2) Table 4.1; and • insulated cavity closers should be installed where appropriate.	L(V2) 10.3

Note: The U-values for roof windows and rooflights are based on the U-value having been assessed in the vertical position. If a particular unit has been assessed in a plane other than the vertical, the standards given in the AD should be adjusted following the guidance given in BR 443.

4.10 Chimneys and fireplaces

In accordance with the Clean Air Act 1993, it is an offence for dark smoke to be emitted from a chimney of any building. A person found guilty of an offence under this section is liable, on summary conviction, to a fine in accordance with the Standard Scale contained in the Criminal Justice Act – at the time of publication the maximum fine for such offences was £5,000!

Reminder: The subsections include details for buildings other than dwellings only where they differ from the requirements for dwellings.

Reminder: Free downloads of all the diagrams and tables referenced in the subsections are available on-line at the Governments' Planning Portal: www.gov.uk/government/collections/approved-documents.

Requirements for dwellings

4.10.1 Types of chimneys

4.10.1.1 Factory-made metal chimneys

* Where a factory-made metal chimney passes through   J 1.43
  a wall, sleeves should be provided.
* Joints between chimney sections should *not* be
  concealed within ceiling joist spaces or within the
  thicknesses of walls.

Following installation of a factory-made metal chimney, it should be a simple measure to withdraw the appliance without having to dismantle the chimney.

* Factory-made metal chimneys should *not* be installed   J 1.45
  near combustible materials.
* Where a factory-made metal chimney passes through
  a cupboard, storage space or roof space, the chimney
  should be no closer to combustible material than
  defined in BS EN 1856-1 (see AD-J Diagram 13).

Nevertheless, encasing the chimney in non-combustible material is recommended.

4.10.1.2 Flueblock chimney systems

There are two types of flueblock chimney systems; one for gas-burning appliances, and the other (often called a 'chimney block system') for solid-fuel-burning appliances.

* Flueblock chimneys:
  * should be constructed using factory-made   J 1.29
    components that are suitable for the intended
    application;
  * should be installed with sealed joints;   J 1.30
  * should only have bends and offsets that have been J 1.30 &
    formed with matching factory-made components.   J 4.16

- Flueblocks that are not intended to be bonded into   J 4.16
  surrounding masonry should be supported and
  restrained in accordance with the manufacturer's
  installation instructions.

- Where a flue pipe or chimney penetrates a fire   J 4.18
  compartment wall or floor, it must not breach the
  fire-separation requirements.

Where a building is to be altered for a different use (such as being converted into flats, which is judged as being a change of use):

- The fire resistance of the walls of existing masonry   J 1.31
  chimneys may need to be improved (see AD-J
  Diagram 12).

4.10.1.3 Masonry and flueblock chimneys

- The thickness of the walls around the flues (excluding   J 2.17
  flue liners) should comply with AD-J Diagram 20.
- Combustible material should **not** be located where it   J 2.18
  could be ignited by the heat dissipating through the
  walls of fireplaces or flues.

AD-J Diagram 21 shows a method of meeting this requirement so that combustible material is at least:

1   200mm from the inside surface of a flue or fireplace recess;
2   40mm from the outer surface of a masonry chimney or fireplace
     recess.

4.10.1.4 Thatched roofs

Thatched roofs can be vulnerable to spontaneous combustion caused by heat being transferred from flues and building up in thick layers of thatch that are in contact with the chimney. To reduce the risk, it is recommended that rigid twin-walled insulated metal flue liners are used within a ventilated (top and bottom) masonry chimney void, provided they are adequately supported and not in direct contact with the masonry.

Further information and recommendations are contained in HETAS Information Paper 1/007, 'Chimneys in thatched properties'.

- The clearances to flue outlets that discharge on,   J 2.12
  or are in close proximity to, roofs with surfaces
  which are readily ignitable (e.g. covered in thatch or
  shingles) should be increased to those shown in AD-J
  Diagram 18.

4.10.2 Structure

A chimney is an architectural ventilation structure made of masonry, clay or metal that isolates hot toxic exhaust gases or smoke produced by a boiler, stove, furnace, incinerator or fireplace from human living areas.

4.10.2.1 End restraint

- The ends of all small single-storey buildings and   A 2C25
  annexes should be bonded or otherwise securely tied
  throughout their full height to a chimney, pier or
  buttressing valley.
- The intermediate chimneys (buttressing walls or
  piers) should provide lateral restraint to the full height
  of the supported wall.
- The length of the buttressing wall should be at least   A 2C26
  1/6 of the overall height of the supported wall and
  be bonded or securely tied to the supporting wall and
  also at the other end to the chimney, buttressing wall
  or pier.
- Floors and roofs should act to transfer lateral forces   A 2C33
  from walls to buttressing walls, piers or chimneys.

4.10.2.2 Maintenance of chimneys

- If a chimney cannot be cleaned directly through the   J 2.16
  appliance, a debris-collecting space should be provided   J 3.38
  within the chimney for emptying and cleaning.
- Where a chimney is not adequately supported by ties   A 2D1
  or securely restrained in any way, its height should
  not be greater than 4.5 times the width – provided
  that the density of the masonry is more than 1,500
  kg/m^3 (see AD-A Diagram 20).

> • The chimneys foundation should project as indicated   A 2E2f
> in AD-A Diagram 22.

4.10.3 Fire safety

How do you ensure a chimney is safe?

1. Examine the firebox: look for any cracks, gaps, or signs of wear in
    the lining of the firebox (the interior of the fireplace).
2. Look for tell-tale smoke stains.
3. Make sure your grate is the right size.
4. Check the chimney.
5. Double-check your fire extinguisher.

Proprietary fire-stopping and sealing systems are available and may be
used. Other fire-stopping materials include:

*   cement mortar;
*   gypsum-based plaster;
*   cement or gypsum-based vermiculite/perlite mixes;
*   glass fibre, crushed rock, blast furnace slag or ceramic-based
    products (with or without resin binders);
*   intumescent mastics.

> • Joints between fire-separating elements should be   B(V1)
> fire-stopped.   9.24a
> • Openings in compartment walls or compartment   B(V1) 7.20
> floors may be made for chimneys, appliance
> ventilation ducts or ducts encasing one or more flue
> pipes, provided they meet the provisions in Section 9
> of AD-B(V1).

4.10.4 Flues

AD-J provides guidance on how to meet the requirements in terms of
constructing a flue or chimney, where each flue serves one appliance only.

> • Chimneys shall consist of a wall or walls enclosing   J 0.4
> one or more flues (see AD-J Diagram 2).

- Flues designed to serve more than one appliance can   J 1.25
  meet the requirements by following the guidance
  in BS 5410-1 (for oi) and BS 5440-1 (for gas-fired
  systems).

Note: Each solid-fuel appliance should have its own flue.

4.10.4.1 Condensates in flues

- Chimneys and flues should provide satisfactory   J 1.26
  control of water condensation such as:
  - insulating flues so that gases within the flue do not
    condense during normal operation;
  - using lining components that are impervious to
    condensates and suitably resistant to corrosion (e.g.
    BS EN 1443, 'W' designation), and by making
    appropriate provisions for draining, avoiding
    ledges, crevices, etc.;
  - making provisions for the disposal of condensate
    from condensing appliances.

4.10.4.2 Concealed flues

- When a flue is routed within a void, access should   J 1.47
  be provided at various strategic points to allow the
  following visual checks at installation and during
  servicing:
  - that the flue is continuous and adequately
    supported throughout its length;
  - that all joints have been correctly assembled and
    sealed;
  - that required gradients of fallback to the boiler
    and other required drain points have been
    provided.

See AD-J Diagram 14 for further description of the requirements.

- Access for concealed flues should permit visual   J 1.147
  inspection (particularly at any joints).
- Concealed flues should not pass through another
  dwelling.
- Access hatches should be at least 300mm × 300mm
  or larger where necessary.

4.10.4.3 *Connecting flue pipes*

- Whenever possible, flue pipes should be   J 1.32
  manufactured from:
  - cast iron (BS EN 1856-2);
  - mild steel (BS1449-1, with a flue wall thickness of
    at least 3mm);
  - stainless steel (BS EN 10088-1, grades 1.4401,
    1.4404, 1.4432 or 1.4436 with a minimum flue
    wall thickness of 1mm);
  - vitreous enamelled steel (BS 6999).
- Flue pipes with spigot and socket joints should be   J 1.33
  fitted with the socket facing upwards.
- Joints should be made gas-tight.

Combustible materials in the building fabric should be protected
from the heat dissipation from flues so that they are not at risk of
catching fire.

- Connecting flue pipes: J 2.14
  - should not pass through any roof space, partition,
    internal wall or floor, unless they pass directly into
    a chimney through either a wall of the chimney or
    a floor supporting the chimney;
  - should be guarded if they are likely to be
    damaged or if the burn hazard they present is not
    apparent;
  - should be located so as to avoid igniting   J 2.15 &
    combustible material (see **AD-J** Diagram 19).   J 1.45

4.10.4.4 *Flue outlets*

- Flue pipes should have the same diameter or equivalent   J 2.4
  cross-sectional area as the appliance's flue outlet.

- Flues should not be smaller than the appliance's   J 2.5
  flue outlet or that recommended by the appliance
  manufacturer.

 Note: For further details concerning sizes of flues, see AD-J Table 1.

- Outlets from flues should allow the dispersal of   J 3.23
  products of combustion and, if a balanced flue, the
  intake of air.

- Flue outlets should be protected where flues are at   J 3.24
  significant risk of blockage.

- Flues serving natural draught open-flued appliances   J 3.25
  should be fitted with outlet terminals if the flue
  diameter is no greater than 170mm.

In areas where nests of squirrels or jackdaws are likely, it is advisable to
fit a protective cage with a mesh size between 6mm and 25mm designed
for solid-fuel use.

- Flue outlets should have a guard if persons could   J 3.26
  come into contact with them or if they could be
  damaged.

- Flues that discharge at low level near boundaries   J 1.52
  should ensure safe flue gas dispersal (see AD-J
  Diagram 34).

4.10.4.5 *Flue systems*

- Flue systems should offer least resistance to the   J 1.48
  passage of flue gases by minimising changes in
  direction or horizontal length.
- Wherever possible, flues should be built so that they
  are straight and vertical ***except*** for the connections
  to combustion appliances with rear outlets, where the
  horizontal section should not exceed 150mm.
- Where bends are essential, they should be angled at
  no more than 45° vertical.

- Provision should be made to enable flues to be swept   J 1.49
  and inspected (see AD-J Diagram 15) on a regular
  basis.

- A flue should **not** have openings into more than one   J 1.50
  room or space except for the purposes of:
  - inspection and/or cleaning; or
  - a chimney fitting an explosion door, draught break,
    draught stabiliser or draught diverter.

- Openings for inspection and cleaning should have an   J 1.51
  access cover that has the same level of gas-tightness
  as the flue system and an equal level of thermal
  insulation.

- After the appliance has been installed, it should be   J 1.50
  possible to sweep the **whole** flue.

4.10.4.6 Height of flues

- Flues should be high enough (normally 4.5m) to   J 2.8
  ensure there is sufficient draught to clear the products
  of combustion.

- The outlet from a flue should be above the roof of   J 2.10
  the building so that the products of combustion can
  discharge freely and will not present a fire hazard,
  whatever the wind conditions (see AD-J Diagram 17).

Note: Chimney heights and/or separations may need to be increased in particular cases where wind exposure, surrounding tall buildings, high trees or high ground could have adverse effects on flue draught.

4.10.4.7 Relining flues

A flue liner is the wall of the chimney that is in contact with the products of combustion (see AD-J Diagram 2).

In certain circumstances, relining is considered 'building work' within the meaning of Regulation 3 of the Building Regulations and must, therefore, **not** be undertaken without prior notification to the Local Authority. If you are in doubt, you should consult your Local Authority's Building Control Department or an Approved Inspector.

- If a chimney has been previously relined using a metal   J 1.39
  lining system **and** the appliance is being replaced,
  then the metal liner should also be replaced, unless it
  has been recently installed and is in good condition.

- Flexible flue liners should **only** be used to reline a   J 1.40
  chimney. They should **not** be used in a new chimney
  as the primary liner.

- Plastic flue pipe systems can be used in some cases,   J 1.41
  provided that they are supplied by or specified by the
  appliance manufacturer.

Flues should be swept to remove deposits before being relined.

4.10.4.8 Re-use of existing flues

Where it is proposed to bring a flue in an existing chimney back into use,
or to use a flue with a different type of appliance or appliance rating, the
flue and chimney should be checked and, if necessary, altered to ensure
that they satisfy the requirements for the proposed use (see J 1.36 for
additional advice).

Note: Oversize flues can be unsafe. A flue may, however, be lined in
order to reduce the flue area to suit the intended appliance.

4.10.4.9 Repair of flues

If the installation and maintenance of a fireplace and/or chimney are
deemed as being a 'material change of use', it is a **mandatory requirement** that the building is brought up to the standards required by ADs
J1 to J3. If renovation, refurbishment or repair amounts to (or involves
the provision of) a new or replacement flue liner, it is considered 'building
work' within the meaning of Regulation 3 of the Building Regulations
and must, therefore, **not** be undertaken without prior notification to the
Local Authority.

Examples of work that would need to be notified include:

- Relining work by the creation of new flue walls and   J 1.34
  insertion of new linings.

- A cast in situ liner that significantly alters the flue's   J 1.35
  internal dimensions.

4.10.4.10 Ventilation

Rooms or spaces intended to contain flueless appliances may need permanent ventilation and purge ventilation (e.g. openable windows), or adjustable ventilation and rapid ventilation.

• Rooms or spaces that contain a decorative fuel effect (DFE) fire should have permanently open air vents.	J 3.11
• Flues should be high enough to ensure sufficient draught to safely clear the products of combustion.	J 3.21

4.10.5 Fireplaces

Note: Examples of length (L) and height (H) for large and unusual fireplace openings are shown in AD-J Diagram 45.

4.10.5.1 General guidance

• Fireplace recesses (see the AD-J Diagram 5 series) are a structural opening (sometimes called a 'builder's opening') formed in a wall or in a chimney breast, from which a chimney leads and which has a hearth at its base.	J 0.4–17

4.10.5.2 Construction of fireplace gathers

To minimise resistance to the proper working of flues, tapered gathers should be provided in fireplaces for open fires or corbelling of masonry, as shown in AD-J Diagram 22.

Alternatively, a suitable canopy (as shown in AD-J Diagram 23) or a prefabricated appliance chamber incorporating a gather may be used.

• Tapered gathers should be provided in fireplaces for open fires.	J 2.21

4.10.5.3 Fireplace recesses

Simple recesses (see AD-J Diagram 5a) are suitable for closed appliances such as room heaters, stoves, cookers or boilers, and can be used for accommodating open fires and freestanding fire baskets. Fireplace recesses are often lined with firebacks to accommodate inset open fires (see AD-J Diagram 5c).

- Fireplaces for open fires should be constructed so that   J 2.30
  they adequately protect the building fabric from fire.
- Fireplace recesses can either: J 2.30b
  - be made from masonry or concrete (AD-J
    Diagram 28); or
  - be a prefabricated factory-made appliance
    chamber made of insulating concrete, having a
    density of between 1,200 kg/m³ and 1,700 kg/m³,
    and with the minimum thickness as shown in AD-J
    Table 4.

4.10.5.4 *Fireplace lining components*

Note: Lining components and/or decorative treatments, fitted around openings, reduce the opening area.

- Fireplace recesses containing inset open fires need   J 2.31
  to be heat-protected and should be lined with either
  lining components or lining the recess with suitable
  firebricks as shown in AD-J Diagram 29.

4.10.5.5 *Hearths*

A hearth is a base intended to safely isolate a combustion appliance from people, combustible parts of the building fabric and soft furnishings. (See AD-J Diagram 6.)

4.10.5.5.1 CONSTRUCTION OF HEARTHS

- Hearths should be constructed so that, whilst in   J 2.22
  normal use, they prevent combustion appliances
  setting fire to the building fabric and furnishings –
  as well as limiting the possibility of people being
  accidentally burnt.
- All hearths shall safely isolate combustion appliances   J 0.4–27
  from people, soft furnishings and the building.

- If the chimney is not independently supported, the hearth should be able to support the weight of the appliance and its chimney. This includes: J 2.23
  - hearths made of non-combustible board/sheet material; or
  - tiles at least 12mm thick; or
  - constructional hearths.
- Constructional hearths should: J 2.24a
  - be planned in accordance with AD-J Diagrams 24 and 24;
  - be made of solid, non-combustible material (such as concrete or masonry that is at least 125mm thick). J 2.24b
- Combustible material should **not** be placed beneath constructional hearths unless: J 2.25
  - there is an air space of at least 50mm between the underside of the hearth and the combustible material; or
  - the combustible material is at least 250mm below the top of the hearth (see AD-J Diagram 25).
- An appliance should be located on a hearth so that: J 2.26
  - it is surrounded by a surface that is free of combustible material;
  - the surface of a superimposed hearth is laid wholly or partly upon a constructional hearth.

The edges of the hearth surface should be marked (e.g. by a change in level) to warn building occupants and to discourage combustible floor finishes such as a carpet from being laid too close to the appliance.

- Combustible material that is placed on or beside a constructional hearth should **not** extend under a superimposed hearth by more than 25mm **or** be closer than 150mm (measured horizontally) to the appliance. J 2.28

Some methods for meeting these provisions are shown in AD-J Diagram 27.

4.10.5.5.2 HEARTHS FOR OIL APPLIANCES

- Oil appliance hearths are required to prevent the   J 4.24
  building catching fire and, while it is **not** a health
  and safety provision, it is customary to top them with
  a tray for collecting spilt fuel.

4.10.5.5.3 WALLS ADJACENT TO HEARTHS

- Walls that are not part of a fireplace recess or a   J 2.32
  prefabricated appliance chamber, but are adjacent
  to hearths or appliances, should protect the building
  from catching fire. (See AD-J Diagram 30.)

4.10.6 Maintenance

4.10.6.1 Access to combustion appliances for maintenance

- Safe access to appliances for maintenance purposes   J 1.60
  should be provided.
- Roof space installations of gas-fired appliances   J 1.60
  should comply with the requirements of BS
  6798:2014.

4.10.6.2 Dry lining around fireplace openings

- Gaps around a fireplace opening (such as a fireplace   J 1.53
  surround, cladding or dry lining), should be sealed to
  prevent any leakage from the fireplace opening into
  the void behind.
- The sealing material should be capable of remaining   J 1.53
  in place despite any relative movement between the
  decorative treatment and the fireplace recess.

4.10.6.3 *Provision of information*

On completion of work:

• A report should be drawn up by the person carrying out the work, to show what materials and components have been used and to confirm that flues have passed appropriate tests.	J 1.54

Guidance on testing is given in Appendix E to AD-J.

• Flues should be checked to show that they are free from obstructions, satisfactorily gas-tight and constructed with materials and components of sizes which suit the intended application.	J 1.55

See Appendix A to AD-J for detailed checklists for the checking and testing of hearths, fireplaces, flues and chimneys.

• Where the building work includes the installation of a combustion appliance, tests should include flue pipes and the gas-tightness of joints between flue pipes and combustion appliance outlets.	J 1.55
• A spillage test should be carried out with the appliance under fire, as part of the process of commissioning.	J 1.55
• Hearths should indicate the area where combustible materials should not encroach.	J 1.56

4.10.6.4 *Notice plates for hearths and flues*

• If a hearth, fireplace (including a flue box designed to accommodate a gas-burning appliance), flue or chimney is installed or extended, a notice plate containing the following information should be permanently posted in the building:
  • the location of the hearth, fireplace, flue box or beginning of the flue;
  • the category of the flue and generic types of appliances that can be safely accommodated;
  • the type and size of the flue (or its liner if it has been relined) and the manufacturer's name;
  • the installation date.
	J 1.57

Notice plates should be securely fixed:

* next to the electricity consumer unit;
* next to the chimney or hearth described;
* next to the water supply stopcock.

4.10.7 Combustion appliances and fuel storage systems

Incomplete combustion occurs when the supply of air or oxygen is poor. Water is still produced, but carbon monoxide and carbon are produced instead of carbon dioxide. The carbon is released as soot, whilst carbon monoxide as a poisonous gas. For this reason, stoves, cookers and boilers should have permanently open air vent(s).

4.10.7.1 Appliances burning solid fuel (air supply)

* Any room or space containing an appliance burning solid fuel (with a rated output up to 50kW) should have a permanent air vent opening of at least the size shown in AD-J Table 1.   J 2.2

4.10.8 Gas-burning devices

The Gas Safety (Installation and Use) Regulations require that:

* Gas fittings, appliances and gas storage vessels must only be installed by a competent person.
* Any person in control of any gas work must ensure that a competent person carries out that work.
* Any person carrying out gas installation, whether an employee or self-employed, **must** be a member of a class of persons approved by the HSE and registered with the Gas Safety Register.

Important elements of the regulations include:

* Precautions must be taken to ensure that all installation pipework, fittings, appliances and flues are installed safely.   J 3.5e

• When any gas appliance is installed, checks are required to ensure compliance with the regulations.	J 3.5f
• All flues must be installed in a safe position.	J 3.5g
• No alteration is allowed to any premises in which a gas fitting or gas storage vessel is fitted that would adversely affect the safety of that fitting or vessel, causing it no longer to comply with the regulations.	J 3.5h

Outlets from flues should be situated externally so as to allow the products of combustion to dispel and, if a balanced flue, the intake of air (see AD-J Diagram 35).

4.10.8.1 Back boilers

• Back boilers should adequately protect the fabric of the building from heat (see AD-J Diagram 36).	J 3.39

4.10.8.2 Gas fires (other than flueless gas fires)

Gas-fired appliances should **only** be located where accidental contact is unlikely and where they can be surrounded by a non-combustible surface which provides adequate separation from combustible materials.

• Gas fires may be installed in fireplaces which have flues designed to serve solid-fuel appliances – **provided** that it can be shown to be safe.	J 3.7

4.10.8.3 Kerosene and gas-oil burning appliances

Kerosene (class C2) and gas-oil (class D) appliances have the following, additional, requirements:

• Open-fired oil appliances should **not** be installed in rooms such as bedrooms and bathrooms.	J 4.2
• Flues should be sized to suit the intended appliance and to ensure sufficient discharge velocity to prevent flow reversal problems.	J 4.4

- The outlet from a flue should be situated externally   J 4.6
  so as to ensure:
  - the correct operation of a natural draft flue;
  - the intake of air if a balanced flue;
  - dispersal of the products of combustion.

AD-J Diagram 41 (and its associated Table 41) indicates typical positioning to meet this requirement.

- Flexible metal flue liners should be installed in one   J.3.37
  complete length without any joints inside the chimney.
- Other than for sealing at the top and the bottom, the   J 4.22
  space between the chimney and the liner should be
  left empty.
- Flues that may be expected to serve appliances   J 4.23
  burning class D oil should be made of materials that
  are resistant to acids.

4.10.9 Carbon monoxide alarms

Carbon monoxide (CO) is a colourless, odourless and tasteless gas that is slightly less dense than air and is toxic to humans and animals when encountered in high concentrations.

Under current Building Regulations, in England and Wales it is now a **mandatory requirement** to fit carbon monoxide alarms where new or replacement fixed solid-fuel appliances are installed in a home.

In Northern Ireland, it a **legal requirement** to fit carbon monoxide alarms in **all** homes where a new or replacement appliance, not used solely for cooking, is installed.

- Carbon monoxide alarms should comply with BS   J 2.35
  EN 50291-2 and be powered by a battery designed
  to operate for the working life of the alarm. Such
  alarms should have a warning device to alert users
  when the working life is due to end.
- The carbon monoxide alarm should be in the same   J 2.36
  room, between 1m and 3m horizontally from the
  appliance, either:
  - on the ceiling at least 300mm from any wall;
  - on a wall, as high up as possible (above any doors
    and windows,) but not within 150mm of the ceiling.

4.10.10 Conservation of fuel and power

4.10.10.1 Secondary heating

A secondary heating appliance may meet part of the demand for space heating.

• When calculating the primary energy rates and emission rates for a dwelling which has a secondary heating appliance:	L(V1) 2.8
• use the Standard Assessment Procedure to calculate the value of heat provided by the secondary heating system;	
• use the efficiency of the secondary heating appliance with its appropriate fuel to calculate the dwelling's primary energy rate and dwelling emission rate.	
• If a continuously burning, manually operated, oil-fired vaporising appliance is provided for secondary heating, then it requires no further control. However, if it is an electrically operated appliance, it will require an integral remote or thermostatic control system.	L(V1) 6.10

Requirements for buildings other than dwellings – Additional requirements

This section provides the details for chimneys and fireplaces in buildings other than dwellings, only where they differ from the requirements already given above.

• Carbon monoxide alarms should comply with BS EN 50291-2 and be powered by a battery designed to operate for the working life of the alarm. Such alarms should have a warning device to alert users when the working life is due to end.	B(V2) 8.31

In accordance with the Clean Air Act 1993, it is an offence for dark smoke to be emitted from a chimney of any building. A person found guilty of an offence under this section is liable, on summary conviction, to a fine in accordance with the Standard Scale contained in the

Criminal Justice Act (at the time of publication, the maximum fine for such offences was £5,000!).

4.10.11 Conservation of fuel and power

• Reasonable provision shall be made for the conservation of fuel and power in buildings by: • limiting heat gains and losses; • providing fixed building services which use no more fuel and power than is reasonable in the circumstances.	B(V2) Schedule 1

4.11 Stairs

Reminder: The subsections include details for buildings other than dwellings only where they differ from the requirements for dwellings.

Subsection	Description	Page
Requirement for dwellings		
4.11.1	Air-circulation systems in houses with a floor more than 4.5m above ground level	395
4.11.2	Structure	396
4.11.3	Fire safety	396
Requirements for buildings other than dwellings – Additional requirements		
4.11.4	Access to and use of buildings	404
4.11.5	Fire safety	408
4.11.6	Protection from falling, collision and impact	417
4.11.7	Resistance to the passage of sound	424

Reminder: Free downloads of all the diagrams and tables referenced in the subsections are available on-line at the Governments' Planning Portal: www.gov.uk/government/collections/approved-documents.

Requirements for dwellings

4.11.1 Air-circulation systems in houses with a floor more than 4.5m above ground level

To avoid the possibility of an air-circulation system allowing smoke or fire to spread into a protected stairway, the following precautions should be taken:

• Transfer grilles should **not** be fitted in any wall, door, floor or ceiling enclosing a protected stairway.	B(V1) 2.9

• Any duct passing through an enclosure to a protected stairway or entrance hall:
  • should be made of rigid steel; and
  • construction with all joints to be fire-stopped.
	B(V1) 3.23

• Ventilation ducts supplying or extracting air directly to or from a protected stairway should not serve any other area. • Any mechanical ventilation system which recirculates air to both stairway and other areas should be designed to shut down on the detection of smoke within the system.	B(V1) 9.7
• Any corridor or lobby next to a stairway should have a smoke vent positioned so that the top edge of the vent is as high as the top of the door to the stairway.	B(V1) 3.50

• Any smoke detected in common corridors or lobbies should cause the vents to simultaneously open on:
  • the storey where the fire is located;
  • the top of the smoke shaft; and
  • the stairway.
	B(V1) 3.51

• The top storey of the stairway should have a vent, with a minimum free area of $1m^2$ to the outside.	B(V1) 3.52
• In single-stair buildings, if a smoke detector in a common part of the building detects a fire, it should automatically cause the vents on the storey where the fire is initiated to open, as well as others located in common parts of the building and at the head of the stairway.	B(V1) 3.53

- In buildings with more than one stair, the control
  system should open the vent at the head of the
  stairway before, or at the same time as, the vent on
  the storey where the fire is located. Other smoke
  vents may be activated manually.

4.11.2 Structure

4.11.2.1 Interruption of lateral support

- Where an opening in a stairway adjoins a wall   A 2C37a
  (and in doing so interrupts the lateral support of the
  wall), the maximum permitted length of the opening
  is 3m.

Other than connections (e.g. via mild steel anchors) either side of the
opening, there should be no other interruption of lateral support.

4.11.2.2 Wall cladding on escape routes

- Where wall cladding is required to provide   A 3.5
  pedestrian guarding for stairs with vertical drops
  of 600mm (or more than the height of two risers),
  account should be taken of the additional imposed
  loading.

4.11.3 Fire safety

Note: As a result of the Grenfell Tower inquiry and the subsequent
changes to the legislation, this information has now changed (but obviously not as much as buildings other than dwellings!), and is now much
more detailed than before.

- Stairs and service shafts connecting compartments   B(V1) 7.21
  should be protected to restrict the spread of fire
  between the compartments.

- Any stair or other shaft passing directly from one
  compartment to another should be enclosed in a
  protected shaft – and for a typical one- or two-storey
  dwelling, the requirement now has more emphasis on:
  - the provision of smoke alarms;
  - the provision of openable windows for
    emergency exit.

B(V1) 7.22
B(V1) 7.24

Sprinklers do not need to be provided in common areas such as stairs, corridors or landings when these areas are fire sterile.

4.11.3.1 Firefighting shafts

- Buildings with firefighting shafts should have fire
  mains (with a maximum hose distance of 45m) in
  both the firefighting stairs and, where necessary, in
  protected stairways.

B(V1) 14.2

- The outlets from fire mains should be located within
  the protected stairway enclosure.

B(V1) 14.4

- The doors of a firefighting lift landing should be a
  maximum of 7.5m from the door to the firefighting
  stair and both the stair and lobby of the firefighting
  shaft should be provided with a means of venting
  smoke and heat.

B(V1) 15.8

Note: AD-B(V1) Tables B3 and C1 detail the minimum performance criteria for firefighting shafts and their doors.

4.11.3.2 Flats

Every flat should have access to alternative escape routes.

- All habitable rooms in a multi-storey flat should have
  access to an internal protected stairway leading to a
  final exit.

B(V1) 3.76

- A multi-storey flat (above 4.5m) without its own
  external entrance at ground level should have a
  protected stairway, sprinkler system and smoke
  alarms (see AD-B(V1) Diagrams.3.5 and 3.6).

B(V1) 3.21

- Alternative exits from a flat should lead to a final exit   B(V1) 3.22
  via a common stair.

4.11.3.3 Helical and spiral stairs

Helical stairs, spiral stairs and fixed ladders may form part of an escape route subject to the following restrictions:

• Any door opening towards a corridor or a stair should be recessed, to prevent its swing interfering with its effective width.	B(V1) 3.96
• Helical and spiral stairs should be designed in accordance with BS 5395-2.	B(V1) 3.86
• Fixed ladders may **only** be used to gain access to unmanned plant rooms or provided where a conventional stair is impractical.	B(V1) 3.85

4.11.3.4 Escape routes with single steps and fixed ladders

Fixed ladders should only be provided where a conventional stair is impractical, such as for access to unmanned plant rooms.

• If a stair is used as a firefighting stair, it should be at least 1,100mm wide between the walls or balustrades, provided that any handrails do not intrude more than a maximum of 100mm on each side.	B(V1) 3.60
• Stairlift guide rails may be ignored, but it should be possible to park the lift's chair or carriage in a position that does not obstruct the stair or landing.	B(V1) D4c

 Note: A stair that is an acceptable width for everyday use will be sufficient for escape purposes.

4.11.3.4.1 EXTERNAL ESCAPE STAIRS

 Note: Where more than one escape route is available from part of a building, then some of these escape routes may be by way of an external stair.

• All doors and glazing leading to the external escape stair should be fire-resisting and fixed shut.	B(V1) 2.17 B(V1) 3.68
• Any part of the fire-resistant external envelope of the building within 1.8m of (and 9m vertically below) the flights and landings of an external escape stair should be of fire-resisting construction (see AD-B (V1) Diagram 2.7).	
• Any part of the building within 1.8m of the escape route should be made as a fire-resisting construction.	
• External escape stairs greater than 6m must be protected from the effects of adverse weather conditions.	
• Access to an external escape stair may be via a flat roof, provided that the roof is clearly defined and guarded by walls and/or protective barriers.	B(V1) 3.30
• In dwellings with more than one stair, the stairs should provide an effective alternative means of escape.	B(V1) 2.4

4.11.3.5 Protected stairway

Cavity barriers should be provided above enclosures to a protected stairway in a dwelling which has a floor more than 4.5m above ground level (see Diagram 2.3).

• All basements should be served by a separate escape stair.	B(V1) 3.71
• Other stairs may connect with the basement if there is a protected lobby or corridor between the stairs and accommodation at each basement level.	B(V1) 3.72
• Every protected stairway should lead to a final exit.	B(V1) 3.81
• The space within a protected stairway should not be used for anything other than a lift well or electricity meters.	B(V1) 3.78
• Where a stair serves an enclosed car park or place of special fire hazard, the lobby or corridor should have a minimum 0.4m^2 of permanent ventilation or be protected by a mechanical smoke-control system.	B(V1) 3.75

- In single-stair buildings, electricity meters should be kept in securely locked cupboards separated from the escape route by fire-resisting construction.   B(V1) 3.79

Note: Class D-s3, d2 plastic rooflights and TP(a) and TP(b) thermoplastic rooflights are suitable for use in protected stairways.

The following services should **not** be within a protected stairway or lobby:

- Gas service and installation pipes and meters.
- Refuse chutes and refuse storage rooms.

- If a protected stairway projects from (is recessed from or at an internal angle to) the adjoining external wall of the building, then the minimum distance between an unprotected area of the building enclosure and an unprotected area of the stair enclosure should be 1,800mm.   B(V1) 3.64

- If a fire could cause the external wall of a protected stairway to heat the stairway, it should be protected by fire-resisting construction (see AD-B(V1) Diagram 3.10).   B(V1) 3.63

- In dwellings with one storey more than 4.5m above ground level, there should be either a protected stairway or an alternative escape route.   B(V1) 2.5

- Dwellings (with upper stories 4.5m above ground level) which are served by just one stair should either have an emergency escape window, external door or direct access to a protected stairway. (See AD-B(V1) Diagrams 3.1, 3.2 and 3.3 for typical examples.)   B(V1) 3.16

- Dwellings with two or more storeys more than 4.5m above ground level (see AD-B(V1) Diagram 2.1d) should (in addition to having a protected stairway or an alternative escape route) have: B(V1) 2.6
  - a sprinkler system throughout; and (ideally)
  - an alternative fire-resisting escape route from each storey that is more than 7.5m above ground level.

4.11.3.6 Galleries

- A gallery should be provided with an alternative   B(V1) 2.15
  exit or an emergency escape window if the gallery
  is a maximum of 4.5m above ground level. (See
  AD-B(V1) Diagram 2.6.)

- Any cooking facilities in a room containing a
  gallery should be remote from the stair to the gallery
  so that it does not interfere with an escape route (see
  AD-B(V1) Diagram 2.6).

4.11.3.7 Lighting and signage

Blocks of flats with a single stair in regular use would not usually require
any fire exit signage.

- Escape stair lighting should be on a separate circuit   B(V1) 3.43
  from the electricity supply to any other part of the   B(V1) 14.2
  escape route.

- Floor identification signs should be located on every   B(V1)
  landing of a protected stairway, protected corridor or   15.14
  lobby that a firefighting lift opens on to.

- Thermoplastic diffusers may be incorporated in   B(V1) 4.16
  ceilings to rooms and circulation spaces, but **not** to
  protected stairways.

4.11.3.8 Loft conversions

- Where the conversion of an existing roof space (e.g.   B(V1) 2.21
  to a two-storey house) means that a new storey is
  going to be added, the stairway must be protected
  with fire-resisting doors and partitions.

4.11.3.9 Means of escape

The design of a building should ensure that there is an alternative way
for people to flee from most situations.

- Where direct escape to a place of safety is not feasible, it should be possible to reach a place of relative safety, such as a protected lobby or common protected corridor; or in multi-storey flats, direct access to a protected internal stairway. B(V1) 3.16

4.11.3.9.1 BASEMENTS

- If a basement contains habitable rooms, it should either have an emergency escape window, external door or a protected stairway leading to a final exit. B(V1) 2.16 B(V1) 3.9

4.11.3.10 *Positioning of smoke and heat alarms*

All dwellings (particularly kitchens) should have a fire detection and alarm system, in accordance with BS 5839-6.

 Smoke alarms should not be fixed over a stair or any other opening between floors.

- All dwellings should have a heat alarm as well as a fire detection and alarm system, in accordance with BS 5839-6. B(V1) 1.1 B(V1) 3.21
- Smoke and heat alarms should have a standby power supply, such as a battery (rechargeable or non-rechargeable) or capacitor. B(V1) 1.4

Where a kitchen area is not separated from the stairway or circulation space by a door, in addition to any smoke alarms in the circulation space(s), there should be an interlinked heat detector or heat alarm in the kitchen.

4.11.3.11 *Passenger lifts*

- Where a passenger lift is provided in a dwelling and it serves any floor more than 4.5m above ground level, it should either be located in the enclosure to the fire protected stairway or be contained in a fire-resisting lift shaft. B(V1) 2.7

- Cavity barriers or a fire-resisting ceiling should be provided above a fire protected stairway enclosure (see AD-B(V1) Diagram 2.3).   B(V1) 2.5

4.11.3.12 Internal stairs

- The door between two rooms which are served by a single window should provide access to the window without passing through the stair enclosure. Nevertheless, both rooms should have access to their own internal stair.   B(V1) 2.3   B(V1) 3.17

Note: To stay within Building Regulations, a domestic staircase needs a rise of between 190mm and 220mm.

4.11.3.13 Single escape stairs

Table AD-B(V2) 2.2 shows the limitations on travel distance which must be applied to buildings with single escape stairs.

- A building may be served by a single escape stair if the building has: B(V1) 3.3
  - a basement with a single escape route;
  - no storey with a floor level more than 11m above ground level and every storey has a single escape route.
- A single escape stair may be used in:
  - small premises;
  - an office building not more than five storeys above the ground storey, provided that the travel distance from every point in each storey does not exceed 18m for escape in one direction only, and every storey more than 11m above ground level has an alternative means of escape (see AD-B(V1) Table 2.1);
  - a factory, provided that the travel distance from every point on each storey does not exceed that given in AD-B(V1) Table 2.1 for escape in one direction only;
  - process plant buildings (with an occupant capacity of not more than 10).

4.11.3.14 Windows

- Windows providing emergency escape should comply   B(V1) 2.10
  with all of the following:
  - a minimum height and width of 450mm;
  - the bottom of the openable area should be a
    maximum of 1,100mm above the floor;
  - windows should be capable of remaining open
    without being held;
  - locks (with or without removable keys) and opening
    stays (with child-resistant release catches) may be
    fitted to escape windows.

If direct escape to a place of safety using an emergency escape window is impracticable, it should be possible to reach a place of relative safety, such as a protected stairway, within a reasonable travel distance.

Requirements for buildings other than dwellings – Additional requirements

The basic requirements for stairs in buildings other than dwellings are the same as for dwellings, with the addition of the following requirements.

4.11.4 Access to and use of buildings

In order to assist people with limited abilities to gain access to and use a building and its facilities, a range of stairs or ramps may be required. Alternatively, lifting devices may be required adjacent to stairs and ramps.

- For common access areas in buildings that contain   K 1.12
  flats, all means of escape routes should have
  a minimum clear headroom of 2m, except in
  doorways.

Although some people will find a stair is easier to use than a ramp, ramps are not necessarily safe and convenient for ambulant disabled people.

- Due to the increased risk of slipping on a ramp in   M2 1.21
  bad weather, steps should be provided as well as a
  ramp.

- Handrails should be set at heights that are convenient for *all* users of the building.   M2 1.36
- Handrails should extend safely beyond the top and bottom of a flight of steps or a ramp.
- A second handrail may be required on stairs in a wide range of building types, particularly in schools, for use by children and people of short stature.

- Handrails should protrude no more than 100mm into a ramped or stepped access.   M2 1.37

4.11.4.1 Construction of steps

- The rise and going should be measured in accordance with AD-K Diagram 1.1.   K 1.2

- Risers should *not* be open.   K 1.6

- Step nosings should: K 1.7
  - be made apparent;
  - be a minimum of 55mm wide on both the tread and the riser;
  - not protrude over the tread below.

If the soffit beneath a stair is less than 2m above floor level, then the area beneath the stair should be protected with either guarding or low-level cane detection, or a barrier.

- In flights of stairs, all of the following should be provided: K 1.15
  - a minimum width between enclosing walls, strings or upstands of 1,200mm;
  - a minimum width between handrails of 1,000mm;
  - where the flight is more than 2m wide, divide it into flights a minimum of 1,000mm wide.

- Single steps should be avoided.   K 1.18
- The maximum number of risers on flights between landings should be:
  - utility stairs – 16 risers;
  - general access stairs – 12 risers (exceptionally 16 in small premises).

- An unobstructed length of 1,200mm should be provided on each landing.   K 1.23
- Doors should not swing across landings.

- Suitable continuous handrails should be provided (dimensions are shown in AD-K Diagram 1.12) on each side of the flights and on each side of the landings.   K 1.35

- Handrails should be provided in accordance with all of the following: K 1.36
  - if a second (lower) handrail is provided, the vertical height from the pitch line of the steps to the top of the second handrail should be 600mm;
  - flights and landings of ramped or stepped flights should have a continuous handrail;
  - handrails should not project into an access route;
  - handrails should contrast visually with the background and not be highly reflective;
  - handrail surfaces should be slip-resistant and not excessively cold or hot to touch;
  - handrail ends should be finished in a way that reduces the risk of clothing being caught.

- Guarding should be provided at the sides of flights and landings with two or more risers.   K 1.40

- Where stairs or ladders are used to access maintenance areas: K 1.42
  - if access is required a minimum of once per month, follow the provisions for private stairs in dwellings or for industrial stairs and ladders in BS 5395-3;
  - if access is required less frequently than once a month, portable ladders may be used.

- Stairs should **not** be directly in line with an access route.   M2 3.50
- A going of at least 300mm is preferred for mobility-impaired people.

4.11.4.2 Stairlifts and other lifting devices

- Internal stairs should always be provided as well as a lifting device.   M2 3.19

- A wheelchair platform stairlift may be considered in an existing building provided its installation does not conflict with requirements for means of escape.   M2 3.23

- The illumination in a passenger lift car, lifting platform or wheelchair platform stairlift should minimise glare, reflection, shadows or pools of light and dark.   M2 3.26

- Wheelchair platform stairlifts should: M2 3.44 / M2 3.24d / M2 3.49
  - only be considered for conversions and alterations where it is not practicable to install a conventional passenger lift or a lifting platform;
  - travel up the string of a stair;
  - not be installed where their operation restricts the safe use of the stair by other people;
  - only be installed in exceptional circumstances in an existing building where it is to serve an intermediate level or a single storey;
  - conform to the requirements of the Supply of Machinery (Safety) Regulations 1992, SI 1992/3073;
  - provide the required clear width of the flight of stairs and landings for means of escape when the wheelchair platform is in the parked position;
  - operate at a speed not exceeding 0.15m/s;
  - have continuous pressure controls;
  - have minimum clear dimensions of 800mm wide and 1,250mm deep;
  - be fitted with clear instructions for use;
  - provide access with an effective clear width of at least 800mm;
  - have controls designed to prevent unauthorised use.

- Wheelchair platform stairlifts:
  - may be more suitable for use in small areas with a unique function, e.g. a small library gallery, a staff rest room or a training room;   M2 3.45

• may not be suitable for users with certain disabilities, e.g. those easily fatigued;	M2 3.46
• are operated by continuous pressure controls, commonly a joystick (but another means of continuous pressure control may be required);	M2 3.47
• are only suitable where users can be instructed in their safe use and where management supervision can be ensured.	M2 3.48

4.11.5 Fire safety

These requirements have changed quite dramatically as a result of the Grenfell Tower inquiry, and now additional provisions apply for protected shafts (that are protected stairways), protected stairways and stairs which are also firefighting stairs.

If the escape stair is:

- the only stair in a building with more than three storeys;
- within a basement storey;
- an external escape stair;
- a firefighting stair;
- or it serves any storey that has a floor level more than 18m above ground or access level;

then the flights and landings of escape stairs should be constructed of materials of class A2-s3, d2 or better. (Further guidance is available in AD-K.)

4.11.5.1 General guidance

When designing a building, there should be an alternative way for people to flee from most situations. Owing to their nature, the fire safety requirements relating to stairs in buildings other than dwellings are quite comprehensive and differ from those for dwellings.

Sprinklers should be provided within the individual flats; they are not required in common areas such as stairs when these areas are fire sterile.

As people in wheelchairs may not be able to use stairways without assistance, it may be necessary to provide refuges on escape routes, assistance down (or up) stairways, or suitable lifts.

4.11.5.2 Stairways

Stairways (particularly those being accessed by wheelchair users) should ensure that:

- The combustible content of such stairways is restricted.
- A stairway or shaft that passes directly from one compartment to another is enclosed in a protected shaft.
- Protected stairways provide a virtually 'fire sterile' area leading to a place of safety outside the building.
- Flames, smoke and gases are controlled by fire-resisting structures and/or by a smoke-control system.
- If a basement storey contains a habitable room, then either an external exit door or window or a protected stairway leading from the basement to a final exit should be provided.

4.11.5.3 Access lobbies

- Escape stairs should have the added protection of a protected lobby/corridor when: B(V2) 3.34
  - the stair is the only one serving a building which has more than one storey above or below the ground storey (except for small premises);
  - the stair serves any storey at a height greater than 18m;
  - the building is designed for phased evacuation;
  - the stair is a protected stairway that leads to a final exit.
- A single escape stair may serve a basement which only has one escape route (see AD-B(V2) Table 2.1).   B(V2) 3.3

4.11.5.4 Basements

- The fire resistance of basement storeys should be that specified for basements in AD-B(V2) Table 4.   B(V2) 18.2

Heat and smoke from basement fires vented via stairs can inhibit access for firefighting personnel.

4.11.5.4.1 PROTECTED STAIRWAYS

- In basements (and enclosed car parks), the lift should be within the enclosure of a protected stairway. Otherwise, the lift should be approached only via a protected lobby or protected corridor.   B(V2) 5.38

4.11.5.4.2 PROVISION OF SMOKE OUTLETS

- Each basement space should have one or more smoke outlets.   B(V2) 18.2
- Smoke outlets should be evenly distributed around the perimeter of the space they serve, at high level, either in the ceiling or wall, and discharge to the open air.   B(V2) 18.5

- If basement storeys are fitted with a sprinkler system, a mechanical smoke extraction system may be provided as an alternative to natural venting.   B(V2) 18.11

 These smoke outlets can also be used by the fire and rescue service to let cooler air into the basement (see Diagram 18.1).

4.11.5.5 Compartmentation

- Stairs and interconnecting compartments should restrict the spread of fire between the compartments.   B(V2) 8.6
- Any stair that passes directly from one compartment to another should be enclosed in a protected shaft.   B(V2) 8.32

4.11.5.6 Escape routes

An escape route can pass through a protected lobby of one stair to reach another; however, the escape route should be a minimum of 1,800mm wide from the foot of the stair to a place of safety.

4.11.5.7   Escape stairs

• The width of escape stairs should: • be no less than the minimum widths given in AD-B(V2) Table 3.1; • not reduce at any point on the way to a final exit; • not exceed 1,400mm in stairs taller than 30m, unless a central handrail is provided.	B(V2) 3.10
• In multi-storey buildings, each storey should have access to more than one stair.	B(V2) 2.4
• If a storey has more than one escape stair, it should not be necessary to pass through one stair to reach another, unless the second stair has a protected lobby.	B(V2) 2.14
• A stair should not form part of the primary circulation route between different parts of the building at the same level unless it is a protected stairway.	B(V2) 2.15
• If an exit route from a stair also acts as an escape route, its width may need to be increased.	B(V2) 3.12
• The minimum stair width depends on the number of stairs provided and the escape strategy. • If the maximum number of people needing to use escape stairs is unknown, the anticipated usage can be calculated using the floor space factors shown in Appendix D.	B(V2) 3.13
• Single steps on escape routes should be prominently marked.	B(V2) 3.26
• Helical stairs and spiral stairs may form part of an escape route provided they are designed in accordance with BS 5395.	B(V2) 3.27

 Note: If these types of stairs are intended to serve members of the public, stairs should be type E (public) stairs.

• Fixed ladders should not be provided as a means of escape for members of the public.	B(V2) 3.28
• Any door opening outwards onto a corridor or stair should be recessed so that its swing does not decease the effective width of an escape route.	B(V2) 5.13

 Note: All access doors to the stair should be fire-resisting and self-closing.

- Unless a smoke-control system (designed in accordance with BS EN 12101-6) has been used, protected lobbies and corridors should be provided at all storeys above ground (except the top storey) if: B(V2) 3.34
  - the stair is the only one serving a building;
  - the stair serves any storey at a height of 18m or more above ground level;
  - the building is designed for phased evacuation;
  - the stair is a firefighting stair.

- A protected lobby should be provided between an escape stair and a place of special fire hazard. B(V2) 3.35

- An escape stair that is part of the only escape route from an upper storey should not continue down to a basement storey. B(V2) 3.40
- The basement storey should be served by a separate escape stair.

- Where multiple escape stairs serve the upper storeys, only one needs to end at ground level. B(V2) 3.41

Other stairs may connect with the basement storeys if there is a protected lobby or a protected corridor between the stairs and the accommodation at each basement level.

4.11.5.8 *External escape stairs*

An external escape stair (see Diagram AD-B(V2) 3.4) should meet the following requirements:

- Where there is more than one escape route available, some escape routes may be via an external escape stair, provided that: B(V2) 3.31
B(V2) 3.32
  - there is at least one internal escape stair from every part of each storey;
  - the route is not intended for use by the public in an 'assembly and recreational' environment;
  - the route serves only office or residential staff accommodation in a 'residential (institutional)' building.

• Stairs more than 6m high should be protected from adverse weather.	B(V2) 3.32
• Glazing in areas of fire-resisting construction should be fixed shut (and fire resistant) (minimum E 30).	
• Access to an external escape stair may be via a flat roof, provided the flat roof is not part of a building intended for use by members of the public.	B(V2) 3.33 & 2.31

4.11.5.9 *Escape lighting*

• In offices, industrial, storage and other non-residential buildings, shops and commercial premises, and car parks, escape lighting is required on stairs that are in a central core or that serve storey(s) more than 18m above ground level.	B(V2) Table 5.1
• Escape stair lighting should be on a separate circuit from the electricity supply to any other part of the escape route.	B(V2) 5.26

4.11.5.10 *Fire mains and access for firefighting personnel*

• Buildings with firefighting shafts should have fire mains in both the firefighting shafts and, where necessary, in protected escape stairs.	B(V2) 16.2
• If a firefighting shaft is provided, outlets from fire mains should be within the protected stairway or protected lobby.	B(V2) 16.4
• Facilities for fire and rescue, such as firefighting lifts, firefighting stairs and firefighting lobbies, are combined in protected firefighting shafts.	B(V2) 17.1

4.11.5.11 *Lifts*

Note: Complete guidance on the design and use of evacuation lifts is given in Annex G to BS 9999.

• Lift wells should be contained in the enclosure of a protected stairway when passing from one compartment to another.	B(V2) 8.32

• Evacuation lifts should be clearly identified.	B(V2) 3.8
• At each storey, the lifts should be approached through a protected lobby.	B(V2) 3.21

• Lifts should not be used when there is a fire in the building, unless the following conditions are met:
  • they are appropriately sited and protected;
  • they contain safety features to ensure they remain usable during a fire.
	B(V2) 5.32

• A building with a storey more than 18m above the fire and rescue service vehicle access level should have one or more firefighting shafts containing a firefighting lift.	B(V2) 17.2

If a firefighting lift is provided, it can be used to evacuate disabled people. (See AD-B(V2) Diagram 7.1 for a complete description of the requirements for the inclusion of a firefighters lift in a firefighters shaft.)

4.11.5.12 Live/work units

Where a flat serves as a workplace for its occupants:

• The maximum travel distance to the flat entrance door or an alternative means of escape (not a window) from any part of the working area should not exceed 18m.
• Any windowless accommodation should have escape lighting which illuminates the route if the main supply fails.

Note: Standards for the installation of this sort of escape lighting are provided in BS 5266-1.

4.11.5.13 Multi-storey buildings

• In multi-storey buildings, there should be a sufficient number of adequately sized and protected escape stairs. (See AD-B(V2) Table 8.1 for details of maximum floor areas.)	B(V2) 2.3
• Every part of each storey must have access to more than one stair.	

4.11.5.14 Refuges

Note: Refuges are designed to be relatively safe waiting areas for short periods, and where a refuge is in a stairway, any signs should be accompanied by a blue mandatory sign worded '**Refuge – keep clear**'.

• Refuges and evacuation lifts should be clearly identified.	B(V2) 3.8
• A refuge should be provided for each protected stairway affording egress from each storey, except storeys consisting exclusively of plant rooms.	B(V2) 3.4
• The following are examples of satisfactory refuges: • an enclosure: e.g. a compartment (see AD-B(V2) Diagram 3.1), protected lobby, protected corridor or protected stairway (see AD-B(V2) Diagram 3.2); • an area in the open air: e.g. a flat roof or balcony which is sufficiently protected (or remote) from fire risk and has its own means of escape.	B(V2) 3.5
• Refuges should be: • a minimum of 900mm × 1,400mm; • accessible by someone in a wheelchair.	B(V2) 3.6
• Where a refuge is a protected stairway: • the wheelchair space should not reduce the width of the escape route; • access to the wheelchair space should not obstruct the flow of persons escaping.	B(V2) 3.4
• Refuges should have an emergency voice communication (EVC) system complying with BS 5839-9.	B(V2) 3.7
• Where refuge is formed by compartmentation, two fire doorsets in the partition are necessary (see AD-B(V2) Diagram 3.1).	B(V2) 3.5

Disabled people should not be left alone indefinitely in a refuge.

4.11.5.15 Student accommodation

• Student accommodation should be designed for independent use by ambulant disabled users.	M2 4.4
• Wheelchair users and ambulant disabled people should be able to wash or bathe either independently or with assistance from others.	M2 5.19
• In student accommodation, a wheelchair-accessible toilet should be available for disabled visitors.	M2 4.19

4.11.5.16 Phased evacuation

Note: In a phased evacuation, the first people to be evacuated are those with reduced mobility and those on the storey most immediately affected by the fire.

• Where a building is designed on the basis of phased evacuation, the stairways should be approached through a protected lobby or protected corridor at each storey (except the top).	B(V2) 3.21
• The minimum width of stairs for phased evacuation is given in AD-B(V2) Table 8.	B(V2) 3.22
• A building designed for phased evacuation should have an internal speech communication system, for communication between a control point at fire and rescue service access level and a fire warden, on every storey.	B(V2) 3.21
• The minimum width of stairs designed for phased evacuation should be in accordance with AD-B(V2) Table 3.3.	B(V2) 3.22

4.11.5.17 Ventilation

In buildings other than dwellings, more sophisticated automatic control systems such as occupancy sensors (using local passive infrared detectors) or indoor carbon dioxide concentration sensors (using electronic carbon dioxide detectors) can be used as an indicator of occupancy level and, therefore, body odour.

• There should be some means of ventilating common corridors/lobbies and stairs – such as mechanical ventilation.	B(V2) 3.35
• A separate ventilation system should be provided for each protected stairway (extracted air should not be recirculated).	B(V2) 10.6

• A protected shaft containing a protected stairway may contain either:
  • a duct for pressurising the protected stairway to keep it smoke free; and/or
  • a duct to ventilate the protected stairway.
	B(V2) 8.36

• An as-built plan of the building should be provided for ventilation systems with a smoke-control function, including mode of operation and control systems.	B(V2) 19.3

Note: Guidance on the design of smoke-control systems is available in BS EN 12101-6.

4.11.6 Protection from falling, collision and impact

4.11.6.1 Access for maintenance

In a dwelling, the maximum pitch of a stair shall be 42°, rise 150mm and 220mm, with any going between 220mm and 400mm.

4.11.6.2 Alternating tread stairs

Alternating tread stairs may only be used in a loft conversion.

• Alternating tread stairs may be used as shown in AD-K Diagram 1.1 in a loft conversion when the stair is for access to only one habitable room and the construction of the alternate tread stair complies with the following:
  • alternating steps are uniform with parallel nosings;
  • all treads have slip-resistant surfaces;
  • tread sizes over the wider part of the step conform to AD-K Table 1.1;
  • a minimum clear headroom of 2m is provided.
	K 1.2
	K 1.30

4.11.6.3 Construction of steps

• The rise and going should be measured in accordance with AD-K Diagram 1.1.	K 1.2
• Risers should not be open.	K 1.6
• Steps should have level treads, according to AD-K Table 1.1.	K 1.5

• The overlap treads of steps with open risers (especially those that are likely to be used by children under five years old) should:
  • be made apparent;
  • be a minimum of 55mm wide on both the tread and the riser;
  • not protrude over the tread below.
	K 1.7

For common access areas in buildings that contain flats:

• Step nosings should be apparent (with a suitable tread nosing profile as shown in AD-K Diagram 1.2) and risers should not be open.	K 1.10

If the soffit beneath a stair is less than 2m above floor level, then the area beneath the stair should be protected with either guarding or low-level cane detection, or a barrier.

• In flights of stairs, all of the following should be provided:
  • a minimum width between enclosing walls, strings or upstands of 1,200mm;
  • a minimum width between handrails of 1,000mm;
  • where the flight is more than 2m wide, divide it into flights a minimum of 1,000mm wide.
	K 1.15

• Single steps should be avoided.
• The maximum number of risers on flights between landings should be:
  • utility stairs – 16 risers;
  • general access stairs – 12 risers (exceptionally 16 in small premises).
	K 1.18

• An unobstructed length of 1,200mm should be provided on each landing.	K 1.23
• Doors should not swing across landings.	

• Suitable continuous handrails should be provided (dimensions are shown in AD-K Diagram 1.12) on each side of the flights and on each side of the landings.	K 1.35

• Handrails should be provided in accordance with all of the following: **K 1.36**
  • if a second (lower) handrail is provided, the vertical height from the pitch line of the steps to the top of the second handrail should be 600mm;
  • flights and landings of a ramped or stepped flights should have a continuous handrail;
  • handrails should not project into an access route;
  • handrails should contrast visually with the background and not be highly reflective;
  • handrail surfaces should be slip-resistant and not excessively cold or hot to touch;
  • handrail ends should be finished in a way that reduces the risk of clothing being caught.

• Guarding should be provided at the sides of flights and landings with two or more risers.	K 1.40

• Where stairs or ladders are used to access maintenance areas: **K 1.42**
  • if access is required a minimum of once per month, follow the provisions for private stairs in dwellings or for industrial stairs and ladders in BS 5395-3;
  • if access is required less frequently than once a month, portable ladders may be used.

4.11.6.4 Fixed ladders

In dwellings, a fixed ladder may be used – with fixed handrails on both sides – but only for access to a loft conversion that contains one habitable room.

Retractable ladders may **not** be used as means of escape.

4.11.6.5 Guarding and handrails

• Guarding should be provided at the sides of flights and landings where there is a drop of more than 600mm.	K1 1.41
• For stairs that are likely to be used by children under five years old, the construction of the guarding shall be such that a 100mm sphere cannot pass through any openings in the guarding and children will not easily be able to climb the guarding (see AD-K Diagram 1.3).	K1 1.39 K3.1
• Guarding should also be provided to safeguard:	

- the edges of any part of a floor, gallery, balcony or roof;
- any light well, basement or similar sunken area next to a building;
- a vehicle park.

4.11.6.5.1 DESIGN OF GUARDING

• Guarding should be provided in accordance with the height shown in AD-K Diagram 3.1.	K 3.2
• Any wall, parapet, balustrade or similar obstruction may be used as guarding.	
• Guarding must be capable of resisting, as a minimum, the loads given in BS EN 1991-1-1 (together with its UK Annex and PD6688-1-1).	

Further guidance on the design of barriers and infill panels can be found in BS 6180.

4.11.6.5.2 HANDRAILS FOR STAIRS

• If the stairs are 1,000mm or wider, there should be a handrail (900mm to 1,100mm from the pitch line or floor) on both sides. (See AD-K Diagram 1.11.)	K 1.34

4.11.6.5.3 HEADROOM FOR STAIRS

For all buildings:

• On the access between levels, the minimum headroom shown in AD-K Diagram 1.3 should be provided.	K 1.11

For loft conversions in dwellings:

• Where there is not enough space to achieve the height shown in AD-K Diagram 1.3, the reduced headroom shown in AD-K Diagram 1.4 may be provided.	K 1.13

4.11.6.6 *Length of flight of stairs for all buildings*

• If stairs have more than 36 risers in consecutive flights, a minimum of one change of direction between flights shall be made, as shown in AD-K Diagram 1.6.	K1 1.17

4.11.6.7 *Landings for stairs*

• Landings shall be provided at the top and bottom of every flight (see AD-K Diagram 1.6).	K1 1.20

A landing may include part of the floor of the building and it should be kept clear of permanent obstructions – such as a door which might swing across the landing at the bottom of the stairs (see Diagram 1.6) or doors to cupboards that could cause an obstruction (see Diagram 1.7).

4.11.6.8 *Spiral and helical stairs*

• The design of spiral and helical stairs shall be in accordance with BS 5395-2.	K 1.28

This requirement is **not** to be confused with BS EN ISO 5395-2:2013, which is for garden equipment, combustion engine powered lawnmowers and pedestrian-controlled lawnmowers!

4.11.6.9 Steepness of stairs – rise and going

- The rise and going shall meet the requirements   K 1.25
  shown in AD-K Diagram 1.1 and Table 1.1.
- The maximum pitch of a stair shall be 42°, rise
  150mm and 220mm, with any going between
  220mm and 400mm.

4.11.6.10 Ramps

4.11.6.10.1 CONSTRUCTION OF RAMPS

If the change in level is less than 300mm, a ramp should be provided instead of a single step.

- If the soffit beneath any ramp is less than 2m above   K 2.7
  floor level, the area beneath the ramp should be
  protected with some sort of barrier (e.g. guarding,
  low-level cane detection, etc.).

4.11.6.10.2 DESIGN OF RAMPS

For all buildings, ramps and landings should be designed in accordance with AD-K Diagram 2.2.

For ramps that are likely to be used by children under five years old, the construction of the guarding shall be such that a 100mm sphere cannot pass through any openings in the guarding and children will not easily climb the guarding.

4.11.6.10.3 STEEPNESS OF RAMPS

- The relationship between the gradient of a ramp   K 2.3
  and its going between landings should be as shown
  in AD-K Diagram 2.3.

4.11.6.10.4 INSTALLATION OF RAMPS

• Ramps should be readily apparent or clearly signposted.	K 2.2

• A ramp surface should be:
  • slip-resistant, especially when wet, and a colour that will contrast visually with that of the landings;
  • similar to the landing.
	K 2.4

• A kerb on the open side of any ramp or landing should be a minimum of 100mm high and contrast visually with the ramp or landing.	K 2.5
• Where the change of level is 300mm or more, in addition to the ramp, there should be two or more signposted steps. • If the change of level is less than 300mm, a ramp should be provided instead of a single step.	K 2.6
• If the soffit beneath any ramp is less than 2m above floor level, the area beneath should be protected with guarding.	K 2.7
• A ramp that provides access for people should have a minimum width between walls, upstands or kerbs of 1,500mm.	K 2.9
• Ramps should be kept clear of permanent obstructions.	K 2.10
• A handrail should be provided on both sides of the ramp.	K 2.11

• All of the following should be provided:
  • ramps that are less than 1,000mm wide should have a handrail on one or both sides;
  • ramps that are 1,000mm or wider should have a handrail on both sides;
  • ramps that are 600mm or less in height do not require handrails;
  • the top of the handrails should be at a height of 900mm to 1,000mm above the surface of the ramp;
  • handrails should give firm support and allow a firm grip;
	K 2.12
	K 2.13

- handrails may form the top of the guarding if you can match the heights;
- at the foot and head of a ramp, there should be landings which are a minimum of 1,200mm long and are clear of any door swings or other obstructions;
- any intermediate landings should be a minimum of 1,500mm long and clear of any door swings or other obstructions;
- all landings should be level or have a maximum gradient of 1:60 along their length;
- if it is not possible to see from one end of the ramp to the other, or the ramp has three flights or more, then intermediate landings a minimum of 1,800mm wide and a minimum of 1,800mm long should be provided as passing places.

4.11.6.11 *Tapered treads*

• Consecutive tapered treads shall use the same going.	K 1.26
• If a stair consists of straight and tapered treads, the going of the tapered treads shall not be less than the going of the straight treads.	K 1.27

4.11.6.12 *Width of flights of stairs*

If the flight is more than 2m wide, divide it into flights a minimum of 1,000mm wide, as shown in Diagram 1.5.

• In exceptional circumstances where severely sloping plots are involved (and a stepped change of level within the entrance storey is unavoidable), the minimum stair width within the entrance storey of a dwelling is 900mm.	K 1.16

4.11.7 *Resistance to the passage of sound*

Common internal parts of buildings which contain flats or rooms for residential purposes shall be designed in such a way that reverberation around these common parts of the building is minimised.

4.11.7.1 Piped services

Piped services (excluding gas pipes) and ducts that pass through separating floors should be surrounded with sound-absorbent material for their full height, and enclosed in a duct above and below the floor.

4.11.7.2 Reverberation

Requirement E3 states that 'Buildings shall be designed and constructed so as to restrict the transmission of echoes'. The guidance notes provided in AD-E (Section 7.6) cover two methods (Method A and Method B) for determining the amount of additional absorption that is required for corridors, hallways, stairwells and entrance halls that provide access to flats and rooms used for residential purposes.

Method A is applicable to stairs and requires the following to be observed:

• The ceiling area should be covered with the additional absorption.	E 7.10
• The underside of intermediate landings, the underside of the other landings, and the ceiling area on the top floor should all be covered with the additional absorption.	E 7.11
• The absorptive material should be equally distributed between all floor levels.	E 7.12

Method B is not suited to stairwells as it is primarily intended for corridors, hallways and entrance halls.

4.11.7.3 Sound insulation testing

The person completing the building work should arrange for sound insulation testing to be carried out.

• Impact sound insulation tests should be carried out without a soft covering (e.g. carpet, foam-backed vinyl, etc.) on the stair floor.	E 1.10
• Testing should **not** be carried out between living spaces, corridors, stairwells or hallways.	E 1.8

4.11.7.4 *Stair treatments*

The 'stair treatment' consists of a stair covering an independent ceiling with absorbent material.

• The underside of the stair within the cupboard should   E 4.37 be lined with plasterboard (minimum mass per unit area of 10kg/m^2) together with an absorbent layer of mineral wool (minimum density of 10kg/m^3). • The cupboard walls should be built from two layers of plasterboard (or equivalent), each sheet with a minimum mass per unit area of 10kg/m^2. • A small, heavy, well-fitted door should be fitted to the cupboard.

If there is no cupboard under the stair, an independent ceiling should be constructed below the stair.

• An independent ceiling with absorbent material (see   E 4.37
AD-E Diagram 4.5) requires: E 4.27
  • at least two layers of plasterboard with staggered
    joints, with a minimum total mass per unit area of
    20kg/m^2;
  • an absorbent layer of mineral wool laid on the
    ceiling, with a minimum thickness of 100mm, and
    a minimum density of 10kg/m^3.
• The ceiling should be supported by one of the
following methods:
  • independent joists fixed only to the surrounding
    wall;
  • independent joists fixed to the surrounding walls
    with additional support provided by resilient
    hangers attached directly to the existing floor base.

Stairs that separate a dwelling from another dwelling (or part of the same building) shall comply with the following:

• Resist the transmission of impact sound (e.g. footsteps   E2 and furniture moving). • Resist the flow of sound energy through walls and floors. • Minimise the level of airborne sound. • Resist flanking transmission from stairs connected to the separating wall.

- All new stairs constructed within a dwelling that is   E 0.9
  purpose-built or formed by a material change of use
  shall meet the laboratory sound insulation values set
  out in AD-E Table 2.

4.12 Windows

Any alterations to windows, from a flat window to a bay or bow window, may require Planning Permission and should be referred to the Local Planning Office.

 Reminder: The subsections include details for buildings other than dwellings only where they differ from the requirements for dwellings.

Subsection	Description	Page
Requirements for dwellings		
4.12.1	Access to and use of buildings	428
4.12.2	Fire safety	428
4.12.3	Site preparation and resistance to moisture	431
4.12.4	Conservation of fuel and power	432
4.12.5	Resistance to the passage of sound	435
4.12.6	Ventilation	436
4.12.7	Drainage and waste disposal	439
4.12.8	Combustion appliances and fuel storage systems	439
4.12.9	Protection from falling, collision and impact	440
Requirements for buildings other than dwellings – Additional requirements		
4.12.10	Access to and use of buildings	441
4.12.11	Structure	442
4.12.12	Fire safety	443
4.12.13	Ventilation	444
4.12.14	Protection from falling, collision and impact	445
4.12.15	Conservation of fuel and power	447

 Reminder: Free downloads of all the diagrams and tables referenced in the subsections are available on-line at the Governments' Planning Portal: www.gov.uk/government/collections/approved-documents.

Requirements for dwellings

4.12.1 Access to and use of buildings

In order to ensure the prevention of unauthorised access to dwellings (including flats), since 1st October 2015 current Building Regulations require test evidence showing that all new building windows have passed the security aspects of PAS 24.

• The main doors for entering a dwelling (usually the front door) should have a door view unless other means exist to see callers, such as clear glass within the door, a window next to the doorset or some form of remote technological security device.	Q 1.4
• Ground floor, basement and other easily accessible windows (including easily accessible rooflights) should be secure windows.	Q 2.1
• Windows should be made to a design that has been shown by test to meet the security requirements of British Standards publication PAS 24, and fitted in accordance with manufacturer's instructions.	Q 2.2

4.12.2 Fire safety

Note: All dwellings should be provided with smoke alarms and openable windows for emergency egress.

4.12.2.1 Basements

• Basements in dwellings which contain a habitable room should have a suitable external window or protected stairway for emergency egress.	B(V1) 2.16

4.12.2.2 Cavity barriers

• Cavity barriers should be provided around openings such as windows.	B(V1) 8.3

4.12.2.3 Emergency escape windows and external doors

Windows should be designed so that they will remain in the open position without having to be held open by the person making their escape.

- Windows provided for emergency escape should comply with the following conditions: B(V1) 2.10
  - the person escaping is able to reach a place free from danger from fire;
  - the route through the window may be at an angle rather than straight through;
  - the bottom of the openable area should be not more than 1,100mm above the floor;
  - there should be an unobstructed openable area of at least 0.33m^2 (450mm high and 450mm wide).
- Glazing in areas that are designed to be fire resistant should be fixed shut.   B(V1) 2.17

Note: See Diagram AD-B(V1) 2.5.

4.12.2.4 Galleries

- If a gallery floor is not more than 4.5m above ground level, it may be provided with an emergency egress window. (See AD-B Diagram 2.6.)   B(V1) 2.15

4.12.2.5 Live/work units

- If a flat is being used as a workplace for occupants **and** people who do not live on the same premises, then an alternative means of escape (not a window) **must** be provided.   B(V1) 3.24

4.12.2.6 Conversions

- Loft-room occupants should be able to gain access to a first-floor-level escape window.   B(V1) 2.23

4.12.2.7 Means of escape

Note: All dwellings should be provided with smoke alarms and openable windows for emergency egress.

* All habitable rooms (except kitchens): B(V1) 2.1
  * in the ground storey, should either open directly onto a hall leading to either the entrance or a suitable exit; or be provided with a window or door;
  * in the upper storey(s) of a dwelling (not more than B(V1) 2.2 4.5m above ground), should have an emergency escape window (or external door), or direct access to a protected stairway.

Note: Two rooms may be served by a single window provided that any door between them does not pass through the stair enclosure (see AD-B(V1) Diagram 2.1).

* An inner room is permitted on a storey that is a B(V1) 2.11 maximum of 4.5m above ground level, provided that it has an emergency escape window.

4.12.2.8 Replacement windows

A replacement window is a controlled service and replacement window openings should provide the same potential for escape as the window it replaces.

Repair work to windows does **not** fall within the definition of building work.

4.12.2.9 Thermoplastic materials

It is important to ensure that the amount of thermal radiation that will fall from a dwelling's window openings onto a neighbouring building is not enough to start a fire in the other building.

* Rigid thermoplastic material may be used to glaze B(V1) 4.13 external windows to rooms, but **not** external windows to circulation spaces.

- Rooflights may be constructed of thermoplastic   B(V1) 4.14
  material if the lower surface is rigid and the size and
  location of the rooflights follow the limits shown in
  AD-B(V1) Diagram 4.2 and Table 4.2.

4.12.3 Site preparation and resistance to moisture

Building Regulations relating to site preparation and resistance to contaminants and moisture will need to be met once the preparation work starts on the site.

4.12.3.1 Cracking of external walls

Vertical cracks usually occur in foundation walls when there is insufficient support below.

- Cracks in masonry external walls (caused by hot   C 5.18
  weather or subsidence after prolonged droughts) will
  enable the penetration of rain and this needs to be
  taken into account when designing a building.

The junctions between cladding and window openings are particularly vulnerable to moisture ingress.

4.12.3.2 External walls – resistance to surface condensation and mould growth

An exterior wall typically forms part of a building envelope, separating the accommodation inside from that outside.

- Junctions between elements and openings such as   C 5.36
  windows should resist surface condensation and
  mould growth.

Note: Window junctions should be designed to Accredited Construction Details (see BRE IP17/01).

4.12.3.3 Joints between doors and windows

When two individual windows or a door and window are joined, the seam between the frames of the two units is called the mull, which is short for mullion.

• Joints between doors (and walls) with window frames should resist the penetration of precipitation.	C 5.29
• Sill elements that do not form a complete barrier to rain and snow should be provided with a damp-proof course to direct moisture towards the outside, particularly at a lintel.	C 5.30
• Checked rebates should be used in all windows and reveals (see AD-C Diagram 13) in areas subject to driving rain.	C 5.32

4.12.3.4 Installing energy-efficiency measures

• Building work should not reduce the ventilation already available to a dwelling unless it can be demonstrated that after the work is completed, it out meets minimum requirements.	F(V1) 3.6 & 3.7

Note: Appendix D to AD-F(V1) provides a checklist for determining the ventilation provision in an existing dwelling.

4.12.4 Conservation of fuel and power

To be introduced by 2025, the Future Homes Standard will require new-build homes to be future-proofed with low carbon heating and world-leading levels of energy efficiency. With effect from June 2022, AD-L (*Conservation of fuel and power*) and AD-F (*Ventilation and overheating*) have been updated and republished. In addition, two new ADs, *Overheating* (AD-O) and *Infrastructure for the chrging of electric veicles* (AD-S), will also becom law in June 2022.

4.12.4.1 U-values

• U-values should be assessed for the whole fabric element (e.g. in the case of a window, the combined performance of the glazing and the frame), using the methods and conventions in BR 443.	L(V1) 4.1
• The U-value of a window should either be calculated using the actual size and configuration of the window; or • calculated for a standard window that is 1.23m wide × 1.48m high, using the hot-box method described in BS EN ISO 12567-1.	L(V1) 4.2
• For windows and roof windows, U-values should be calculated based on a vertical position; and for rooflights, on a horizontal position.	L(V1) 4.5

• The U-values for:
  • a window in a new dwelling, U = 1.6W/(m²·K);
  • a new window in an existing dwelling, U = 1.42W/(m²·K).
	L(V1)
	Table 4.1
	L(V1)
	Table 4.2

Note: If windows cannot meet these requirements because of the need to maintain the character of the building, then the fittings should not exceed a centre pane U-value of $1.2W/(m^2 \cdot K)$, or single glazing should be supplemented with low-emissivity secondary glazing.

4.12.4.2 Overheating

It must be remembered that whilst AD-O provides guidance on window openings for removing excess heat from residential buildings, and AD-B gives guidance on the size of escape windows, AD-O also covers the **security** implications when installing large openings for removing excess heat. (The actual window-locking systems for doors and windows in dwellings also needs to conform to the security requirements provided in AD-Q).

• Louvered shutters, window railings and ventilation grilles should **not**:
  • allow body parts to become trapped;
  • allow the passage of a 100mm diameter sphere;
  • taper in a way that allows finger entrapment (!).
• Looped cords must be fitted with child safety devices.
	O 3.11

4.12.4.3 *Removal of excess heat*

- Excess heat can obviously be removed from the   O 2.10
  residential building by opening windows – but if the   O 3.1
  dwelling is in a noisy area, this is not the ideal choice.
  In those circumstances, installing ventilation louvres
  in external walls or using mechanical ventilation and
  cooling will be just as effective.

4.12.4.4 *Energy performance certificates*

When a dwelling is erected, the person carrying out the work must provide an energy performance certificate to the owner of the building, as well as a notice to the BCB that a certificate has been given.

4.12.4.5 *Extensions*

4.12.4.5.1 CONSERVATORIES AND PORCHES

- Where a dwelling is extended by adding a   F(V1) 3.21
  conservatory or porch that exceeds 30m², the
  general ventilation rate could be achieved by using
  background ventilators.

Note: If it is a new conservatory or porch that exceeds 30m², then purge ventilation might be more appropriate.

4.12.4.5.2 AREA OF WINDOWS

In most circumstances, the total area of windows, roof windows and doors in an extension should not exceed 25% of the floor area of the extension plus the total area of any previous windows or doors which (as a result of the extension works) no longer exist.

4.12.4.6 *Solar gain*

Note: Whilst solar gains are beneficial in winter (i.e. to offset the demand for heating), they can contribute to overheating in the summer.

- Solar gains in summer should (or could) be limited by   O 2.7
  any of the following means:
  - fixed shading devices (shutters, external blinds,
    overhangs or awnings);
  - glazing design (size, orientation, g-value and depth
    of the window reveal);
  - building design (balcony location);
  - shading from adjacent permanent buildings,
    structures or landscaping.

 It should be remembered that although solar gain will be reduced with these measures, there is every likelihood that there will be an increase in the use of electric lighting, which will have an impact on the predicted CO_2 emissions.

4.12.4.7 Insulation

- Windows should be installed so that the thermal   L(V1)
  integrity of the insulated pane is maintained and: 4.20 &
  - window designs minimise thermal bridging;   4.21
  - tightly installed, fully insulated and continuous
    cavity closers are used;
  - insulated plasterboard should be used in reveals to
    abut jambs and should be considered within reveal
    soffits.
- Window frames should be taped with air-sealing tape L(V1)
  to surrounding structural openings.   4.21 h

4.12.5 Resistance to the passage of sound

When conducting tests for sound, windows should be closed.

- Where a window head is near to the existing ceiling, E 4.28
  the new independent ceiling may be raised to form a
  pelmet recess, as shown in AD-E Diagram 4.4.
- Test reports (with windows closed) should contain the E App B.4
  dimensions and position of any windows in external
  walls.

4.12.6 Ventilation

There shall be adequate means of ventilation offered for people in the building.

To meet this requirement, ventilation may be delivered through natural ventilation, mechanical ventilation or a combination of both, and windows should ensure that:

- Sufficient quantities of outdoor air for occupants' health is provided.
- The entry of external air pollutants is minimised.
- Indoor air pollutants are rapidly diluted.
- Low levels of noise from ventilating fans are produced.

• There should be reasonable access for maintaining ventilation systems.	F(V1) 1.8
• Fans operating in normal mechanical background ventilation mode are not overly noisy.	F(V1) 1.6
• If purge ventilation is going to be used, account has been taken of the likely amount of outside noise from openable windows.	F(V1) 1.7
• Extract ventilation used to extract water vapour and indoor air pollutants from areas such as kitchens and bathrooms is either intermittent or continuous operation.	F(V1) 1.9

Note: Sufficient information (provided in a clear manner, for a non-technical audience) concerning overheating and maintenance requirements must be given to owners so that appliances can be used effectively.

4.12.6.1 Noise

• Whilst open windows can be made secure by using fixed or lockable louvred shutters or installing a fixed or lockable window grille or railing, in locations where external noise is a problem, it is quite normal for windows to be closed during sleeping hours (11pm to 7am).	O 3.7 O 3.2

For methods of overcoming noise problems, see: Association of Noise Consultants, *Acoustics, Ventilation and Overheating: Residential Design Guide* (2020).

4.12.6.2 Purge ventilation

A system for purge ventilation should be provided in each habitable room that is capable of extracting at least four air changes per hour per room directly to the outside.

• Openings (e.g. windows or doors) or a mechanical extract ventilation system are both acceptable for purge ventilation.	F(V1) 1.28
• Where purge ventilation is delivered through openings in a habitable room, the minimum opening areas shall be as listed in AD-F(V1) Table 1.4.	F(V1) 1.29

Note: Depending on the design of the dwelling and the external climate, it may be possible to exceed four air changes per hour, but it should be remembered that hinged or pivot windows with an opening angle of less than 15° are not suitable for purge ventilation.

4.12.6.3 Natural ventilation with background ventilators and intermittent extract fans

Note: This guidance is only suitable for less airtight dwellings.

• In a room with no openable window, an intermittent extract fan should be provided.	F(V1) 1.51
• Intermittent extract fans may be used for all wet rooms such as kitchens, utility rooms, bathrooms and sanitary accommodation.	F(V1) 1.47
• Purge ventilation using openings (e.g. windows or doors as shown in AD-F(V1)Table 1.4,) or a mechanical extract ventilation system, should be provided in each habitable room.	F(V1) 1.26

Note: Extract rates for sanitary accommodation can be met by using windows to act as purge ventilation.

• Background ventilators are intended to normally be left open (see AD-F(V1) Table 1.7 for details regarding the minimum equivalent area of background ventilators).	F(V1) 1.52–1.58

Note: A window with a night latch position is not adequate for background ventilation because of the risk of draughts, security and the difficulty of measuring the equivalent area.

- Automatic controls (such as a humidity control) for intermittent extract should have a manual override to allow the occupant to turn the extract ventilation on or off.

F(V1) 1.50

4.12.6.4　Addition of a habitable room (not including a conservatory) to an existing dwelling

- Ventilation requirements can be met if background ventilators are used.

F(V1) 3.17

4.12.6.5　Ventilation of a habitable room through another room or a conservatory

- In an internal habitable room that does not have openable windows, ventilation can be achieved through either another habitable room or a conservatory.

F(V1) 1.40–1.44

4.12.6.6　Replacing windows

- If existing windows have background ventilators, the replacement windows should also include background ventilators that are not less than the background ventilators included in the original windows, and that contain either manual or automatic controls for the occupant.

F(V1) 3.14

- As replacing the windows is likely to increase the airtightness of the dwelling, care should be taken to ensure that the ventilation in the dwelling is no worse than it was before the work was carried out.

F(V1) 3.15

4.12.6.7　Airtightness in existing dwellings

- When carrying out work in existing dwellings, installing windows, roof windows or rooflights, the controlled fitting should be well fitted and reasonably draught-proof.

L(V1) 4.23

• New and replacement windows, roof windows and rooflights should:
  • be draught-proofed;
  • meet the minimum standards given in AD-L(V1) Table 4.2;
  • have insulated cavity closers where appropriate.
	L(V1) 10.3

• If a window is enlarged or a new one is created, the area of windows or roof windows should not exceed 25% of the total floor area of the dwelling.	L(V1) 10.5

The term 'controlled fitting' refers to the entire unit of a window, roof window, rooflight or door, including the frame. Replacing glazing or a window (or even a door) in its existing frame is **not** considered as providing a controlled fitting and, therefore, does not need to meet the energy-efficiency requirements.

4.12.7 Drainage and waste disposal

Note: Rainwater should be carried from the roof of the building.

• Eaves drop systems should prevent water from entering windows.	H3 1.13
• External storage areas for waste containers should be away from windows.	H6 1.12

4.12.8 Combustion appliances and fuel storage systems

A combustion appliance is an apparatus where fuel is burned to generate heat for space heating, water heating, cooking or other similar purposes. Typical combustion appliances are boilers, warm air heaters, water heaters, fires, stoves and cookers.

4.12.8.1 Carbon monoxide alarm location

• Carbon monoxide alarms should be located as high up as possible (above any windows) but **not** within 150mm of the ceiling.	J 2.36

4.12.8.2 Locations of flue outlets near windows

Flue outlets should be protected with terminal guards if persons could come into contact with them or if they could be damaged.

• A flue should **not** penetrate a roof within 2,000mm horizontally of a dormer window or other opening.	J Diagram 35
• If a flue outlet discharges within reach from the ground or a window, it should be designed to prevent the entry of any matter that could obstruct the flow of flue gases.	J 3.26
• For flues in proximity to roof windows, the minimum separation distances in AD-J Diagram 35 should be applied.	J 4.7

4.12.9 *Protection from falling, collision and impact*

4.12.9.1 Glazing in critical locations

• In critical locations (see AD-K Diagram 5.1) all buildings should: • permanently protect glazing; • choose glazing that is in small panes and is robust; • ensure that glazing, if it breaks, will break safely.	K 5.2

For the purposes of this recommendation, a 'small pane' is an isolated pane or one of a number of panes held in glazing bars less than 6mm thick and with a maximum area of $0.5m^2$.

• If glazing in a critical location is protected by a permanent screen (see AD-K Diagram 5.4), then the glazing itself does not need to comply with the requirements of AD-K.	K 5.8

4.12.9.2 Pedestrian guarding

Provision shall be made to prevent people moving in or about the building from colliding with open windows or skylights.

- 800mm of guarding is required (where it is necessary   K 3.1
  for safety) around the edges of any part of a window
  or other opening.

This does not apply to windows in loft extensions.

Requirements for buildings other than dwellings – Additional requirements

The basic requirements for windows in buildings other than dwellings are the same as for dwellings, with the addition of the following requirements.

The general rules about window openings (especially those used as an exit) are that they must meet the following requirements:

- Width – not less than 450mm.
- Height – not less than 450mm.
- Clear openable area – not less than 0.33m².
- Cill height – the bottom of the openable area should be no more than 1,100mm above the floor area.

4.12.10 *Access to and use of buildings*

- Potential hazards on access routes adjacent to   M2 1.5
  buildings (such as open windows) should be avoided
  so that people with impaired sight or hearing are not
  injured.

- Openable windows and window controls in sleeping   M2 4.24
  accommodation should be:
  - located between 800 and 1,000mm above the floor;
  - easy to operate without using both hands
    simultaneously.

4.12.10.1 *Bedrooms*

- For wheelchair users, an accessible and adaptable   M4(2)
  bedroom should provide a minimum clear access   2.24c
  route, 750mm wide, from the doorway to the window.

4.12.10.2 Habitable rooms – living, kitchen and eating areas

• Similar to bedrooms, any glazing to the principal window of the principal living area should start a maximum of 850mm above floor level or at the minimum height necessary for guarding to the window (also see AD-K Diagram 5.1 for further advice on critical areas).	M4(2) 2.24c

4.12.10.3 Services and controls

To assist people who have reduced reach, unless at least one window in the principal living area is fitted with a remote opening device (located between 450mm and 1,200mm above floor level), then:

• The handles to all other windows need to be located between 450mm and 1,400mm above floor level.	M4(2) 2.30

4.12.11 Structure

The stability of any building can be affected by ground conditions, which should be investigated and assessed to ensure that the ground can safely support the building. In addition, the building must be constructed so that the combined dead, imposed and wind loads are sustained and transmitted by it to the ground.

4.12.11.1 Small single-storey non-residential buildings and annexes

• A single leaf door is permitted in a small single-storey building.	A 2C38ii
• There should be no other openings within 2m of a wall containing a major opening.	
• The total size of all openings in a wall (that does not contain a major opening) should not exceed 2.4m².	
• The size and location of openings should be in accordance with AD-A Diagram 17.	
• There should not be more than one opening between piers.	
• Unless there is a corner pier, the distance from a door to a corner should not be less than 390mm.	

4.12.12 Fire safety

Windows are not acceptable as final exits.

4.12.12.1 General guidance

Note: Windows and glazing frames are not considered to be 'a wall' and a ceiling does not include window frames or frames in which glazing is fitted.

4.12.12.2 Cavity barriers

• Cavity barriers should be provided at the edges of cavities, including around window openings.	B(V2) 9.3
• Cavity barriers made around openings may be formed by a window frame, provided that the frame is constructed of steel (minimum 0.5mm thick) or timber (minimum 38mm thick).	B(V2) 9.14

4.12.12.3 Escape routes

• Fire-resisting windows should be fixed shut and fire resistant in the vicinity of an escape stair (see AD-B(V2) Diagram 3.4).	B(V2) 3.32
• Escape lighting is required in all windowless accommodation in an office, industrial, storage or commercial premises, shop, car park, or other non-residential premises (see AD-B(V2) Table 5.1).	B(V2) 5.21

4.12.12.4 Resisting fire spread

• The amount of thermal radiation that falls on a neighbouring building from window openings and other unprotected areas in the building on fire should not be enough to start a fire in the other building.	B(V2) B4

• In a building with a storey 18m or more in height, the insulation material (such as window spandrel panels) used in the construction of an external wall should be class A2-s3, d2 or better (see AD-B(V2) Diagram D6).	B(V2) 12.6
• Window frames and glass are not defined as an 'element of structure' and so do not comply with Regulation 7(2).	B(V2) 12.16d
• Window spandrel panels and infill panels are defined as an 'element of structure' and must comply with Regulation 7(2).	

4.12.12.5 *Thermoplastic materials*

• TP(a) rigid thermoplastic materials may be used to glaze external room windows, but not windows to external circulation spaces.	B(V2) 6.14

For more details, see AD-K.

4.12.12.6 *Venting of heat and smoke from basements*

• A basement storey or compartment containing rooms with windows does not require a smoke outlet.	B(V2) 18.2

4.12.13 Ventilation

4.12.13.1 *Carbon dioxide (CO_2) monitoring*

• CO_2 monitors should be placed at breathing height and away from windows, doors or ventilation openings where practicable.	F(V2) 1.23

4.12.13.2 *Providing ventilation*

• For a room with no openable window, the extract ventilation should operate both while the room is occupied and for a minimum of 15 minutes after occupants have left the room.	F(V2) 1.31

- When other building work is carried out that will affect the ventilation of the existing building (such as replacing windows or doors), then the ventilation of the building should either meet the standards in the relevant AD or be no less satisfactory than before the work was carried out.   F(V2) 3.2

4.12.13.3 Replacing existing windows with background ventilators

- If the existing windows have background ventilators, then any replacement windows should also include background ventilators.   F(V2) 3.5
- New background ventilators should not be smaller than the original background ventilators that were in the window, and they should be controllable either automatically or by the occupant.
- Replacing the windows is likely to increase the airtightness of the dwelling.   F(V2) 3.6

Note: If ventilation is not provided via a mechanical ventilation with heat recovery system, then increasing the airtightness of the building may reduce the beneficial ventilation in the building. It is important to ensure that ventilation is no worse than it was before the work was carried out.

The most cost-effective way to improve ventilation is to incorporate background ventilators in replacement windows.

4.12.14 Protection from falling, collision and impact

For work relating to projecting windows, skylights and ventilators, you should consider Regulation 15(2) of the Workplace (Health, Safety and Welfare) Regulations 1992.

4.12.14.1 Critical locations

Critical locations include areas where two parts of a building are at the same level but separated by transparent glazing which can make people think they can directly walk from one part to the other. There are two methods to indicate glazing, permanent and alternative methods, as shown in AD-K Diagram 5.1.

• Permanent manifestation to make glazing apparent.	K 7.4
• Alternative indications of glazing include items such as mullions, transoms, door framing or large pull or push handles (see AD-K Diagram 7.1).	K 7.4

4.12.14.2 *Location of controls*

• The controls to operate windows and skylights should have one of the following:
  • controls located as shown in AD-K Diagram 8.1;
  • a safe manual or electrical means of remote operation.
	K 8.1

4.12.14.3 *Prevention of collision with open windows*

• People should be prevented from colliding with open windows, skylights or ventilators.	K 5.1
• Where parts of windows, skylights and ventilators project inside or outside a building, this should be indicated and guarded as shown in AD-K Diagrams 6.1 and 6.2.	K 6.1

4.12.14.4 *Robustness*

Some glazing materials (such as annealed glass) gain strength through thickness; others such as polycarbonates or glass blocks are inherently strong.

• The maximum dimensions for annealed glass of different thicknesses for use in large areas (e.g. fronts to shops or showrooms, or in offices, factories and public buildings) with four edges supported are in AD-K Diagram 5.5.	K 5.5

- In an impact test, a breakage is considered 'safe' if it creates one of the following: K 5.X
  - just a small clear opening, with detached particles no larger than the specified maximum size;
  - disintegration into small detached particles;
  - broken glazing in separate pieces that are not sharp or pointed.

4.12.14.5 Safe access methods

- If it is not possible to safely clean a glazed surface standing on the ground, or a floor or other permanent stable surface, then either: K 9.1
  - provide windows of a size and design that allows the outside to be cleaned safely from inside the building (see AD-K Diagram 9.1); or
  - provide access ladders, as shown in AD-K Diagram 9.2.

Note: If none of the above methods are possible, scaffolding will have to be used.

4.12.15 Conservation of fuel and power

The basic requirements for windows are the same as for dwellings, with the addition of the following requirements caused by the adoption of Regulation 25B and the envisaged 'nearly zero-energy building' for all new buildings.

4.12.15.1 General requirements

- New or replaced windows, roof windows and rooflights should meet the standards in AD-L(V2) Table 4.1. L(V2) 11.7

- Replacement windows, roof windows or rooflights (but excluding display windows) should have a U-value higher than: L(V2) D4
  - for windows and roof windows – $3.30 \text{W}/(\text{m}^2 \cdot \text{K})$;
  - for rooflights – $3.80 \text{W}/(\text{m}^2 \cdot \text{K})$.

4.12.15.2 Continuity of insulation

• All windows, roof windows and rooflights should be insulated and draught-proofed to at least the same extent as in the existing building.	L(V2) 10.13

4.12.15.3 Extensions

A 'controlled fitting' in the context of a window, roof window or rooflight refers to a whole unit, i.e. including the frame.

• If windows, roof windows and rooflights are replaced, the following must be taken into account: • units should be draught-proofed; • units should meet the minimum standards shown in AD-L(V2) Table 4.1; • insulated cavity closers should be installed where appropriate.	L(V2) 10.3

Note: Where the extension is a conservatory or porch, the windows **between** the building and the extension should be insulated and draught-proofed.

Replacing the glazing whilst retaining an existing frame is not providing a controlled fitting, and so such work is not notifiable.

4.12.15.4 Limiting fabric standards

• The U-value of a window should be assessed using the methods and conventions set out in BR 443, based on the combined performance of the glazing and the frame of a window.	L(V2) 4.4
• Using the hot-box method shown in BS EN ISO 12567-1 (for windows) and BS EN ISO 12567-2 (for roof windows), the actual U-value of a window's glazing can be calculated for: • the smaller of the two standard windows defined in BS EN 14351; • the standard window configuration set out in BRE 443; • the specific size and configuration of the actual window.	L(V2) 4.4

- As shown in AD-L(V2) Table 4.1, the limiting   L(V2) 4.6
  U-values for new or replacement elements for new
  and existing buildings is:
  - for windows in buildings similar to dwellings –
    $1.6W/(m^2{\cdot}K)$ (or window energy rating Band B);
  - for all other windows, roof windows, rooflights and
    curtain walling – $1.6W/(m^2.K)$.

Note: There is no limit on design flexibility for display windows and similar glazing, but their impact on CO_2 emissions must be taken into account in calculations.

4.12.15.5 *Thermal elements*

Note: Thermal bridges should be avoided at the edges of window openings.

4.12.15.6 *Protection from falling, collision and impact*

Note: For work relating to projecting windows, skylights and ventilators, you should also consider Regulation 15(2) of the Workplace (Health, Safety and Welfare) Regulations 1992.

4.12.15.6.1 PREVENTION OF COLLISION WITH OPEN WINDOWS

- People moving in or about a building should be   K 5.1
  prevented from colliding with open windows,
  skylights or ventilators.
- Where parts of windows, skylights and ventilators
  project inside or outside a building, this should be
  indicated and guarded as shown in AD-F Diagrams
  6.1 and 6.2.

Note: The projecting part of any window in spaces which are used infrequently (and then only for maintenance purposes) should be clearly marked rather than using guarding measures.

4.13 Doors

Unless specified otherwise, all doors should be fire doors.

Reminder: The subsections include details for buildings other than dwellings only where they differ from the requirements for dwellings.

Note: ADs P (*Electrical safety*) and Q (*Security*) **only** apply to dwellings.

Reminder: Free downloads of all the diagrams and tables referenced in the subsections are available on-line at the Governments' Planning Portal: www.gov.uk/government/collections/approved-documents.

Requirements for dwellings

4.13.1 Access to and use of buildings

The threshold to a building should be an accessible threshold. Where a step into the dwelling is unavoidable, the rise should not exceed 150mm and should be aligned with the outside face of the door threshold.

A ramped approach should have a landing 1,200mm long (minimum) which is clear of the swing of any door.

4.13.1.1 Principal communal entrance

The principal door of a dwelling should have a minimum clear opening width of 775mm between the inside face of door (when open) and the inside edge of door frame or stop. (See AD-M Diagram 1.1.)

• The principal door of a dwelling should have a minimum clear opening width of 775mm between the inside face of door (when open) and the inside edge of door frame or stop. (See AD-M Diagram 1.1.)	M4(1) 1.9
• The principal communal entrance should comply with all of the following:	M4(1) 1.9 M4(2) 2.14
• have a level landing a minimum of 1,500mm wide and 1,500mm long directly outside the entrance and clear of the swing of any door;	**(Optional requirement)** M4(3) 3.14
• an entrance door (or gate) with a clear opening width of 850mm (minimum), when measured in accordance with AD-M Diagram 2.2;	**(Optional requirement)**
• a minimum 300mm nib should be provided to the leading edge of the door (or gate) and the extra width should be maintained for a minimum distance of 1,200mm beyond it;	
• the reveal on the leading side of the door (usually the inside) has a maximum depth of 200mm;	
• the threshold is an accessible threshold;	

- lobby or porch doors should be a minimum of 1,500mm apart and there should be a minimum of 1,500mm between door swings;
- door entry controls should be mounted 900–1,000mm above floor level.

4.13.1.2 Communal passenger lifts

- Communal passenger lifts should comply with all of the following: M4(1) 1.14
    - provision of a clear landing directly in front of the lift door at every floor level;
    - a minimum clear opening width of 800mm for doors;
    - a dwell time of five seconds before the doors begin to close;
    - the system should be capable of being overridden by a door re-activating device.

4.13.1.3 Wheelchair-accessible dwellings

The following are additional optional requirements for wheelchair-accessible dwellings (also see AD-M4(1) Diagram 13):

- A clear turning circle 1,500mm in diameter should be provided inside the entrance area, behind the entrance door when closed.
- A minimum 200mm nib is provided to the following edge of the door (or gate) and the extra width created by this nib is maintained for a distance of a minimum 1,800mm beyond it.
- The door is located reasonably centrally within the thickness of the wall while ensuring that the depth of the reveal on the leading face of the door (usually the inside) is a maximum of 200mm.
- Where power-assisted opening is provided, the opening force of the door is more than 30N from 0° to 30° or more than 22.5N from 30° to 60° of the opening cycle.

4.13.2 *Combustion appliances and fuel storage systems*

• The carbon monoxide alarm should be located in the same room as the appliance, above any doors but not within 150mm of the ceiling.	J 2.36
• No permanently open vent is needed if the room containing a flueless gas appliances has a door direct to outside.	J Diagram 33
• A flue may have openings into more than one room or space to enable an explosion door to be fitted.	J 1.50

4.13.2.1 *Provisions where there is a risk of oil pollution*

• Where bund walls of a fuel storage system are part of the wall of a building enclosing the tank, any door through the walls should be above bund level.	J 5.10c

4.13.3 *Conservation of fuel and power*

4.13.3.1 *Airtightness in existing dwellings*

Note: New fabric elements that are used in existing dwellings should meet the limiting standards in AD-L(V1) Table 4.2.

• When carrying out work in existing dwellings, care should be taken to reduce unwanted heat loss through air infiltration from (in particular) doors, which should be well fitted and reasonably draught-proof.	L(V1) 4.23
• If new and replacement doors are part of an entirely new fitting, the whole of that fitting has to be: • draught-proofed; • meeting the minimum standards given in AD-L(V1) Table 4.2; • using insulated cavity closers where appropriate.	L(V1) 10.3

> • If a door is enlarged or a new one is created, the L(V1) 10.5
> combined area of windows, roof windows, rooflights
> and doors should not exceed 25% of the total floor
> area of the dwelling.

> **Note:** If the combined area of windows and doors
> exceeds 25% of the total floor area of the dwelling,
> the energy efficiency of the dwelling will need to be
> improved.

When carrying out work on an existing dwelling, care should be taken to reduce unwanted heat loss through thermal bridging.

4.13.4 Fire safety

4.13.4.1 Means of escape

All habitable rooms should be provided with suitable means for emergency egress from each storey via doors or windows.

A communicating door between rooms should be provided so that it is possible to gain access to the escape window without passing through the stair enclosure, and (when fitted) manual call points for fire alarm systems should be adjacent to exit doors.

4.13.4.2 Alarms – smoke and heat alarms and detectors

Alarm units designed for wall-mounting must be fitted above the level of doorways.

Where there is no door between a kitchen area and a stairway or circulation space by a door, the kitchen should have a heat detector or heat alarm as well as smoke alarms in the circulation space(s).

> • Smoke and heat alarms should have a standby B(V1) 1.4
> power supply, such as a battery (rechargeable or non-
> rechargeable) or capacitor.

4.13.4.3 Basements

A basement space should have one or more smoke outlets.

• Basements without smoke outlets may be vented to other spaces indirectly, by opening connecting doors.	B(V1) 16.2

• Basement storeys containing habitable rooms should have a fire escape route via one of the following:
  • an emergency escape window or external door;
  • a protected stairway leading from the basement to a final external protected door.
	B(V1) 2.16

4.13.4.4 Air-circulation systems

• Transfer grilles should not be fitted to a stair enclosure door.	B(V1) 2.9a B(V1) 3.23

4.13.4.5 Cavities and openings in compartments

Cavities in the construction of a building provide a ready route for the spread of smoke and flame.

• Cavity barriers should be provided around openings (such as doors), and at the junction between an internal cavity wall and a door assembly, forming a fire-resisting barrier.	B(V1) 8.3
• Openings in a compartment wall common to two or more buildings may be made for a fire doorset to provide a means of escape.	B(V1) 7.19

• Doors should contain vision panels:
  • where doors on escape routes divide corridors;
  • where doors are hung to swing both ways.
	B(V1) 3.97

Note: Glazing in doors should comply with AD-B(V1) Table B5.

• Openings in a protected shaft may be made for a fire doorset that provides a means of escape.	B(V1) 7.29

4.13.4.6 Electrical safety

• Connecting an electric gate or garage door to an existing isolator switch is not notifiable work, but installing a new circuit from the consumer unit to the isolator is notifiable.	P 2.7b

4.13.4.7 Escape routes

 The only projections allowed on escape routes below 2m are door frames.

• All doors on escape routes should open:
   • in the direction of escape;
   • a minimum of 90°;
   • with a swing that does not reduce the effective width of any escape route across a landing.
	B(V1) 3.90
	 B(V1) 3.95

• A fire doorset can be used to separate the dead-end portion of a common corridor from the rest of the corridor (see AD-B(V1) Diagrams 3.7 and 3.8).	B(V1) 3.36
• If two rooms only have access a single escape window, the door between these rooms should not pass through a stair enclosure to gain access to this window.	B(V1) 3.17

• Doors to external escape stairs (see AD-B(V1) Diagram 3.11) should be:
   • fire resistant;
   • fitted with a self-closing device;
   • fitted wiith fire-resisting glazing.
	B(V1) 3.68

• Single steps on escape routes should be prominently marked.	B(V1) 3.84
• A single step on the line of a doorway is acceptable; however, final exits should not present a barrier for disabled people.	
• Where the route to a final exit does not include stairs, a level threshold and, where necessary, a ramp should be provided.	
• A common corridor connecting two or more storey exits should be connected with a fire doorset fitted with a self-closing device.	B(V1) 3.35

- All habitable rooms (excluding kitchens):   B(V1) 2.1
    - should either have an opening leading to a final   B(V1)3.15
    exit or to an emergency escape door;

    - 4.5m above ground level with only one stair,   B(V1) 2.2
    should either have an emergency escape window   B(V1) 3.16
    or external door leading to a protected stairway.

- Basement storeys containing habitable rooms should   B(V1) 2.16
either have an emergency escape window or an
external door.

- Where automatic doors are placed across an escape   B(V1) 3.98
route:
    - their width and their failsafe system should be
    capable of providing an outward opening from any
    open position;
    - they will have a monitored failsafe system to open
    the doors if the mains electricity supply fails.
- As a precaution against a power failure, non-
automatic swing doors should be provided
immediately adjacent to the revolving or automatic
door, or turnstile.

Doorways providing access to a means of escape should be clearly
marked by an exit sign (in accordance with BS ISO 3864-1 and BS
5499-4).

4.13.4.7.1 DOORS ON ESCAPE ROUTES

- Doors on escape routes should not be fitted with a   B(V1) 3.91
lock, latch or bolt fastening.
- Doors on escape routes should only be fitted with
simple fastenings that are:
    - easy to operate;
    - operable from the side approached by people
    escaping;
    - operable without a key; and
    - operable without requiring people to manipulate
    more than one mechanism.

- If a secure door is operated by a code, keypad, swipe   B(V1) 3.92
  card, proximity card or biometric data, etc., and
  if the door provides escape in either direction, a
  security override mechanism should be installed on
  both sides of the door.

- Self-closing fire doors may be held open by: B(V1) C6
  - a fusible link;
  - an automatic release mechanism activated by a fire
    detection and alarm system;
  - a door closing delay device.

- Rolling shutters across a means of escape should only   B(V1) C9
  be released by a heat sensor, such as a fusible link or
  electric heat detector, in the immediate vicinity of the
  door.

4.13.4.7.2 SINGLE STEPS ON ESCAPE ROUTES

- Single steps on escape routes should be prominently   B(V1) 3.84
  marked.
- A single step on the line of a doorway is acceptable;
  however, final exits should not present a barrier for
  disabled people.
- Where the route to a final exit does not include stairs,
  a level threshold and, where necessary, a ramp should
  be provided.

4.13.4.8 Doors in compartment walls

The only opening in a compartment wall common to two or more
buildings should be for a fire escape door which should have the same
fire resistance as that of the wall.

Basic advice should be provided to occupants on how to maintain and
use fire doors and other fire equipment.

4.13.4.9 Fire doorsets

- All fire doorsets should be classified in accordance with BS EN
  13501-2 and tested in accordance with BS EN 1634-1.

- Self-closing fire doorsets should be positioned so that smoke will not affect access to more than one stairway.
- Glazing in any fire-resisting doorsets should be fire-resisting and fixed shut.
- Basic advice should be provided on how to maintain and use fire doorsets and other fire equipment.

Fire doorsets should be classified in accordance with BS EN 13501-2.

• All fire-resisting doors should have the appropriate performance given in AD-B Appendix B(V1) Table C1.	B(V1) C1
• With the exception of lift doors (which are tested from the landing side only), a smoke leakage test should be completed from each side of a doorset, separately. (See BS EN 1634-1 for further guidance.)	B(V1) C3
• All fire doors should be fitted with a self-closing device except for fire doors within flats and dwelling houses.	B(V1) C5
• Two fire doors may be fitted in the same opening provided that each door is capable of closing the opening.	B(V1) C7

Note: It should be noted that often fire doorsets do not provide any significant insulation.

• The hinge on which a fire door is hung should be made entirely from materials that have a minimum melting point of 800°C.	B(V1) C10
• All fire doorsets should be marked with the appropriate fire safety sign.	B(V1) C11 B(V1) C12
• Other than doors to and within flats and dwellings, all fire doorsets should be marked on both sides, except for fire doorsets to cupboards and service ducts, which should be marked on the outside.	

Note: Limitations on the use of uninsulated glazing in fire doorsets are given in AD-B(V1) Table B5.

- The design, construction, installation and   B(V1) C14
  maintenance of fire doorsets constructed with non-
  metallic doors should comply with BS 8214.
- Guidance on timber fire-resisting doorsets may
  be found in the 'Timber Fire-Resisting Doorsets'
  standard published by the Timber Research and
  Development Association (TRADA).
- Guidance for metal doors is given in 'Code of
  Practice for Fire-Resisting Metal Doorsets' published
  by the Door and Shutter Manufacturers' Association
  (DSMA).

4.13.4.10 Loft conversions

- If a new storey is created above 4.5m, fire-resisting   B(V1) 2.21
  doors should be provided; and where the layout is
  open plan, new partitions should be provided to
  enclose the escape route (see AD-B(V1)
  Diagram 2.2). Alternatively:
  - sprinkler protection could be provided to the   B(V1) 2.23
    open-plan areas;
  - a fire-resisting partition and door could be used to
    separate the ground storey from the upper storeys;
  - cooking facilities could be separated from the
    open-plan area with fire-resisting construction.

 In cases where it would be undesirable to replace an existing door, owing
to its historical or architectural nature, the possibility of upgrading it to
meet modern fire requirements should be investigated.

4.13.4.11 Flats

- Self-contained flats in student residences with their   B(V1) 1.11
  own entrance door should have a separate automatic
  fire detection system.
- Fire detection and alarm systems in sheltered housing   B(V1) 1.12
  flats should connect to a central monitoring point or
  alarm receiving centre.

- Every flat entrance door that is separated from the common stair, protected lobby or protected corridor (see AD-B(V1) Diagram 3.7) should have a maximum travel distance of 30m to a protected exit, as shown in AD-B(V1) Table 3.   B(V1) 3.27

- There should be no more than 30m between the flat entrance door and a protected or final exit.   B(V1) 3.27

- Bedrooms should be separated from living accommodation by fire-resisting construction and fire doorsets.   B(V1) 3.19

4.13.4.11.1 ALTERNATIVE EXITS FROM FLATS

- Any alternative exit from a flat should be remote from the main entrance door to the flat and lead to a final exit via a door: B(V1) 3.22
  - to an access corridor/lobby or common balcony;
  - to a common stair;
  - to an escape route over a flat roof;
  - to an external stair.

4.13.4.11.2 LIVE/WORK UNITS

- In flats that serve as a workplace for both occupants and people who do not live on the premises, the maximum travel distance between any part of the working area and the flat entrance door or an alternative means of escape (but not a window) should be *no more* than 18m.   B(V1) 3.24

4.13.4.12 *Firefighting shafts*

In low-rise buildings without deep basements, access for firefighting personnel is typically achieved by providing the fire service with vehicle access and means of escape.

A building with a storey more than 18m above the fire and rescue service vehicle access level should have one or more firefighting shafts, each containing a firefighting lift.

- The doors of a firefighting lift landing should be a   B(V1) 15.8
  maximum of 7.5m from the door to the firefighting
  stair (see AD-B(V1) Diagram 15.1).
- The stair and lobby of a firefighting shaft should have
  a means of venting smoke and heat.

4.13.4.13 Garages

- Fire doors serving an attached or integral garage   B(V1) 5.7
  should be fitted with a self-closing device.
- If a door is provided between a dwelling and the
  garage (see AD-B(V1) Diagram 5.1):
  - the floor of the garage should be laid to allow fuel
    spills to flow away from the door to the outside;
  - the door opening should be positioned at least
    100mm above garage floor level.

4.13.5 Protection from falling, collision and impact

4.13.5.1 Doors on landings

A landing may include part of the floor of the building and should be
kept clear of permanent obstructions.

- Doors to cupboards and ducts that open over a   K 1.21
  landing at the top of a flight of stairs, as shown in   K 1.24
  AD-K Diagram 1.7, should be kept shut or locked
  shut when under normal use.
- Doors that swing across a landing at the bottom
  of a flight of stairs should comply with AD-K
  Diagram 1.8.

4.13.5.2 Glazing in critical locations

For doors and door side panels, the risk is greatest for glazing between
floor and shoulder level when near to door handles and push plates, especially when normal building movement causes doors to stick.

- Glazing material installed in a door or in a door side   K 5.4
  panel which has a pane width exceeding 900mm
  should satisfy the requirements of Class 2 of BS EN
  12600 or Class B of BS 6206.

4.13.6 Security

AD-Q (*Security*) is a fairly new requirement for England and Wales and
came into effect in October 2015.

4.13.6.1 General guidance

All easily accessible doors (including garage and communal entrance
doors) that provide access into a dwelling, or into a building containing a
dwelling, should be secure doorsets. Where access to the dwelling can be
gained via an interconnecting doorset from the garage, then either the
garage doorset (pedestrian and vehicular) or the interconnecting doorset
should be a secure doorset.

4.13.6.2 Design of secure doorsets

- Secure doorsets should either: Q 1.2
  - meet the security requirements of British
    Standards publication PAS 24;
  - be designed and manufactured in accordance
    with AD-Q Appendix B.
- The main entrance door should have a door viewer,   Q 1.4
  unless other means exist to see callers (such as clear
  glass within the door or a window next to the doorset.
- Doorsets should also have a door chain or door
  limiter.

In some situations, a door chain or limiter is not appropriate; for example
where a warden may need emergency access to residents in sheltered
housing.

Alternative caller-identification measures, such as electronic audio-visual
door entry systems or CCTV can be used to identify visitors.

4.13.6.3 Provision of essential information

• Basic information on fire doorsets fitted with a self-closing device and other doors equipped with relevant hardware is required.	B(V1) 17.3
• Records should include details of passive fire-safety measures, including fire doorsets fitted with a self-closing device and other doors equipped with relevant hardware (such as electronic security locks).	B(V1) 17.6

4.13.6.4 Material change of use and change to energy status

If there is a material change of use and/or a change to energy status:

• New or replaced doors should meet the limiting standards in AD-L(V1) Table 4.2.	L(V1) 11.7
• Any door that separates a conditioned space from an unconditioned space, that has a U-value higher than $3.30\text{W}/(\text{m}^2\cdot\text{K})$, should be replaced.	

4.13.6.5 U-values

U-values should be assessed using the methods and conventions in BR 443.

• The required U-value for a new door (including a glazed door) in a new dwelling is $1.6\text{W}/(\text{m}^2\cdot\text{K})$ – see AD-L(V1) Table 4.1.	L(V1) 4.7
• U-values should be assessed for the combined performance of the door, any glazing and the frame.	L(V1) 4.1
• The U-value of a door can be calculating by using:	L(V1) 4.4
• the actual size and configuration of the door; or	
• a standard door 1.23m wide × 2.18m high; or	
• the default value from the Standard Assessment Procedure Table 6e.	
• New thermal elements should meet the standards in AD-L(V1) Table 4.2.	L(V1) 10.7

Note: If fully glazed external pedestrian doors cannot meet the requirements of AD-L(V1) Table 4.2 (because of the need to maintain the character of the building), then either the fittings should not exceed a centre pane U-value of $1.2\,W/(m^2{\cdot}K)$ or, if there is any single glazing, it should be supplemented with low-emissivity secondary glazing.

4.13.6.6 Limiting solar gains

• In order to limit solar gains, the maximum glazing in doors (particularly conservatories) should be limited as indicated in AD-O Tables 1.1 and 1.2.	O 1.6

4.13.6.7 Noise

• In locations where external noise may be an issue, there is every likelihood that openings will be closed during sleeping hours (11pm to 7am), which in turn will cause overheating.	O 3.2
• Doors and windows that can be opened wider than 100mm may form part of the overheating mitigation strategy and to counteract some of the effect of noise, open doors can be made secure by using fixed or lockable louvred shutters or a fixed or lockable grille or railing.	O 3.9 O 3.7

4.13.6.8 Removing excess heat

• Excess heat can be removed from residential buildings by opening windows and doors, and the effectiveness of this method is improved by cross-ventilation.	O 2.10 O 3.1

Note: The minimum free areas for buildings (or parts of buildings) with or without cross-ventilation are detailed in AD-O Tables 1.3 and 1.4.

4.13.7 Site preparation and resistance to moisture

4.13.7.1 Cladding

• Care should be taken with the junctions between cladding and door openings to prevent moisture ingress.	C 5.24

4.13.7.2 Joints between doors and walls

• The joint between walls and door and window frames should:
  • not be damaged by precipitation;
  • resist the penetration of precipitation to the inside of the building.
	C 5.29

• Damp-proof courses should be provided to direct moisture towards the outside (particularly under doors which do not form a complete barrier to the transfer of rain and snow).	C 5.30
• In areas of the country subject to driving rain, rebates should be used in all door reveals, and the frame should be set back behind the outer leaf of masonry, which should overlap it.	C 5.32

Note: Insulated finned cavity closers may be used instead of rebates.

4.13.7.3 External walls – resistance to surface condensation and mould growth

• Junctions between the elements and details of openings, such as doors in an external wall, should be designed to Accredited Construction Details (available to download from: www.planningportal. gov.uk).	C 5.36

4.13.7.4 *Rainwater drainage*

• Eaves drop systems should be designed to prevent rainwater from entering doorways.	H 1.13

4.13.7.5 *Air permeability testing*

• Older houses are unlikely to achieve an air permeability of less than 5.0m³/(h.m²) at 50 Pa, unless the building fabric has been substantially upgraded, such as:
   • fitting all external doors with integral draught seals and letterbox seals;
   • providing internal and external sealing around external doors and window frames.
	J App F.2

4.13.8 *Resistance to the passage of sound*

4.13.8.1 *Corridor walls and doors*

• All doors should have good perimeter sealing (including the threshold) and:
   • a minimum mass per unit area of 25kg/m²; or
   • a minimum sound reduction index of 29 dB Rw (measured according to BS EN ISO 140-3:1995 and rated according to BS EN ISO 717-1:1997).
	E 2.26
	E 4.20
	E 6.6

• Noisy parts of the building should preferably have a lobby, double door or high performance doorset to contain the noise.	E 2.27, E 4.21 & E 6.7

Note: Sound insulation will be reduced by the presence of a door.

4.13.8.2 *Cupboards under the stairs*

• A small, heavy, well fitted door should be used on a cupboard under all, or part, of a stair.	E 4.37

4.13.8.3 Insulation

- Doors should be installed in such a way that: L(V1) 4.14
  - the thermal integrity of the insulated plane is maintained;
  - door units should be located with an overlap between the inner face of the unit and the inner face of the external leaf of 50mm;
  - fully insulated and continuous cavity closers should be used;
  - insulation should be installed within the threshold zone.

4.13.8.4 Lightweight doors

Lightweight doors provide a lower standard of sound insulation than walls. One method of improving sound insulation is to use a doorset.

4.13.8.5 Tests for soundproofing

- When conducting soundproof tests, doors should be   E 2.16
  closed and kitchen units, cupboards, etc., on all walls E 2.17
  should have their doors open and be unfilled.

Although not specifically required in test reports, it may be useful to have a description of the building including dimensions and position of any doors in external walls.

4.13.9 Ventilation

AD-F sets minimum standards for purge ventilation for rapidly diluting indoor air pollutants and extracting water vapour where necessary in habitable rooms in dwellings. However, AD-O may require a higher standard to enable excess heat to be removed. In this case, the higher of the two standards should be followed.

- Openable doors can provide purge ventilation in   F1 1.28
  habitable rooms.
- To ensure ventilation throughout the dwelling, doors   F 3.27
  should be undercut by 10mm (if the floor is fitted) or
  20mm (if the floor finish is not fitted).

- The minimum height × the width of an external   F1 1.30
  door (such as a patio door) should be 1/20 of the
  floor area (see AD-F1 Table 1.4 for details).

If a room contains more than one external door (or a combination of
one external door and at least one openable window), the areas of all the
opening parts may be added together to achieve at least 1/20 of the floor
area of the room.

4.13.9.1 Purge ventilation

Purge ventilation is used to remove high concentrations of pollutants
(such as fumes from painting) and water vapour; and a system for purge
ventilation should be provided in each habitable room.

- Purge ventilation should be capable of extracting   F(V1) 1.27
  at least four air changes per hour (4 ach) per room
  directly to the outside.
- If purge ventilation is delivered through openings in a   F(V1) 1.28
  habitable room (e.g. doors or windows), the minimum
  total area of the openings should be 1/20 of the floor
  area of the room (see AD-F(V1) Table 1.4).
- It may be possible to achieve four air changes per   F(V1) 1.30
  hour with smaller openings, depending on design or
  external climate, in which case expert advice should
  be sought.

4.13.9.2 Ventilation of a habitable room or conservatory through another room

- A habitable room which does not have any outside   F(V1)
  windows may be ventilated through either another   1.40–1.44
  habitable room or a conservatory provided that:
  - the habitable room or conservatory has openings
    to the outside which provides both purge and
    background ventilation; and
  - there is a permanent opening between the two
    rooms with a minimum area of 1/20 of the
    combined floor area of the two rooms. (See
    AD-F(V1) Diagram 1.3.)

4.13.9.3 Sanitory conveniences

- A door should separate a place containing a sanitary
  convenience (together with its hand-washing facilities)
  from any place used for the preparation of food.

  G 4.10
  G 4.17

4.13.9.4 Energy efficiency

Energy-efficiency measures must be provided that limit the heat loss through the doors, etc., by suitable means of insulation.

Note: The minimum total area of background ventilators in each room should follow the guidance shown in AD-F(V1) Table 1.7.

4.13.9.4.1 INSTALLING ENERGY-EFFICIENCY MEASURES

- Replacing more than 30% of the total existing doors
  is categorised as a major measure.

  F(V1)
  Table 3.1

- When considering the impact on ventilation when
  carrying out works in existing dwellings, AD-F(V1)
  Table 3.1 should be consulted, and these measures
  should then be compared to AD-F(V1) Diagram 3.1
  which categorises the impact on ventilation.

  F(V1) 3.09

AD-F(V1) Table D1 provides a complete checklist for ventilation provision in existing dwellings.

4.13.9.5 Adding a habitable room (not including a conservatory) to an existing dwelling

- If the additional room is connected to an existing
  habitable room which has no windows opening to
  the outside, the requirement for adequate ventilation
  can be met with only background ventilation – unless
  the total background ventilator equivalent area is less
  than 5,000mm².

  F(V1) 3.17

4.13.9.6 Addition of a conservatory or a porch to an existing building

• Where a dwelling is extended by adding a conservatory or porch, the work is exempt from the energy-efficiency requirements provided that the extension is at ground level and the floor area of the extension does not exceed 30m².	F(V1) 3.21

4.13.9.7 Building work on existing dwellings

• When building work is carried out that will affect the ventilation of the existing dwelling (such as replacing a door), then the ventilation of the dwelling should be no be less satisfactory than before the work was carried out.	F(V1) 3.2

Note: The minimum total area of background ventilators in each room should follow the guidance shown in AD-F(V1) Table 1.7.	F(V1) 3.13

4.13.9.8 Dwelling extensions

When carrying out work in existing dwellings, care should be taken to reduce unwanted heat loss through air infiltration. This is particularly important when a dwelling is extended.

• The U-value of replacement fabric in an existing dwelling should: • be no worse than that of the element being replaced; and • meet the limiting standards listed in AD-L(V1) Table 4.2. • The U-value of a replacement door should be calculated using the actual size and configuration of the door.	L(V1) 10.7

- When a dwelling is extended, doors with greater than 60% glazed area, should have a maximum U-value of $1.2W/(m^2 \cdot K)$ and a frame actor of 0.7. — L(V1) Table D1
- The total area of doors (and windows and roof windows) should not be greater than 25% of the floor area of the extension. — L(V1) 10.7
- If an existing dwelling has a total useful floor area of over $1,000m^2$, the overall energy efficiency of the dwelling might need to be improved if any proposed work includes an extension that increases the capacity of any fixed building service. — L(V1) 12.1

4.13.9.9 Historic buildings

Particular issues relating to building work in historic buildings will warrant sympathetic treatment. In this case, advice from experts could be beneficial, particularly when restoring the historic character of a building that has been subject to previous inappropriate alteration (e.g. replacement doors).

4.13.9.10 Major renovation

Major renovation means the renovation of a building where more than 25% of the surface area of the building envelope undergoes renovation.

Note: When assessing whether the area proportion constitutes a major renovation of a building, the surface area of the whole of the external building envelope (including doors) should be taken into account.

Requirements for buildings other than dwellings – Additional requirements

The basic requirements for doors in buildings other than dwellings are the same as for dwellings, with the addition of the following requirements (which are particularly relevant for ambulant disabled people).

Note: Unless specified otherwise, all doors should be fire doors.

The Building Regulations state that a fire door or window should have an unobstructable area, which needs to be at least 0.33m² and at least 450mm high and wide.

For access doors to comply with Building Regulations, they must have a step no greater than 15mm and have a clear opening of 775mm.

4.13.10 *Access to and use of buildings*

• Wheelchair users should have adequate space to stop and pass through doors without having to reverse.	M2 1.23
• The landing at the foot and head of a ramp should be clear of any door swings.	M2 1.26h
• Intermediate landings should be clear of any door swings.	M2 1.26i
• No doors should swing across landings.	M2 1.33e
• Doors in frequent use should have a level threshold.	M2 2.6
• All door entry systems should be accessible.	M2 2.7f

4.13.10.1 *Doors to accessible entrances*

• Doors to the principal entrance should be accessible to all.	M2 2.8
• Entrance doors may be either manually or power operated under manual or automatic control.	
• Entrance doors should be capable of being held closed when not in use.	
• An automatic sliding door may be used.	M2 2.10
• Once open, all doors to accessible entrances should be wide enough to allow unrestricted passage.	M2 2.11
• Self-closing doors to accessible entrances should ideally be power operated.	M2 2.13
• The effective clear width through a single leaf door should be in accordance with AD-M Table 2 and Diagram 9.	

4.13.10.2   Glass doors, screens and entrance lobbies

• The presence of a door should be apparent when it is shut and when it is open.	M2.2.23
• The minimum length of the lobby is related to the size and projection of the chosen door. (See AD-M2 Diagram 10.)	M2 2.29a–c
Note: Where both doors of a lobby are automatic sliding doors, the length can be reduced.	M2 2.27
• Any reception point should be easily identifiable from the entrance doors.	M2 3.6b

4.13.10.3   Internal doors

• If internal doors are required, the use of self-closing devices should be minimised.	M2 3.7
• Low-energy powered door systems may be used in locations not subject to frequent use or heavy traffic.	
• The presence of doors, whether open or closed, should be apparent.	M2 3.8
• Internal doors should comply with the following:	M2 3.10

- • Internal doors should comply with the following:
  - • fire doors may be held open with an electro-magnetic device;
  - • fire doors, particularly to individual rooms, may be fitted with swing-free devices that close when activated by smoke detectors or a fire alarm system (or when the power supply fails);
  - • door-opening furniture should contrast visually with the surface of the door;
  - • the opening force at the leading edge of the door is not more than 30N from 0^0;
  - • there should be an unobstructed space of at least 300mm on the ***pull*** side of the door;
  - • door frames should contrast visually with the surrounding wall;
  - • glass doors should be clearly defined with appearance and indication;

- fully glazed doors should be clearly differentiated from any adjacent glazed wall or partition;
- low-energy powered swing door systems should be able to operate in a manual or power-assisted mode.

• The effective clear width through a door should be in accordance with AD-M2 Table2 and Diagram 8.	M2 3.10
• Doors from unisex wheelchair-accessible toilets may project into a corridor (if it is not a major access or an escape route), provided the corridor is 1,800mm wide at that point – but these doors should be recessed.	M2 3.14h

4.13.10.4 *Manually operated non-powered entrance doors*

• A space alongside the leading edge of a door should be provided to enable a wheelchair user to reach and grip the door handle.	M2 2.15
• Door furniture on manually operated non-powered doors should be easy to operate by people with limited manual dexterity.	M2 2.16

• Manually operated non-powered entrance doors will satisfy the requirement if:
  • the opening force at the leading edge of the door is not more than 30N;
  • latches can be operated with one hand using a closed fist, e.g. a lever handle;
  • door-opening furniture contrasts visually with the door surface and is not cold to the touch.
	M2 2.17

4.13.10.5 *Powered entrance doors*

• Revolving doors are not considered accessible.	M2 2.20

• Powered entrance doors should:
  • ensure swing doors that open towards people approaching the doors have visual and audible warnings of their automatic operation;
  • have a controlled sliding, swinging or folding action;
	M2 2.21

- have a safety stop that is activated if the doors begin to close when a person is passing through;
- revert to manual control or failsafe in the open position in the event of a power failure;
- not project into any adjacent access route when open;
- have manual controls that are operable with a closed fist;
- contrast visually with the background against which they are seen.

4.13.10.6 Sanitary accommodation

- WC cubicle doors should be operable by people with limited strength or manual dexterity.   M2 5.3
- Doors to cubicles should be capable of being opened if a person has collapsed against them while inside the cubicle.
- All doors to WC cubicles and wheelchair-accessible unisex toilets should open outwards.

- WC compartment doors, and doors to wheelchair-accessible unisex toilets, changing rooms or shower rooms should be fitted with light action privacy bolts.   M2 5.4
- Doors to wheelchair-accessible unisex toilets, changing or shower rooms should have an emergency release mechanism.
- Doors should not obstruct emergency escape routes when they are open.

- Doors to compartments for ambulant disabled people should be outward opening and have a horizontal closing bar fixed to the inside face.   M2 5.14c

4.13.10.7 Wheelchair-accessible bedrooms

The maximum permissible opening force of the door from access corridor to a wheelchair-accessible bedroom is 30N, and the width and actual accessibility of the entrance should be in accordance with AD-M2 Diagram 2.

• A proportion of wheelchair-accessible bedrooms should have a connecting door to an adjacent bedroom to enable a wheelchair user to visit (or be assisted by) companions in other bedrooms.	M2 4.21
• For people with limited manual dexterity, electronic card-activated locks for bedroom entrance doors and lever taps in sanitary accommodation should be provided.	M2 4.23
• Swing doors for built-in wardrobes and other storage systems should open through 180°.	M2 4.24b
• Handles on hinged and sliding doors should be easy to grip and operate, and should contrast visually with the surface of the door.	M2 4.24c
• A balcony should have a door which complies with AD-M2 Table 2. • The door threshold to the balcony should be level.	M2 4.24p

4.13.10.8 *Wheelchair-accessible lifts*

• Lift door systems should allow time for people to enter or leave the lift without coming into contact with closing doors.	M2 3.30
• Lift cars and lifting platforms may be provided with opposing doors to allow a wheelchair user to leave without reversing out.	M2 3.33 M2 3.41
• Power-operated horizontal sliding doors should provide an effective clear width of at least 800mm (nominal).	M2 3.34e
• Doors should be fitted with timing devices and re-opening activators.	M2 3.34f

4.13.11 Conservation of fuel and power

• Thermal bridges should be avoided around door openings.

• Doors which are incorporated in an extension should be insulated and draught-proofed to at least the same extent as in the existing building.

• Where doors are to be provided, their performance should be no worse than:

- pedestrian doors with a U-value of $2.2W/(m^2.K)$;
- pedestrian doors where the door has more than 60% of its external face area glazed with a U-value of $1.8W/(m^2.K)$;
- high-usage entrance doors with a U-value of $3.5W/(m^2.K)$.

Full details for calculating the U-value of doors in different situations is contained in BS EN 14351-1.

Note: New or enlarged pedestrian doors should not exceed the values listed in AD-L2B Table 2.

4.13.11.1 *Carbon dioxide (CO₂) monitoring*

• CO_2 monitors **must** be placed at breathing height and away from doors or ventilation openings whenever possible.	F(V2) 1.23 App C

4.13.11.2 *Providing ventilation*

• Extract ventilation should operate in rooms without openable doors (e.g. a storeroom) while the room is occupied and for at least 15 minutes after the occupants have left the room.	F(V2) 1.31
• If building work (such as replacing a door) is carried out, the building's ventilation system should be unaffected and certainly not less satisfactory than before the work was carried out.	F(V2) 3.2

4.13.12 Structure

Note: In some non-residential buildings, final exit doors have security locks that are used only when the building is empty.

4.13.12.1 *U-values*

Note: Any door that separates a conservatory or porch from the building that has been retained or, if removed, replaced by another door, is exempt from the following requirements, if the floor area of the extension does not exceed 30m².

• If a pedestrian door intended for a new and existing building cannot meet the requirements of AD-L(V2) Table 4.1, then it should have a U-value of 1.2W/(m²·K) and single glazing should be supplemented with low-emissivity secondary glazing.	L(V2) 4.6
• Limiting U-values in new and existing buildings and air permeability in new buildings (as stated in AD-L(V2) Table 4.1) are: • for pedestrian doors – 1.6W/(m²·K); • for vehicle access and similar large doors – 1.3W/(m²·K); • for high-usage entrance doors – 3.0W/(m²·K).	L(V2) 10.9 c

The U-value of a door should be assessed by calculating:

* The actual size of the door.
* The actual configuration of the door (normally 1.23m wide × 2.18m high).

4.13.12.2 *Consequential improvements*

• Consequential improvements may include replacing existing doors that have a U-value higher than 3.30W/(m²·K).	L(V2) Table D1 Table D2

4.13.13 *Material change of use*

• If there is a material change of use (or a change to energy status) to existing doors, roof doors or rooflights and they separate a room from an unconditioned space (or the external environment) and they have a U-value higher than 3.30W/(m²·K) – then they should be replaced to meet the limiting standards shown in AD-L(V2) Table 4.1.	L(V2) 11.7

4.13.13.1 Small single-storey non-residential buildings

- The size and location of openings should be in
  accordance with **AD-A** Diagram 17.
  A 2C38ii
- A single leaf door is permitted in a small single-storey
  building.
- Unless there is a corner pier, the distance from a door
  to a corner should not be less than 390mm.

4.13.14 Fire safety

Unless specified otherwise, all doors should be fire doors that fully comply
with BS 476.

4.13.14.1 Fire safety information

- An as-built plan of the building should be available,
  showing fire doorsets (with or without a self-closing
  device) and all of the other doors and their usage.
  B(V2) 19.3
- Additional information for complex buildings should
  include all passive fire-safety measures, including
  fire doorsets fitted with a self-closing device and
  other doors equipped with relevant hardware (e.g.
  electronic security locks).
  B(V2) 19.6
- Self-closing fire doors should be fitted with an
  automatic release mechanism.
  B(V2) 2.15

Note: Manual call points for an electrically operated fire warning system
should be sited adjacent to exit doors.

4.13.14.2 Cavity barriers

Cavities that have been made during the construction of a building provide a ready route for the spread of smoke and flame; consequently, they
need to be restricted by cavity barriers (which can be formed by a door
frame if the frame is constructed of steel or timber).

• Cavity barriers (see AD-B(V2) Diagram 9.1) should be provided: • to divide cavities; • to close the edges of cavities (especially around doors);	B(V2) 9.2
• at the junction between an internal cavity wall and every compartment door assembly which forms a fire-resisting barrier.	B(V2) 9.3
• Openings in cavities (or a compartment wall common to two or more buildings): • should be limited to those containing a fire doorset with the same fire resistance as the wall;	B(V2) 8.30 & 8.31
• may be made for fire doorsets which have at least 30 minutes fire resistance.	B(V2) 9.17

4.13.14.3 *Means of escape*

The number of escape routes and exits that should be provided depends on the number of occupants in the room, tier or storey, and the limits on travel distance to the nearest exit, as given in AD-B(V2) Table 2.1.

 Fire resistance test criteria for doors and standards of performance are summarised in AD-B(V2) Table C1.

• Doors on escape routes should either be fitted with a lock, latch or bolt fastening, or a simple fastening that is easy to operate without a key and openable from the side approached by people escaping.	B(V2) 5.7
• Doors on escape routes should be readily openable. (For further guidance, see AD-B(V2) Appendix C.)	B(V2) 5.6
• Doors of fire-resisting construction (minimum RE 30) should be provided within 1,800mm of an escape route's place of safety (such as the foot of the stairway).	B(V2) 3.32
• A door above an external escape stairway does not require fire resistance. (See AD-B(V2) Diagram 3.4.)	B(V2) 3.33
• Every doorway providing access to a means of escape, other than exits in ordinary use, should be marked by an exit sign in accordance with BS ISO 3864-1 and BS 5499-4.	B(V2) 5.28

• Escape routes should have a minimum clear headroom of 2m.	B(V2) 5.16
• External escape stairs should have fire-resisting doors which are fitted with self-closing devices. (See AD-B(V2) Diagram 3.4.)	B(V2) 3.33
• Revolving or automatic doors that are placed across escape routes should either be: • outward facing, failsafe automatic doors (with a monitored failsafe system to open the doors if the mains electricity supply fails); or • non-automatic swing doors immediately adjacent to the revolving or automatic door.	B(V2) 5.15
• A corridor that is more than 12m long, connects two or more storey exits, and provides access to an alternative escape route should be divided by fire doorsets fitted with self-closing devices.	B(V2) 2.26

4.13.14.4 *Protection of escape routes*

• Fire doorsets are required in protected shafts (see AD-B(V2) Diagram 8.3).	B(V2) 8.3
• The doors at both ends of a service area should be self-closing fire doorsets. (See AD-B(V2) Diagram 2.4.)	B(V2)
• The use of uninsulated glazed elements in fire-resisting doors is described in AD-B(V2) Section 13.	B(V2) 13
• Doors on escape routes (both within and from the building) should be readily openable.	B(V2) 5.6
• A lock, latch or bolt fastening should not be fitted to doors on escape routes.	B(V2) 5.9

In locations such as hotel bedrooms, locks may be fitted that are operated from the outside by a key and from the inside by a knob or lever.

• A secure door operated by a code, combination, swipe or proximity card, etc., should be capable of being overridden from the side approached by people making their escape.	B(V2) 5.7

- Electrically powered locks should return to the   B(V2) 5.8
  unlocked position on operation of the fire alarm
  system, loss of power, system error or on activation
  of a manual door release unit.
- There should be a manual door release unit on **both**
  sides of a door that provides escape in either direction.
- Doors on escape routes from rooms with an occupant   B(V2) 5.9
  capacity of more than 60 in places of assembly, shop
  and commercial buildings should either **not** be fitted
  with a lock, latch or bolt fastening, or be fitted with
  panic fastenings in accordance with BS EN 1125.

The door of any doorway or exit should be hung to open in the direction
of escape.

- Doors on escape routes should be hung to open not   B(V2) 5.15
  less than 90° with a swing that is clear of any change   B(V2) 5.12
  of floor level (a single step on the line of a doorway is
  acceptable).

- Doors that open towards a corridor or a stairway   B(V2) 5.13
  should be sufficiently recessed to prevent its swing
  from encroaching on the effective width of the
  stairway or corridor.

- Vision panels are required where doors on escape   B(V2) 5.14
  routes sub-divide corridors, or where any doors swing
  both ways.

4.13.14.5 Residential care homes

- All bedrooms in a care home should have fire-   B(V2) 2.42
  resisting doors.

- Self-closing doors should not present an obstacle to B(V2) 2.45
  the residents of the building.
- Door-closing devices for fire doors should take
  account of the needs of residents.
- Door hardware should be appropriate (e.g. bedrooms
  with free-swing door closers and circulation spaces
  with hold-open devices).

- Where a sprinkler system is provided, fire doors to   B(V2) 2.46
  bedrooms do not need to be fitted with self-closing
  devices.

4.13.15 Protection from falling, collision and impact

• If, during normal use, any door (other than a fire escape door) swings out by more than 100mm towards an access route, it should be protected as shown in AD-K Diagram 10.2. In addition:	K 10.2
• doors should not swing across landings;	K 1.23
• the foot and head landings of a ramp should be greater than 1,200mm long and clear of any door swings;	K 2.13
• intermediate landings should be greater than 1,500mm long and clear of any door swings or other obstructions.	

4.13.15.1 Critical locations

Critical locations (see AD-K Diagram 5.1) often include large, uninterrupted areas of transparent glazing in the internal or external doors of shops, showrooms, offices, factories, and public or other non-domestic buildings. People moving in or around a building might not see door glazing in these critical locations and can collide with it. In order to avoid this happening:

• The presence of a door should be apparent when it is shut and when it is open.	M2 2.23
• Glazing should be immediately apparent.	K 7.3
• An alternative method of glazing, such as mullions, transoms, door framing or large pull or push handles should be considered (see AD-K Diagram 7.1).	

Glass doors and/or glazed screens can also be provided at two levels (see AD-K Diagram 7.2).

• Glass doors/glazed screens should:	K 7.4
• contrast visually with the background;	
• contain a logo or sign or a decorative feature;	
• be clearly marked with a high-contrast strip at the top and on both sides.	
• Where glass doors may be held open, they should be protected with guarding.	

4.13.15.2 *Doors and gates*

Doors and gates should be constructed so that:

• They include vision panels towards the leading edge of the door, as shown in AD-K Diagram 10.1.	K 10.1a
• Upward-opening doors and gates are fitted with a device to stop them falling.	K 10.1c
• Power-operated doors and gates include safety features (e.g. a power switch operated by a pressure-sensitive door edge).	K 10.1d
• The stop switch is readily identifiable and accessible.	
• In the event of a power failure, the facility for manual or automatic opening is retained.	

4.14 Access routes

Reminder: The subsections include details for buildings other than dwellings only where they differ from the requirements for dwellings.

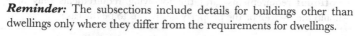

Subsection	Description	Page
Requirements for dwellings		
4.14.1	Access to and use of buildings	486
4.14.2	Site preparation and resistance to contaminants and moisture	488
4.14.3	Drainage and waste disposal	489
4.14.4	Fire safety	491
4.14.5	Ventilation	494
4.14.6	Sanitary appliances	494
4.14.7	Combustion appliances and fuel storage systems	495
4.14.8	Protection from falling, collision and impact	496
4.14.9	Conservation of fuel and power	498
4.14.10	Infrastructure for the charging of electric vehicles	499

Subsection	Description	Page
Requirements for buildings other than dwellings – **Additional requirements**		
4.14.11	Disabled access	501
4.14.12	Fire safety	502

Reminder: Free downloads of all the diagrams and tables referenced in the subsections are available on-line at the Governments' Planning Portal: www.gov.uk/government/collections/approved-documents.

Requirements for dwellings

4.14.1 Access to and use of buildings

In 2016, AD-M was updated and now wheelchair-user dwellings are **'optional** requirements' and the following requirements only apply should the **owner** of the property desire this to be the case.

4.14.1.1 Approach to a dwelling

Note: The folowing provisions do not apply beyond the curtilage of the development.

• Approach routes should be level, gently sloping, or, where necessary, ramped.	M 1.4
• The approach route should be safe and convenient for everyone.	M 1.3

On steeply sloping plots, a stepped approach can be used.

• Where a driveway forms all, or part of, the approach route, an additional allowance of at least 900mm wide should be provided so that a wheelchair user can pass a parked car.	M 1.6
• A ramped approach should have a minimum clear width of 900mm.	M 1.7
• Where it is not possible to achieve step-free access to any private entrance, a stepped approach is acceptable if every flight with three or more risers has a suitable handrail to one side.	M 1.8

The principal communal entrance door of the building containing the dwelling should ensure that:

• The entrance door has a minimum clear opening width of 775mm.	M 1.9
• The entrance threshold is an accessible threshold. • The ground surface (or entrance flooring) does not impede wheelchairs.	M 1.14
• To provide easy access between rooms on the entrance storey, a stepped change of level within the entrance storey should be avoided where possible.	M 1.16

For private entrances where a step into the dwelling is unavoidable, the rise should be no more than 150mm.

• To enable access into habitable rooms and to a WC in the entrance storey, door and hall widths should comply with all of the requirements shown in AD-M Diagram 1.2.	M 1.15

• To enable easy access to a WC:
  • the door should open outwards and have a clear opening width in accordance with AD-M Table 1.1;
  • there should be clear space to access the WC in accordance with AD-M Diagram 1.1.
	M 1.17

4.14.1.2 Communal lifts and stairs

• Where a lift is provided, it should be suitable for a wheelchair user.	M 1.10
• Where the dwelling is on an upper floor and **does not** have lift access, the stair provided should be a general access stair.	M 1.12
• Where the dwelling is on an upper floor and **does** have lift access, the stair provided may be a utility stair.	

4.14.1.3 New dwellings

• Reasonable provision should be made in a new dwelling for most people, including wheelchair users, to approach and enter the dwelling and to access habitable rooms and sanitary facilities on the entrance storey.	M4(1)

4.14.2 Site preparation and resistance to contaminants and moisture

Note: Those parts of the land associated with the building should be improved (where necessary) to meet the requirements of the Building Regulations.

4.14.2.1 Accessible thresholds

• Where an accessible threshold is provided (see AD-C Figure 6.13.3), the fall should be between 1 in 40 and 1 in 60 in a single direction away from the doorway, and the sill (leading up to the door threshold) should have a maximum slope of 15°.	C 5.33

4.14.2.2 Sub-soil drains

• If an active sub-soil drain is cut during excavation, and if it passes under the building, it should have access points outside the building.	C 3.3

If floor levels need to be nearer to the ground (i.e. to provide level access), sub-floor ventilation can be provided through offset (periscope) ventilators.

4.14.3 Drainage and waste disposal

4.14.3.1 General requirements

• Buildings should **not** be constructed over a manhole or inspection chamber or other access fitting on a sewer.	H4 1.3
• Access points to sewers (serving more than one property) should be in places where they are accessible and apparent for use in an emergency.	
• Air admittance valves (used for sanitary pipework) should be accessible for maintenance and should be removable to give access for clearance of blockages.	H1 1.33
• Special measures for controlling rodents having access to buildings must be controlled.	H1 2.22

4.14.3.2 Access for clearing blockages

• Access points should be provided only if blockages could not be cleared without them, and these should be removable to give access for clearance of blockages.	H1 2.13
• If a trap forms part of an appliance, the appliance should be removable.	H1 1.6
• All other traps should be fitted directly after the appliance and should be removable or be fitted with a cleaning eye.	
• Pipes should be laid to even gradients and any change of gradient should be combined with an access point.	H1 2.19
• Rodding points should be provided to give access to any lengths of discharge pipe which cannot be reached by removing traps or appliances with internal traps.	H1 1.25 H1 1.34
• Rainwater drainage should be provided to paved areas within the curtilage of a building.	H3 0.1

4.14.3.3 Greywater and rainwater storage tanks

- Rainwater drainage should be provided to paved areas within the curtilage of a building.   H3 0.1
- Greywater and rainwater tanks should be provided with access for emptying and cleaning.   H2 1.70c
- Access covers should be of durable quality (having regard to the corrosive nature of the tank contents) and should be lockable or otherwise engineered to prevent personnel entry.

4.14.3.4 Septic tanks

- Septic tanks should be:
  - provided with access for emptying and cleaning;   H2 1.24
  - sited within 30m of vehicle access and where they can be emptied and cleaned.   H2 1.17
- The inlet and outlet pipes of a septic tank should be provided with access for sampling and inspection in accordance with BS 5395-3.   H2 1.23

4.14.3.5 Cesspools siting

- Cesspools should be sited within 30m of vehicle access and where they can be emptied and cleaned.   H2 1.60
- Cesspools should have no openings except for the inlet, access for emptying and ventilation.   H2 1.62

 The inlet of a cesspool should be provided with access for inspection.

4.14.3.6 Refuse storage

- Access should be provided from the building to the place for storing refuse, and from the place of storage to the collection point.   H6 1.3
- Dwellings should have access to where movable, individual or communal waste containers can be stored.   H6 1.2

• In high-rise domestic developments, chutes (if provided) should have close fitting access doors at each storey.	H6 1.18
• Secure containers with close fitting lids should be used in order to prevent access by vermin.	H6 1.16
• Storage areas should not interfere with pedestrian or vehicle access to buildings.	H6 1.12
• The collection point should be reasonably accessible to the size of vehicles typically used by the waste collection authority.	H6 1.11

 Note: It is unlawful to obstruct access provided for removal of waste without the consent of the Local Authority.

4.14.4 Fire safety

 A building should be capable of being easily and rapidly evacuated in case of fire.

4.14.4.1 Access for fire appliances and firefighters

 Note: If additional means of escape or provisions for firefighting access are planned, then you should seek advice from the fire and rescue service as early as possible.

• There should be access for fire appliances to buildings and facilities in and around the building to assist firefighters in the saving of life of people. (See AD-B(V1) Diagram 13.1 and Table 13.1.)	B(V1) 13.1–13.4
• Firefighting shafts should serve all storeys through which they pass and any basement storeys more than 10m below the fire and rescue vehicle access level.	B(V1) 15.4
• Buildings 18m or more above the fire and rescue service vehicle access level, and with a floor area of more than 900m², should have two firefighting shafts.	B(V1) 15.3 & 15.5
• In any building, the hose laying distance should be a maximum of 60m from the fire main outlet in a firefighting shaft (see AD-B(V1) Diagram 15.3).	B(V1) 15.7

4.14.4.1.1 FIREFIGHTERS ESCAPE ROUTES

• In low-rise buildings without deep basements, access for firefighting personnel is typically aimed at providing fire service vehicle access and means of escape.	B(V1) 15.1
• A building with a storey more than 18m above the fire and rescue vehicle access level should have one or more firefighting shafts that contains a firefighting lift (see AD-B(V1) Diagram 15.1).	B(V1) 15.2

4.14.4.2 Fire mains

• Access to fire mains should be provided. • There should be vehicle access for a pump appliance to within 45m of all points within the dwelling.	B(V1) 14.7
• Buildings taller than 50m without firefighting shafts should be provided with fire mains where fire service vehicle access is not provided.	B(V1) 14.3

If mains water is not available, then another source of water (such as water from a spring, river, canal or pond) will have to be available for access (minimum of 45,000 litres) on a hardstanding with a pumping appliance.

4.14.4.3 Final exit

• Dwellings more than 4.5m above ground level should have a protected stairway that provides access to at least two ground level final exits that are separated from each other by fire-resisting construction and a fire doorset.	B(V1) 2.5

4.14.4.4 Means of escape

Note: Passenger lifts in dwellings are **not** considered acceptable as a means of escape unless they are suitably designed and located in either the enclosure to a protected stairway or a fire-resisting lift shaft.

• Upper storeys that are served by only one stairway should either have an emergency escape window, external escape door or direct access to a protected stairway.	B(V1) 2.1
• Every doorway or exit providing a means of escape should be marked by an exit sign in accordance with BS ISO 3864-1 and BS 5499-4.	B(V1) 3.45
• Access to an external escape stair may be via a flat roof. (A fixed ladder may be provided if a conventional stairwway proves impractical – such as access to a plant room.)	B(V1) 3.69 B(V1) 3.85

4.14.4.5 Flats

Note: In dwellings, a fixed ladder may be used – with secure handrails on both sides – for access to a loft conversion that contains one habitable room.

• In multi-storey flats, all habitable rooms (excluding kitchens) should have direct access to a protected internal stairway leading to an exit from the flat.	B(V1) 3.16
• In blocks of flats fitted with fire mains, access should be provided for a pumping appliance to within 18m of each fire main inlet connection point. Inlets should be on the face of the building.	B(V1) 13.5

All habitable rooms should have direct access to a protected landing.

• In dwellings with two or more storeys that are more than 4.5m above ground level (and the alternative escape route can only be accessed via a protected stairway, upper storey or a landing within the protected stairway enclosure), then a separate, fire resistant, alternative escape route should be provided which is separated from the lower storeys.	B(V1) 2.6

4.14.4.5.1 LOFT ESCAPE

• Occupants of a loft room should have access to a first-storey escape window.	B(V1) 2.23

4.14.1.6 Floor identification

• Floor identification signs should be located on every landing of a protected stairway, a protected corridor, lobby, or open-access balcony into which a firefighting lift opens.	B(V1) 15.14
• In multi-storey flats with two or more entrances, the flat number has only to be indicated on the normal access storey.	B(V1) 15.16

4.14.5 Ventilation

Note: If a unit contains both living accommodation and space for commercial purposes (e.g. a workshop or office), the whole unit should be treated as a dwelling.

• In a central plant room, adequate space should be provided for the maintenance of the plant.	F(V1) 1.8
• There should be reasonable access for maintenance for the purpose of changing filters, replacing defective components, and cleaning ductwork, etc., in new dwellings.	

Note: Ventilation devices should be fitted with accessible controls.

4.14.6 Sanitary appliances

• Hand-washing facilities should be located in a room containing a sanitary convenience.	G 5.10
• A sanitary appliance used for personal washing fitted with a macerator and pump may also be connected to a small-bore drainage system with access to washing facilities discharging directly to a gravity system.	

• A WC fitted with a macerator and pump may be connected to a small-bore drainage system connected to a discharge stack if there is also access to a WC discharging directly to a gravity system.	G 4.24

The cistern should be accessible for maintenance, cleaning and replacement.

4.14.7 Combustion appliances and fuel storage systems

4.14.7.1 Metal chimney and flues

• Flues may pass through communal areas, including purpose-designed ducts where inspection access is provided. • Flues should **not** pass through another dwelling. • Access hatches should be at least 300mm × 300mm.	J 1.47
• Joints between chimney sections should not be concealed within ceiling joist spaces or within the thicknesses of walls without proper access being provided.	J 1.43
• Where a flue is routed within a void, access should be provided at strategic locations to allow the flue to be visually checked and confirmed.	J 1.47

AD-J Diagram 14 shows an acceptable approach to providing access to a horizontal flue located within a ceiling void.

• Provision should be made to enable flues to be swept and inspected.	J 1.49
• Access covers should have the same level of gas-tightness as the flue system and an equal level of thermal insulation. • Openings for cleaning the flue should allow easy passage of the sweeping brush.	J 1.51

4.14.7.2 *Location and support of cylinders*

• There should be a permanent means of safe access to appliances for maintenance.	J 1.60
• Provisions should enable fuel storage cylinders to stand upright, be secured by straps or chains against a wall outside the building in a well-ventilated position at ground level, where they are readily accessible.	J 5.20

4.14.8 Protection from falling, collision and impact

• The minimum headroom for access between levels should be 2m, as shown in AD-K Diagram 13.	K 1.11
• Windows should be designed so that people can safely clean the outside safely from inside the building (see AD-K Diagram 9.1).	K 9.1a
• Dwellings (and common access areas within buildings that contain flats) with ramps which are *less* than 1,000mm wide, should be provided with a handrail on one or both sides. Ramps *more* than 1,000mm wide should be provided with a handrail on both sides.	K 2.12
• Barriers should be provided at any edges which are level with (or above) the floor, ground or any other route used for vehicles (see AD-K Diagram 4.1).	K 4.1

• Pedestrian guarding should be provided at:
  • the edges of any part of a floor, gallery, balcony or roof (including rooflights and other openings);
  • any light well, basement or similar sunken area next to a building;
  • in vehicle parks.
	K 3.1

Any part of a floor of a gallery or balcony, indeed any other place to which people have access, should be provided with balcony guarding for escape purposes.

- If a building is likely to be used by children under five years old, the guarding should *not* have horizontal rails, should stop children from easily climbing it, and its construction should prevent a 100mm sphere being able to pass through any opening of that guarding.   K 3.3

4.14.8.1 Access routes – hazards

- A door that swings out by more than 100mm towards an access route shall be protected as shown in AD-K Diagram 10.2.   K 10.2

- A fixed ladder, with fixed handrails on both sides, may be used for access to a loft conversion containing one habitable room when there is not enough space for a stair.   K 1.32

4.14.8.2 Access stairs

- The rise for access stairs should be between 150mm and 170mm, and the going between 250mm and 400mm.   K 1.3

- Common access stairs in buildings containing flats should provide a stair with steps that use a suitable tread nosing profile (as shown in AD-K Diagram 1.2.) and use risers which are not open.   K 1.10

For common access areas of buildings that contain flats:

- Guarding should be provided at the sides of flights of stairs and landings when there are two or more risers.   K 1.40

- Handrails shall not project into an access route (see AD-K Diagram 1.13).   K 1.36

4.14.9 *Conservation of fuel and power*

• Boarded areas should be provided above the insulation to give access for maintenance.	L(V1) 4.15f
• Roof insulation should be installed when the eaves are still accessible.	L(V1) 4.17i Table C1
• When a boiler or hot water storage vessel is replaced, any accessible pipes in the dwelling should be insulated.	L(V1) 4.25
• Operating and maintenance instructions should be left in an accessible format.	L(V1) 9.1d

4.14.9.1 *Overheating*

This AD-O has interactions with ADs B, F, J, L, K, M and Q. Where different requirement figures are given, the most stringent of them apply.

When determining the free area available for ventilation during sleeping hours, only the proportion of openings that can be opened securely should be considered to provide useful ventilation. This particularly applies for easily accessible bedrooms.

4.14.9.2 *Live/work units*

• The commercial part of a building can revert to residential use as long as **all** of the following apply:	O 0.6 O 1.4 O 1.5
• there is direct access between the commercial space and the residential accommodation;	
• the commercial space and the residential accommodation are within the same thermal envelope;	
• the residential accommodation comprises a substantial proportion of the total area of the unit.	

4.14.10 Infrastructure for the charging of electric vehicles

While AD-S provides guidance on the installation and location of electric vehicle charging points, access requirements must be considered when locating them. Accessible parking spaces must meet requirements – and in these particular cases, you should follow the guidance in AD-M.

Note: For your convenience, basic information concerning electric vehicle charging points that is applicable to **access routes** has been included here. Subsection 4.17 'Electrical safety' contains a full explanation of the requirements of AD-S concerning the infrastructure for the charging of electric vehicles.

4.14.10.1 Charging points

- Electric vehicle charging points must meet the   S 44J
  following minimum standards. They must be:
  - capable of providing a reasonable power output for each parking space for which it is intended to be used;
  - run on a dedicated circuit;
  - compatible with all vehicles which may require access to it.

- Underground cable ducts should meet BS EN 61386-   S 6.10
  24 and the following requirements:
  - cable routes should be laid as straight as possible and with suitable access points, so as to enable cables to be pulled through at a later date;
  - the termination points of cable ducts should be sited where maintenance access is unrestricted.

Connection points identified for future use should leave space to install electric vehicle charging points and sufficient vehicle access for installing, maintaining and using these charging points (as shown in AD-S Figure NEW S2).

4.14.10.1.1 NEW DWELLINGS

- All new residential buildings with associated parking   S S1
  spaces must have access to electric vehicle charging
  points.
- Where associated parking spaces are provided for   S 1.1 & 1.2
  a new residential building, the number of electric
  vehicle charging points will depend on the number of
  dwellings that the car park serves.
- If the number of associated parking spaces for the   S 1.3
  new residential building is more than 10 and more
  than the number of dwellings, then cable routes must
  be provided for all associated parking spaces which do
  not have access to an electric vehicle charging point.

4.14.10.1.2 DWELLINGS RESULTING FROM A MATERIAL CHANGE
             OF USE

- Where one or more dwellings with associated parking   S S2
  result from a building that is undergoing a material
  change of use, then at least one associated parking
  space with access to an electric vehicle charging point
  should be available for the use of each dwelling.

- If an associated parking space is exempt from the   S 2.3
  requirement to install an electric vehicle charging
  point, the number of associated parking spaces with
  access to either cable routes or an electric vehicle
  charging point should be greater than either the
  number of associated parking spaces or the number
  of dwellings that the car park serves.

The following building types may receive special   S 2.4 & 2.7
consideration regarding installing electric vehicle   S 2.5 & 2.6
charging points:
  - buildings of architectural and historical interest
    and that are identified as non-designated heritage
    assets;
  - buildings of architectural and historical interest
    within National Parks, Areas of Outstanding Natural
    Beauty, registered historic parks and gardens,
    registered battlefields and World Heritage Sites;

- Listed Buildings;
- buildings in Conservation Areas;
- scheduled monuments.

• If there are not enough associated parking spaces outside of the covered car park to meet the requirement to install electric vehicle charging points, then cable routes should be installed for associated parking spaces within the covered car park.	S 2.18

Note: The total number of associated parking spaces which have access to either cable routes or an electric vehicle charging point should be the greater of the number of dwellings in the residential building or the number of associated parking spaces.

AD-S Diagram 6.2 shows a possible arrangement of electric vehicle charging points for accessible parking spaces.

Requirements for buildings other than dwellings – Additional requirements

This section provides the details for access routes in buildings other than dwellings, only where they differ from the requirements already given above.

4.14.11 Disabled access

• Separate pedestrian access routes should be provided.	M2 1.13h
• Access routes should not be too narrow, and entrances should have a surface that people are able to travel along easily without the risk of tripping or falling.	M2 1.2 M2 1.9
• Potential hazards on access routes adjacent to buildings should be avoided.	M2 1.5
• Doors that open towards a corridor which is a major access route or an escape route should be recessed.	M2 3.14g
• Powered doors should not project into any adjacent access route.	M2 3.14
• Handrails should not project into an access route.	M2 1.37e

• Floor, wall and ceiling surface materials and finishes should help visually impaired people appreciate the boundaries of rooms or spaces and identify access routes.	M2 4.33
• Each lifting device should have an unobstructed manoeuvring space (of 1,500mm × 1,500mm) or a straight access route (900mm wide) in front of it.	M2 3.28a
• Stepped access routes to audience seating should be provided with fixed handrails.	M2 4.12
• The surface of a parking bay designated for disabled people should allow the safe transfer of a passenger or driver to a wheelchair, and transfer from the parking bay to the access route into the building without undue effort.	M2 1.15
• A buff-coloured blister surface should identify an uncontrolled crossing point across the vehicular route.	M2 1.13h

4.14.12 Fire safety

4.14.12.1 General requirements

Any measures that restrict access to or within the building should not adversely affect fire safety provisions.

• Access to refuse storage chambers should **not** be adjacent to escape routes, final exits, or near to windows of flats.	B(V2) 5.45
• Doors providing access to an external escape stair should fire-resisting and self-closing.	B(V2) 2.26
• Escape routes should have adequate artificial lighting.	B(V2) 5.25
• Every part of each storey in a multi-storey building should have access to more than one stair.	B(V2) 2.4
• Final exits should not present an obstacle to users.	B(V2) 5.22
• Final exits should provide direct access to a street, passageway, walkway or open space.	B(V2) 5.21
• Roller shutters across a means of escape should **only** be released by a heat sensor and should **not** be initiated by smoke detectors or a fire alarm system, unless the shutter forms part of a boundary to a smoke reservoir.	B(V2) App C9

• A building (or part of a building) designed for phased evacuation should have an internal speech communication system.	B(V2) 3.21a

4.14.12.2 Access and facilities for the fire service

• Reasonable provision shall be made to enable fire appliances to gain access to be used near the building.	B(V2) B5
• Facilities must enable access into and within the building for firefighting personnel to both search for and rescue people and fight fire.	
• Vehicle access routes and hardstandings should comply with the guidance of AD-B(V2) Table 15.2.	B(V2) 16.8

Note: Turning facilities (a hammerhead or turning circle) should be provided in any dead-end access route that is more than 20m long.

• For buildings fitted with dry fire mains, fire appliances will need access within 18m of each fire main inlet connection point (inlets on the face of the building) and these inlets should be visible from the parking position of the appliance.	B(V2) 15.4 B(V2) 16.10
• For buildings fitted with wet fire mains, access for a pumping appliance should be within 18m of (and visible from) the fire main and the inlet to replenish the suction tank for the fire main in an emergency.	B(V2) 15.5

Note: Access routes for buildings over 11m tall should comply with the guidance provided in AD-B(V2) Diagram 15.2.

• Where access is provided for high-reach appliances, overhead obstructions (e.g. cables and branches) should be avoided in the zone shown in AD-B(V2) Diagram 15.3 and the guidance listed in AD-B(V2) Table 15.	B(V2) 15.9
• Buildings that do not have fire mains should provide access for vehicles in accordance with AD-B(V2) Table 15.1.	B(V2) 15.1

• There should be vehicle access for a pump appliance to small buildings within 45m of all points within every point of the footprint of the building.	B(V2) 15.1
• Every elevation to which there is vehicle access should have a suitable door (no less than 750mm wide) that provides access to the interior of the building.	B(V2) 15.1 3

4.14.12.3 Flats

Flats should have access to alternative escape routes.

• All habitable rooms in the upper storey of a multi-storey flat should have direct access to an internal protected stairway leading to a final exit.	B(V2) B1 2.4
• A corridor more than 12m long that provides access to alternative escape routes should be sub-divided by self-closing fire doors.	B(V2) 2.26
• A dead-end portion of a corridor that provides access to a point from which alternative escape routes are available should be separated by self-closing fire doors.	B(V2) 2.28
• Fixed ladders should not be used as a means of escape for members of the public.	B(V2) 3.28

Note: Ladders should only be provided where a conventional stair is impractical, such as for access to plant rooms which are not normally occupied.

• Flats should have access to alternative escape routes.	B(V2) 2.20
• Refuges should provide an area accessible to a wheelchair of at least 900mm × 1,400mm in which a wheelchair user can await assistance.	B(V2) 3.6
• If the wheelchair space is within a protected stairway, the wheelchair space should not obstruct the flow of persons escaping (see AD-B(V2) Diagram 2.1).	

Note: In taller blocks of flats, fire and rescue service facilities are required within the building.

4.14.12.4 Ventilation

• Access should be provided to allow inspection, testing and maintenance of both the fire damper and its actuating mechanism.	B(V2) 10.18
• Each compartment in a basement should have direct access to venting, without having to open doors, etc., into another compartment.	B(V2) 18.2
• If the natural smoke outlet terminates at a point that is not readily accessible, it should be kept unobstructed and should only be covered with a non-combustible grille or louvre.	B(V2) 18.8

Note: Stairs or ladders that are used in industrial buildings to access areas for maintenance purposes should meet with the requirements of BS 5395-3.

4.15 Corridors and passageways

Reminder: The subsections include details for buildings other than dwellings only where they differ from the requirements for dwellings.

Subsection	Description	Page
Requirements for dwellings		
4.15.1	Disabled access to and use of dwellings	506
4.15.2	Fire safety	506
4.15.3	Resistance to the passage of sound	509
4.15.4	Glazing in corridors	510
4.15.5	Electrical safety	510
Requirements for dwellings other than buildings – Additional requirements		
4.15.6	Disabled access	510
4.15.7	Fire safety	512

Reminder: Free downloads of all the diagrams and tables referenced in the subsections are available on-line at the Governments' Planning Portal: www.gov.uk/government/collections/approved-documents.

Requirements for dwellings

4.15.1 Disabled access to and use of dwellings

In 2016, AD-M was updated and now wheelchair-user dwellings are **'optional** requirements' and the following requirements only apply should the **owner** of the property desire this to be the case.

• Any door opening towards a corridor or a stair should be recessed to prevent its swing encroaching on the effective width.	B(V1) 3.96

To assist access circulation in dwellings:

• Obstructions (such as a radiator) should not occur opposite or close to a doorway. • Corridors should be a minimum of 750mm width at any point. (See AD-M Table 1.1.) • The widths of clear passageways should comply with AD-M Diagram 1.2.	M4(1) 1.15

4.15.2 Fire safety

In corridors, lobbies or stairways, the combustible content should be restricted.

4.15.2.1 Escape routes

Ancillary accommodation should not be located in, or entered from, a protected lobby or protected corridor forming the only common escape route on that storey.

• All escape routes should have adequate artificial lighting.	B(V1) 3.41

If the mains electricity power supply fails, escape lighting should illuminate the route (including external escape routes).

• Escape routes should have a minimum clear headroom of 2m.	B(V1) 3.38

• Escape route floor finishes should minimise their slipperiness when wet.	B(V1) 3.39
• Any sloping floor or tier should have a pitch of not more than 35° to the horizontal.	B(V1) 3.40
• If the stair serves offices, stores or other ancillary accommodation, they should be separated from the stair by a protected corridor or lobby that has a minimum 0.4m² of permanent ventilation or is protected by a mechanical smoke-control system.	B(V1) 3.28
• In divided corridors with cavities, fire-stopping should be provided to prevent alternative escape routes being affected by fire and/or smoke.	B(V1) 8.6

4.15.2.2 Flats

 Emergency exits should be provided within individual flats.

• Provided that the flat is on a storey served by a single common stair or is at the dead end of a common corridor served by two (or more) common stairs, a single escape route is acceptable from a flat entrance door, provided that the travel distance from flat entrance door to a common stair or stair lobby:	B(V1) 3.27
• in one direction, is less than 7.5m;	B(V1) 3.32
• in more than one direction, is less than 30m.	
• Common corridors should be protected corridors. • Any wall between a flat and the corridor should be a compartment wall.	B(V1) 3.34
• A common corridor connecting two or more storey exits should be divided with a fire doorset fitted with a self-closing device (see AD-B(V1) Diagram 3). • Associated screens must be fire resisting. • Doors should be constructed so that smoke does not affect access to more than one stair.	B(V1) 3.35

Note: A fire doorset fitted with a self-closing device and (if required) a fire-resisting screen should separate the dead-end portion of a common corridor from the rest of the corridor (see AD-B(V1) Diagrams 3.7a, 3.7b and 3.8c).

4.15.2.3 Floor identification

• Floor identification signs should be located on every landing of a protected corridor, lobby and stairway into which a firefighting lift opens.	B(V1) 15.14

4.15.2.4 Lifts

• In basements and enclosed car parks, the lift should be within the enclosure of a protected stairway. Otherwise, the lift should be approached only via a protected corridor (or protected lobby).	B(V1) 3.102 B(V2) 5.38
• If a lift delivers into a protected corridor (or protected lobby) serving sleeping accommodation, and also serves a storey containing a high fire risk (such as a kitchen, communal areas, stores, etc.), then the lift should be separated from the high fire-risk area(s) by a protected lobby or protected corridor.	B(V1) 3.103

4.15.2.5 Protected stairways

• Stairs may connect with the basement storeys if there is a protected corridor or lobby between the stairs and accommodation, at each basement level.	B(V1) 3.72
• Where a stair serves an enclosed car park or place of special fire hazard, the corridor (or lobby) should have a minimum 0.4m² of permanent ventilation or be protected from the ingress of smoke by a mechanical smoke-control system.	B(V1) 3.75
• Any protected exit corridor should have the same standard of fire resistance and lobby protection as the stair it serves.	B(V1) 3.81

4.15.2.6 Smoke vents

Ventilation (natural or mechanical) offers additional protection to the corridors and lobbies to control smoke, particularly in common escape routes.

- Other than small single-stair buildings, all corridors and/or lobbies next to stairs should have a smoke vent positioned as high as practicable.   B(V1) 3.50

The top edge of the vent must be as high as the top of the door to the stair.

- The free area of smoke vents should be a minimum of 1m² from the corridor or lobby into the shaft.   B(V1) 3.51
- Smoke vents in common escape routes should:
  - be located on an external wall with a minimum cross-sectional area of 1.5m²;
  - discharge into a vertical smoke shaft at roof level (at least 0.5m above any surrounding structures within 2m of it horizontally).

Note: If smoke is detected in the common corridor or lobby, both the vents on the storey where the fire is located and those at the top of the smoke shaft and to the stair should open simultaneously, and the vents from the corridors or lobbies on all other storeys should remain closed.

4.15.2.7 Vision panels in doors

- Where doors on escape routes divide corridors or are hung to swing both ways, they should contain vision panels.   B(V1) 3. 97

See AD-M (Vision panels in doors across accessible corridors) and AD-K (Safety of glazing) for further details.

4.15.3 Resistance to the passage of sound

Sound insulation testing should not be carried out between living spaces and corridors, stairwells or hallways.

4.15.3.1 Corridor walls

Separating walls should be used to provide sound insulation between corridors and rooms for residential purposes or between dwelling houses.

Sound insulation will be reduced by the presence of a door.

4.15.4 Glazing in corridors

• An uninsulated glazed screen may be incorporated in the enclosure to a protected corridor entered from the stair.	B(V1) 7.24
• Glazed screens alongside a corridor should be provided with all of the following:	K 7.4

- • Glazed screens alongside a corridor should be provided with all of the following:
  - • a sign at two levels, as shown in AD-K Diagram 7.2;
  - • signs that contrasts visually with the background seen through the glass (both from inside and outside);
  - • signs in the form of a logo or symbol (minimum 150mm high) or another decorative feature.
- • Glazed doors beside or part of a glazed screen shall be clearly marked with a high contrast strip at the top and on both sides.
- • Where glass doors may be held open, they shall be protected with guarding.

4.15.5 Electrical safety

Common access areas in blocks of flats such as corridors and staircases shall comply with the requirements of AD-P for electrical installations.	P 2.2b

Requirements for buildings other than dwellings – Additional requirements

The basic requirements for corridors and passageways in buildings other than dwellings are the same as for dwellings, with the addition of the following requirements.

Note: The minimum width of a passage is measured as the space available between any fixed obstructions.

4.15.6 Disabled access

The following recommendations primarily apply to **newly** erected buildings or a change of use of buildings.

4.15.6.1 Design considerations

• Corridors and passageways should be wide enough to allow people with buggies, or carrying cases, or people on crutches to pass others on the access route.	M2 3.11
• Wheelchair users should have access to adjacent rooms and spaces, be able to pass other people and, where necessary, turn through 180°.	
• There should be a visual contrast between walls and ceilings and between walls and floors.	M2 3.12
• Good acoustic design should be employed that is neither too reverberant nor too absorbent.	M2 3.13

4.15.6.2 General provisions

Corridors and passageways should comply with the following:

• Columns, radiators and fire hoses should not project into the corridor.	M2 3.14
• There should be an unobstructed width along their length of at least 1,200mm.	
• Corridors should have an unobstructed width of at least 1,800mm at reasonable intervals to allow wheelchair users to pass each other.	
• The floor should be level or predominantly level (see AD-M2 Diagram 3 and Table 1).	
• Any sloping section should extend the full width of the corridor or the exposed edge should be clearly identified and protected by guarding.	
• Doors opening towards a major access route or an escape route should be recessed, except those to minor utility facilities (e.g. small storerooms).	
• Patterns that could be mistaken for steps or changes of level should be avoided.	
• Floor finishes should be slip-resistant.	

Note: In school buildings where there are lockers within the corridor, the preferred corridor width dimension is 2,700mm.

4.15.6.3 Internal lobbies

• Internal lobbies should allow a wheelchair user or a person pushing a pram or buggy to move clear of one door before attempting to open the second door. In addition:	M2 3.15
• the length and width of internal single swing doors should be as shown in AD-B2 Diagram 10;	M2 3.16
• glazing within the lobby should not create distracting reflections;	
• junctions of floor surface materials at the entrance to the lobby area should not create a potential trip hazard;	
• columns and ducts, etc., that project into the lobby by more than 100mm should be protected by a guard rail.	

4.15.6.4 Vertical circulation within the building

• A ramp should be provided on any internal circulation route where a change of level is unavoidable.	M2 3.20
• All bedrooms should have an effective clear width of the door from the access corridor that complies with AD-M2 Table 2.	M2 4.24a M2 4.24j
• The door from the access corridor to a wheelchair-accessible bedroom should comply with the relevant provisions for internal doors as shown in AD-M2 Table 2 and Diagram 9.	

4.15.7 Fire safety

Note: Sprinklers should be provided within individual flats, but they are not required in the common areas such as corridors or landings when these areas are fire sterile.

4.15.7.1 Buildings designed for phased evacuation

• Other than the top storey, stairs should be approached through a protected lobby or protected corridor. (See AD-B(V2) Table 3.3.)	B(V2) 3.21 B(V2) 2.21

4.15.7.2 Doors

• Any door opening towards a corridor or a stair should be recessed to prevent its swing interfering with its effective width.	B(V2) X.13
• Where doors on escape routes divide corridors, they should contain vision panels.	B(V2) 5.14

4.15.7.3 Escape lighting

• The following areas require escape lighting: • all common escape routes in residential settings; • internal corridors more than 30m long in office, industrial, storage, other non-residential, shop and commercial settings and car parks; • all escape routes to which the public are admitted in shop and commercial settings, places of assembly and recreation, and car parks.	B(V2) Table 5.3

4.15.7.4 Escape routes

Requirement B1 states:

> *The building shall be designed and constructed so that there are appropriate provisions for the early warning of fire, and appropriate means of escape in case of fire from the building to a place of safety outside the building capable of being safely and effectively used at all material times.*

• Escape routes should be: • suitably located; • sufficient in number and capacity; • satisfactorily lit and exits suitably signed; • able to restrict the spread of fire and remove smoke.	B(V2) B1
• If a corridor is being used as a means of escape, then it should be a protected corridor (see AD-B(V2) Diagrams 2.7 and 2.8).	B(V2) 2.24

• The number of escape routes and exits provided depends on the number of rooms (tiers or storeys) of the occupants and the travel limitations to the nearest exit (see AD-B(V2) Tables 2.1 and 2.2).	B(V2) 2.3
• The width of escape routes and exits depends on the number of people being evacuated (see AD-B(V1) Table 2.3).	B(V2) 2.18
• In buildings designed for phased evacuation, stairs should be approached through a protected lobby or protected corridor.	B(V2) 3.34
• Alarm systems should be installed within parts of the escape routes that comprise circulation areas and circulation spaces, such as corridors.	B(V2) 1.9
• Common corridors in flats should be protected corridors or an automatic fire detection or alarm system should be installed in the whole storey.	B(V2) 2.17
• A common corridor connecting two or more storeys exits should be sub-divided by a self-closing fire door with any associated fire-resisting screen.	B(V2) 2.26
• A corridor that serves as a means of escape should be a protected corridor.	B(V2) 2.24
• All internal corridors more than 30m long require escape lighting.	B(V2) Table 5.1
• A dead-end portion of a common corridor should be separated from the rest of the corridor by a self-closing fire door with, if necessary, a fire-resisting screen.	B(V2) 2.28
• Self-closing fire doors in a primary circulation route should be fitted with an automatic release mechanism.	B(V2) 2.15

Guidance on smoke-control systems using pressure differentials is available in BS EN 12101-67.

4.15.7.5 Exits

Note: Windows are not acceptable as final exits.

• Every protected stairway should lead to a final exit, either directly or via a protected exit corridor.	B(V2) 3.21
• Where a central core has more than one exit, storey exits should be remote from one another and no two exits should be approached from the same undivided corridor.	B(V2) 2.12 Diagram 2.4

Note: The doors at both ends of the area marked 'S' should be self-closing fire doorsets.

4.15.7.6 *Protected stairways, passages, corridors and enclosures*

Protected stairways should discharge either directly to a final exit or via a protected exit passageway to a final exit.

• If the upper storeys of a building have multiple escape stairs, only one needs to end at ground level (the others can go via a protected lobby or a protected corridor between the stairs and basement level).	B(V2) 3.41
• Any protected exit corridor or stair should have the same standard of fire resistance and lobby protection as the stair it serves.	B(V2) 3.36
• If a lift opens into a protected corridor serving sleeping accommodation which also serves a storey containing a high fire-risk area, then the lift should be separated from that area by a protected lobby or protected corridor.	B(V2) 5.39

Note: Corridors that are longer than 2m require sub-dividing doors, which should be fire doorsets fitted with self-closing devices.

• An uninsulated glazed screen may be incorporated in the enclosure to a protected shaft between a stair and a lobby or corridor entered from the stair. (See AD-B(V1) Diagram 8.4 and Table B5.)	B(V2) 8.34

Note: Stairs may connect with the basement storey(s) if there is a protected lobby or a protected corridor between the stair(s) and accommodation at each basement level.

• A protected lobby should be provided between an escape stair and a place of special fire hazard, to protect from the ingress of smoke.	B(V2) 3.35

Protected corridors should be provided for:

- every corridor serving bedrooms;
- every dead-end corridor;
- any corridor common to two or more different occupancies.

• Every corridor more than 12m long which connects two or more storey exits should be sub-divided by self-closing fire doors where necessary, and cavity barriers should be used where appropriate.	B(V2) 2.26

• Dead-end corridors exceeding 4.5m in length should be separated by self-closing fire doors from any part of the corridor which:
  • provides two directions of escape;
  • continues past one storey exit to another (see AD-B(V1) Diagrams 2.10a and 2.10b).
	B(V2) 2.28

• If a cavity exists above the enclosures to a corridor, the possibility of smoke bypassing the enclosure should be restricted by either:
  • fitting cavity barriers across the corridor (see AD-B(V1) Diagram 2.9);
  • enclosing the cavity (on the lower side) by a fire-resisting ceiling.
	B(V2) 2.27

Note: If a corridor is used for a means of escape but is not a protected corridor, then:

- Openings into rooms from the corridor should be fitted with doors (they need not be fire doors).
- Any partitions in corridors which are used as an escape route should be carried up to the soffit of the floor above.

4.15.7.7 Refuges

Note: An enclosure (such as a protected lobby or protected corridor) is considered to be a satisfactory refuge provided that each refuge can provide an area of at least 900mm × 1,400mm in which a wheelchair user can await assistance – **provided** that the wheelchair space does not reduce the width of the escape route.

• If a refuge is sited in a protected stairway (protected lobby or protected corridor), it should not reduce the width of the escape route or obstruct the flow of people escaping.	B(V2) 3.6

4.15.7.8 Residential care homes

• Each bedroom should be enclosed in fire-resisting construction with fire-resisting doors. • Every corridor serving bedrooms should be a protected corridor.	B(V2) 2.42
• A fire in one protected area should not prevent occupants of other areas from reaching a final exit (see AD-B(V2) Diagram 2.11).	B(V2) 2.38

4.15.7.9 Single escape routes and exits

In many cases, there is only one escape route, and this is acceptable if:

• The travel distance to the nearest storey exit is within the limits for routes where escape is possible in more than one direction (see AD-B(V2) Table 2.1).	B(V2) 2.7
• The travel distance does not exceed the limits shown in AD-B(V2) Table 2.1.	

4.15.7.10 Storeys divided into different occupancies

• If a storey is divided into different areas of occupancy, then the means of escape from each occupancy should not pass through any other occupancy. • If a common corridor or circulation space is on an escape route, it should be a protected corridor.	B(V2) 2.17

4.16 Sanitary accommodation, bathrooms and showers

Reminder: The subsections include details for buildings other than dwellings only where they differ from the requirements for dwellings.

Subsection	Description	Page
Requirements for dwellings		
4.16.1	Access and use of buildings	518
4.16.2	Combustion appliances and fuel storage systems	519
4.16.3	Drainage and waste disposal	519
4.16.4	Electrical safety	522
4.16.5	Fire safety	522
4.16.6	Protection from falling, collision and impact	523
4.16.7	Sanitation, hot water safety and water efficiency	523
4.16.8	Smoke alarms	526
4.16.9	Site preparation and resistance to moisture	526
4.16.10	Resistance to the passage of sound	527
4.16.11	Ventilation	527
Requirements for buildings other than dwellings – Additional requirements		
4.16.12	Access to and use of buildings	531
4.16.13	Fire safety	537

Reminder: Free downloads of all the diagrams and tables referenced in the subsections are available on-line at the Governments' Planning Portal: www.gov.uk/government/collections/approved-documents.

Requirements for dwellings

4.16.1 Access to and use of buildings

4.16.1.1 WC facilities

- Wherever possible, the entrance storey should contain WC facilities that enable easy access to a WC (see AD-M Diagram 1.3 and AD-M Table 1.1).   M1 1.17

The availability of WC and sanitary facilities that are accessible and/or wheelchair adaptable is an optional requirement for dwellings.

4.16.2 Combustion appliances and fuel storage systems

• The outlets for flues serving oil-fired appliances with a pressure jet burner should not be closer to vertical sanitary pipework than 300mm.	J Diagram 41
• Open-flued oil-fired appliances should **not** be installed in bathrooms.	J 4.2

4.16.3 Drainage and waste disposal

WCs should have a major flush volume of 5 litres or more.

4.16.3.1 Discharge stacks

• Stacks should discharge to a drain.	H 1.26
• The radius of the bend at the foot of the stack should be as large as possible.	
• Offsets in the 'wet' portion of a discharge stack should be avoided.	H 1.27
• Stacks should have at least the diameter shown in AD-H Table 3 and should not reduce in the direction of flow.	H 1.28
• Urinal stacks should be not less than 50mm.	
• Stacks serving closets with outlets less than 80mm should be not less than 75mm.	
• Stacks serving closets with outlets greater than 80mm should be not less than 100mm.	
• The stack internal diameter should not be less than that of the largest trap or branch discharge pipe.	

Note: Additional information on the maximum flow in a larger building is in AD-H Appendix A. The size of ventilation pipes may be reduced in one and two storey houses.

• Discharge stacks should be ventilated and sub-stacks may be used.	H 1.29
• Ventilated discharge stacks may be terminated inside a building if they are fitted with air admittance valves that comply with BS EN 12380.	H 1.33
• All pipes should be reasonably accessible for repair.	H 1.34

4.16.3.2 Branch discharge pipes

• Branch discharge pipes from a ground floor closet: • may discharge into a stub stack (see AD-H Diagram 5);	H 1.12
• should not discharge if it could cause crossflow into any other branch pipe (see AD-H Diagram 2);	H 1.10
• should not discharge into a stack lower than 450mm above the upturn of the tail of the bend at the foot of the stack in single dwellings of up to 3 storeys (see AD-H Diagram 2);	H 1.11
• should only discharge directly to a drain (see AD-H Diagram 1);	H 1.9
• should terminate between the grating (or sealing plate) and the top of the water seal when discharging to a gully;	H 1.13
• should not have bends in them.	H 1.16
• Condensation from boilers may be connected to sanitary pipework.	H 1.14
• Pipes serving a single appliance should have at least the same diameter as the appliance trap (see AD-H Table 1).	H 1.15
• An unventilated pipe serving more than one appliance should have the minimum diameter shown in AD-H Table 2.	

4.16.3.3 Ventilation of branch pipes

* Branch ventilating pipes: H 1.23
  * which run directly into the outside air, should finish
    at least 900mm above any opening into a building
    closer than 3m (see AD-H Diagram 6);
  * are not required if the length and slope of the   H 1.19
    discharge pipes do not exceed those shown in
    AD-H Table 2 or AD-H Diagram 3;
  * should be connected to the discharge pipe within   H 1.22
    750mm of the trap, and to the ventilating stack
    above the highest 'spill over' level of the appliances
    served (see AD-H Diagram 4);
  * should have a continuous incline from the
    discharge pipe to the point of connection to the
    ventilating stack or stack vent;
  * that are open to outside air should have a wire   H 1.31
    cage or perforated cover, fixed to the end of the
    ventilating pipe.

Note: In areas where rodent control is a problem, cages should be
metallic!

* Branch ventilating pipes to branch pipes serving one   H 1.24
  appliance should be 25mm minimum in diameter.
* Where the branch is longer than 15m (or has more
  than 5 bends), the diameter should be a minimum of
  32mm.

* Rodding points should be provided (above the   H 1.25
  normal spill over level) to discharge pipes which
  cannot be reached by removing traps or appliances
  with internal traps.

4.16.3.4 Pipe gradients and sizes

* Drains carrying foul water should have an internal   H 2.33
  diameter of at least 75mm.
* Drains carrying effluent from a WC should have an
  internal diameter of at least 100mm.

• AD-H Table 6 shows the flattest gradients at which drains should be laid. (Also see AD-H Table A1, which shows approximate flow rates resulting from a typical dwelling.)	H 2.34

 All pipework carrying greywater for re-use should be clearly marked with the word 'GREYWATER'.

4.16.3.5 *Foul drainage from basements*

• Drainage from a basement should be pumped if it contains sanitary appliances and there is a high risk of flooding due to a possible sewer surcharge.	H 2.9
• Where the risk of flooding is considered to be low, an anti-flooding valve should be installed on the drainage from the basement. (See BS EN 13564-3.)	

4.16.4 Electrical safety

• Although a fixed electric heater or mechanical extractor fan in a bathroom is classified as non-notifiable service, work in the space surrounding the bath tap or shower head (as shown in AD-P Diagram 2) falls under the category of notifiable work.	P 2.6

4.16.5 Fire safety

4.16.5.1 *Protected shafts*

• Protected shafts may also include sanitary accommodation and washrooms.	B(V1) 7.22

4.16.5.2 *Inner rooms*

• A bathroom, shower or WC whose only escape route is through another room is acceptable as an inner room.	B(V1) 2.11

As long as the bathroom is separated by fire resisting construction from the adjacent rooms, there is no need for a fire doorset.

4.16.5.3 Protection of openings for pipes

• Openings in cavity barriers (compartment walls or floors) may be used for the passage of pipes.	B(V1) 5.24 B(V1) 7.20
• Openings through a fire-resisting element for pipes should be as few and as small as possible – but above all, fire-stopped.	B(V1) 9.24
• If a pipe is passing through a fire-separating element, unless in a protected shaft, it should have either a tested sealing system (to maintain the fire resistance of the wall, floor or cavity barrier) or have a fire-stopping sleeve around it (see AD-B(V1) Table 9.1 and Diagram 9.2).	B(V1) 9.2 B(V1) 9.3
• If the pipe is made of either lead, aluminium, aluminium alloy, fibre-cement or uPVC (with a maximum nominal internal diameter of 160mm), it may be used as shown in AD-B(V1) Diagram 9.1.	B(V1) 9.5

4.16.6 Protection from falling, collision and impact

• Alternating tread stairs may only provide access to a bathroom in a loft conversion provided it is *not* the only WC in the dwelling.	K 1.29

4.16.7 Sanitation, hot water safety and water efficiency

4.16.7.1 Scale of provision and layout in dwellings

• Dwellings should have at least *one* sanitary convenience and associated hand-washing facility located in their principal/entrance storey.	G 4.7
• A dwelling *must* have at least one bathroom with a fixed bath or shower, and a washbasin.	G 5.6

- Hand-washing facilities should be located in: G 4.9
  - the room containing a sanitary convenience;
  - an adjacent room or place providing the sole
    means of access to the room containing the
    sanitary convenience.

- A place containing a sanitary convenience and/or   G 4.10
  associated hand-washing facilities should be separated
  by a door from any place used for the preparation of
  food – including a kitchen. (See AD-G Diagrams 2
  and 3.)

4.16.7.2 Sanitary conveniences

- Sanitary installations should convey hot water to the   G 3a
  sanitary appliances without waste, misuse or undue   G 2d
  consumption of water.
- A record of the sanitary appliances and white goods
  (such as washing machines and dishwashers) used in
  the water consumption calculation and installed in
  the dwelling shall be provided.

- Water for toilet flushing may be provided from: G 1.6
  - water extracted from wells, springs, boreholes or
    watercourses;
  - harvested rainwater;
  - reclaimed greywater;
  - reclaimed industrial-process water.

All pipework carrying greywater for re-use should be
clearly marked with the word 'GREYWATER'.

- Water intended for sanitary conveniences shall: G 1a–d
  - be reliable and either wholesome, softened
    wholesome or of a suitable quality;
  - convey water to sanitary appliances without waste,
    misuse, undue consumption or contamination of
    wholesome water.

- Where hot and cold taps are provided on a sanitary   G 4.6
  appliance, ***the hot tap should always be on the
  left***.

Note: Even though the requirement only applies to **new** buildings, it is advised that you always ensure that hot water taps are located on the left-hand side of all of your basins and baths because most people are right handed, and so it is intended to stop people (particularly children and poorly sighted people) inadvertently scalding themselves.

• Hot water should be provided to any washbasin, bidet, fixed bath or shower in a bathroom.	G 3.9
• The hot water supply temperature to a bath should be limited to a maximum of 48°C.	G 3.65
• The length of supply pipes between in-line blending valves and outlets should be kept to a minimum, in order to prevent the colonisation of waterborne pathogens.	G 3.68

Blending valves should be compatible with the water source.

4.16.7.3 Chemical and composting toilets

• Chemical toilets or composting toilets may be used where: • suitable arrangements can be made for the disposal of the waste either on or off the site; • the waste can be removed without carrying it through any living space or food preparation areas.	G 4.19
• Composting toilets should **not** be connected to an energy source other than for ventilation or if it is required for the composting process.	G 4.21

4.16.7.4 Discharge to drains

• Any WC fitted with flushing apparatus should discharge to an approved drainage system.	G 4.22
• A discharge pipe should not be connected to a soil discharge stack unless it can be demonstrated that the latter can safely withstand temperatures of the water discharged.	G 3.60
• A urinal fitted with flushing apparatus should discharge through a grating to a discharge stack or a drain.	G 4.23

- A WC fitted with a macerator and pump may be   G 4.24
  connected to a small-bore drainage system that
  discharges to a discharge stack if:
  - there is also access to a WC discharging directly to
    a gravity system;
  - the macerator and pump meet the requirements of
    BS EN 12050-1 or BS EN 12050-3.
- Sanitary appliances used for personal washing should   G 5.9
  discharge through a branch discharge pipe to an
  adequate system of drainage.

4.16.8 Smoke alarms

All dwellings should have a fire detection and alarm system in accordance
with the recommendations of BS 5839-6.

Note: Smoke alarms should **not**, however, be fixed in bathrooms,
showers, or any other place where steam, condensation or fumes could
give false alarms.

- As a minimum, all smoke alarms should: C 1.2
  - be mains operated, in compliance with the relevant
    parts of BS EN 14604 and BS 5839-6;
  - have a standby power supply, such as a rechargeable   C 3.21
    (or non-rechargeable) battery or capacitor.

4.16.9 Site preparation and resistance to moisture

In bathrooms where water may be spilled, any board used as a flooring,
irrespective of the storey, should be moisture resistant.

4.16.9.1 Resistance to damage from interstitial condensation

- Gaps and penetrations for pipes should be filled and   C 6.12
  sealed in areas of high humidity (e.g. bathrooms) to
  avoid excessive moisture transfer to roof voids.

4.16.9.2 Subsoil drainage

• If an active sub-soil drain passes under the building and is cut during excavation, it should either be re-laid, re-routed or re-run.	C 3.3

4.16.10 Resistance to the passage of sound

Fire-stopping should be flexible and prevent rigid contact between pipes and floors.

• Pipes and ducts penetrating a floor separating different flats should be enclosed, as shown in AD-E Diagram 3.6.	E 3.41
• Materials used to construct the enclosure for pipes should have a mass per unit area of 15kg/m² (minimum).	E 3.42

• Piped services (excluding gas pipes) and ducts which pass through separating floors in conversions should:
  • be surrounded with sound-absorbent material;
  • be enclosed in a duct above and below the floor.
	E 4.15

4.16.11 Ventilation

The Building Regulations require that ventilation should extract indoor air pollutants and water vapour from where they are produced, and prevent them from spreading throughout the building.

Without adequate ventilation, mould and internal air pollution might become hazardous to health.

4.16.11.1 Control of ventilation

• Continuously running fans can have a high rate of operation or be equipped with automatic controls for selecting the rate of operation.	F(V1) 1.35
• A manual high-rate control is provided for spaces such as bathrooms and kitchens, whilst automatic controls might include sensors for humidity, occupancy/usage and pollutant release (see AD-F(V1) Table 1.4).	

- Although controls based on humidity sensors can be installed in moisture-generating rooms (e.g. kitchen or bathroom), they should **not** be used for sanitary accommodation where odour is the main pollutant. F(V1) 1.36

4.16.11.2 *Natural ventilation with background ventilators and intermittent extract fans*

- Intermittent extract fans should be fitted in all wet rooms. For kitchens, utility rooms, bathrooms and sanitary accommodation, the extract rates in AD-F(V1) Table 1.1 can be met using an intermittent extract fan. F(V1) 1.47

- If a wet room has no external walls, the intermittent extract fan should extract at four air changes per hour (4 ach). F(V1) 1.48

- For sanitary accommodation, extract rates can be met by following the recommendations associated with purge ventilation. F(V1) 1.49

- Any automatic controls (e.g. humidity control) for intermittent extract should have a manual override to allow the occupant to turn the extract ventilation on or off. F(V1) 1.50

- If a room does not have an openable window, an intermittent extract fan should be provided with controls which enable the fan to operate for at least 15 minutes after the room is vacated. F(V1) 1.51

4.16.11.3 *Extract ventilation*

- All kitchens, utility rooms, bathrooms and sanitary accommodation should have intermittent or continuous extract ventilation to the outside. (See AD-F(V1) Table 1.1 for details of extract rates for these rooms.) F(V1) 1.17

4.16.11.4 Background ventilators for continuous mechanical extract ventilation

- Where continuous mechanical extract ventilation is used: F(V1) 1.64
  - background ventilators should not be in wet rooms;
  - background ventilators should provide a minimum equivalent area of 4,000mm² for each habitable room in the dwelling;
  - background ventilators should provide a minimum total number of ventilators that is the same as the number of bedrooms plus two ventilators.

4.16.11.5 Existing windows without background ventilators

- Whilst replacing the windows is likely to increase the airtightness of the dwelling, if ventilation is not provided via a mechanical ventilation heat recovery system, then the airtightness of the building may reduce beneficial ventilation in the building.   F(V1) 3.15
- To avoid making a dwelling's ventilation provision worse than it was before replacing the windows:
  - provide continuous mechanical extract ventilation; or
  - incorporate background ventilators in replacement windows (provided that they are not wet rooms).

4.16.11.6 Mechanical ventilation with heat recovery ventilation rates

- For dwellings using mechanical ventilation with heat recovery, each wet room should have a minimum continuous mechanical extract ventilation high rate as given in AD-F(V1) Table 1.2.   F(V1) 1.70

Mechanical ventilation with heat recovery systems should be designed to avoid the moist air from the wet room recirculating to the habitable rooms!

4.16.11.7 Addition of a wet room to an existing dwelling

- When a wet room (e.g. kitchen or bathroom, etc.)   F(V1) 3.25
  is added to an existing dwelling, whole dwelling
  ventilation should be extended and extract ventilation
  (see AD-F(V1) Table 1.1) should be provided by:
- continuous extract; or
- intermittent extract; or
- a single room heat recovery ventilator; or
- a background ventilator with at least 5,000mm^2
  equivalent area.

Note: Internal doors should have a 10mm undercut in a 760mm wide
door to allow air to move within the dwelling.

4.16.11.8 Refurbishing a bathroom in an existing dwelling

- If building work is carried out in a kitchen or a   F(V1) 3.30
  bathroom, existing fans (if working correctly) should
  be retained or replaced.
- If there is no ventilation system in the original room,   F(V1) 3.31
  it is not necessary to provide one in the refurbished
  room unless it is likely to make the building less
  compliant with the ventilation requirements of the
  Building Regulations than it was before building work
  started.

If an extractor fan or cooker hood is replaced and the existing cabling is
used, this does not need to be notified to a BCB.

4.16.11.9 Ventilation rates in wet rooms

- Each wet room should have a minimum continuous   F(V1) 1.63
  mechanical extract ventilation high rate as given in
  AD-F(VI) Table 1.2 in Section F(V1) 1.22.

4.16.11.10 Noise

• The average sound pressure level for a ventilator operating under normal conditions in a noise sensitive room (such as a bedroom) should not exceed 30dB; whilst in less sensitive rooms (for example a bathroom), the high rate would be 45dB.	F(V1) 1.7

Requirements for buildings other than dwellings – Additional requirements

This section provides the details for sanitary accommodation, bathrooms and showers for buildings other than dwellings, only where they differ from the requirements already given above.

4.16.12 Access to and use of buildings

Note: Sanitary accommodation should be available to everybody, including for wheelchair users and ambulant disabled people. Ensuite sanitary facilities should also be provided for wheelchair-accessible bedrooms.

4.16.12.1 Ensuite facilities in hotels

• Bath or washbasin taps should be either controlled automatically, or be capable of being operated using a closed fist (e.g. by lever action).	M2 5.4
• Doors to cubicles should be able to be opened if a person inside has collapsed against them.	M2 5.3
• Doors, when open, should not obstruct emergency escape routes.	M2 5.4
• Electronic card-activated locks may be used for bedroom entrance doors.	M2 4.23
• Emergency assistance alarm systems should be fitted in all wheelchair and ambulant disabled WCs, toilets and bathrooms, etc.	M2 5.4

• Emergency assistance alarm systems should:
  • have visual and audible indicators to confirm that an emergency call has been received;
  • have reset controls which should be reachable from a wheelchair, WC and/or shower/changing seat;
  • be distinguishable visually and audibly from the fire alarm.
	M2 5.4

• Fire alarms should emit a visual and audible signal.	M2 5.4
• Heat emitters should be either screened or have their exposed surfaces kept at a temperature below 43°C.	M2 5.4
• Light switches with large push pads should be used.	M2 5.3
• People with limited strength or manual dexterity should be able to operate taps and WC cubicle doors.	

 Lever taps should be used in sanitary accommodation provided for wheelchair-accessible patrons.

• Sanitary fittings and grab bars should contrast visually with background wall and floor finishes.	M2 5.4
• Student accommodation should have a wheelchair-accessible toilet available for disabled visitors.	M2 4.17
• There should be at least as many ensuite shower rooms as ensuite bathrooms.	M2 4.19
• There should be visual contrast between wall and floor finishes.	M2 5.4

• WC compartment doors and doors to wheelchair-accessible unisex toilets, changing rooms or shower rooms should:
  • be fitted with light action privacy bolts;
  • have an emergency release mechanism so that they are capable of being opened outwards, from the outside.
	M2 5.4

• WC cubicles and wheelchair-accessible unisex toilet doors should open outwards.	M2 5.3

4.16.12.2 Provision of toilet accommodation

• Toilet accommodation should be suitable for **all** people who use the building.	M2 5.5

- Wheelchair-accessible unisex toilets should be provided, as well as separate-sex toilets.

- Wheelchair-accessible unisex toilets should *not* be used for baby changing.

 In large building developments, separate facilities for baby changing and an enlarged unisex toilet incorporating an adult changing table are desirable.

4.16.12.3 Wheelchair-accessible unisex toilets

If there is only space for one toilet in a building, it **must** be a wheelchair-accessible unisex type.

• A unisex toilet should be approached *separately* from other sanitary accommodation.	M2 5.8
• Wheelchair users should be able to approach, transfer to and use this sanitary facility.	M2 5.8
• The space should allow wheelchair users to adopt various transfer techniques (independent or assisted).	

If required, horizontal support rails may be used.

• The transfer space alongside the WC should be kept clear to the back wall.	M2 5.8
• A unisex toilet should enable one or two assistants (of either sex) to assist a disabled person.	M2 5.9

A chemical sanitary waste disposal unit may be required in wheelchair-accessible WC accommodation.

• A wheelchair-accessible unisex toilet should be located:	M2 5.10

  - as close as possible to the entrance and/or waiting area of the building;
  - on accessible routes that are direct and obstruction free;
  - in a similar position on each floor of a multi-storey building (allowing right- and left-hand transfer on alternate floors);
  - so that any wheelchair user does not have to travel more than 40m on the same floor;
  - so that they *do not* compromise the privacy of users.

- A wheelchair-accessible unisex toilet should: M2 5.10
  - ensure heat emitters do not restrict the minimum
    clear wheelchair manoeuvring space or the space
    beside the WC used for transfer;
  - ensure that WC pans conform to BS EN 997;
  - have doors that open outward with a horizontal
    closing bar fixed to the inside face;
  - provide the minimum overall heights, dimensions
    and arrangement of fittings as shown in AD-M2
    Diagrams 19 and 20;
  - have an additional drop-down rail on the wall side;
  - have an emergency assistance alarm system;
  - provide an emergency assistance pull cord that is
    easily identifiable and reachable from the WC and
    from the floor close to the WC;
  - provide cisterns with flushing mechanisms
    positioned on the open or transfer side of the
    space, irrespective of handing.
- When more than one unisex toilet is available on a
  single storey, they should have a choice of layouts
  suitable for left- and right-hand transfer.

4.16.12.4 Toilets in separate-sex washrooms

- Ambulant disabled people should be able to use   M2 5.11
  a WC compartment in a **separate**-sex toilet
  washroom, which should have:
  - support rails fitted;
  - sufficient activity space to accommodate people
    with crutches, or impaired leg movements;
  - a WC pan that will accept a variable height toilet
    seat riser.
- An enlarged WC cubicle should be provided for use   M2 5.12
  by people who need extra space and should include:
  - a fold-down table;
  - a minimum manoeuvring space;
  - space clear of any door swing.

In a separate-sex toilet washroom, wheelchair users should have access,
at a lower height than that provided for other users, to both a urinal and
a washbasin.

- WC compartments within separate-sex toilet   M2 5.14
  washrooms should:
  - maintain a 450mm diameter manoeuvring space
    between the swing of an inward opening door, the
    WC pan and the side wall of the compartment;
  - ensure that the minimum dimensions of
    compartments for ambulant disabled people
    comply with AD-M2 Diagram 21;
  - have doors to compartments for ambulant disabled
    people that open outward and are fitted with a
    horizontal closing bar fixed to the inside face;
  - have an enlarged compartment for those who need
    extra space (1,200mm wide) with a horizontal grab
    bar adjacent to the WC, a vertical grab bar on
    the rear wall, and space for a shelf and fold-down
    changing table;
  - ensure any compartment for use by ambulant
    disabled people has a WC pan that conforms to BS
    EN 997;
  - ensure that a wheelchair-accessible compartment
    has the same layout and fittings as the unisex toilet;
  - ensure that a wheelchair-accessible washroom has
    at least one washbasin with its rim set at 720mm to
    740mm above the floor and, for men, at least one
    urinal with its rim set at 380mm above the floor,
    with two 600mm long vertical grab bars with their
    centre lines at 1,100mm above the floor, positioned
    either side of the urinal.

4.16.12.5 Wheelchair-accessible changing and shower facilities

- A self-contained compartment should allow space for   M2 5.16
  a helper.
- Any combined facility should be divided into 'wet'
  and 'dry' areas.
- In large building complexes (e.g. retail parks and large   M2 5.17
  sports centres), one wheelchair-accessible unisex toilet
  should be provided with an adult changing table.

Changing and shower facilities should have:

- An emergency assistance alarm system.   M2 5.18
- A choice of layouts suitable for left- and right-hand transfer when more than one individual changing or shower compartment is available.
- Wall-mounted drop-down support rails and slip-resistant tip-up seats.
- In sports facilities, individual self-contained shower and changing facilities should be available in addition to communal separate-sex facilities.

Note: Facilities for limb storage should be provided.

4.16.12.6   Changing facilities

- The minimum overall dimensions should comply   M2 5.18
  with AD-M Diagram 22.
- The floor of a changing area should be level and slip-resistant when dry or when wet.
- There should be manoeuvring space 1,500mm deep in front of lockers.

4.16.12.7   Shower facilities

- Individual self-contained shower facilities should   M2 5.18
  comply with AD-M Diagram 23.
- The shower curtain should enclose the seat.
- Rails, when they are in a horizontal position, should be able to be operated from the shower seat.
- A shelf should be provided for toiletries (reached from the shower seat or the wheelchair, before or after transfer).
- The floor of the shower and shower area should be slip-resistant and self-draining.
- The markings on the shower control should be logical and clear.
- In commercial developments where showers are provided, at least one of them should be wheelchair accessible.

4.16.12.8 Shower facilities incorporating a WC

• Overall dimensions and the arrangement of fittings in shower facilities incorporating a WC should comply with AD-M Diagram 23.	M2 5.18
• When more than one shower area incorporating a corner WC is provided, there should be a choice of left- and right-hand transfer layouts.	

Wheelchair users and ambulant disabled people should be able to wash or bathe either independently or with assistance from others.

4.16.12.9 Wheelchair-accessible bathrooms

• The minimum overall dimensions for a bathroom incorporating a corner WC should comply with AD-M Diagrams 25 and 26.	M2 5.21
• A choice of layouts for left- or right-hand transfer should be provided when more than one bathroom incorporating a corner WC is available.	
• The bathroom floor should be slip-resistant when dry or wet.	
• The bath should have a transfer seat 400mm deep and equal to the width of the bath.	
• Doors should be outward opening and fitted with a horizontal closing bar on the inside face.	
• An emergency assistance pull cord should be supplied, that is easily identifiable and reachable from the bath or from the floor.	

4.16.13 Fire safety

• A suitable fire alarm (visual and audible) signal should be provided in sanitary accommodation that is used by people with impaired hearing.	B(V2) 1.15
• A bathroom, WC or shower room may be an inner room with an escape route through another room.	B(V2) 2.5
• Sanitary accommodation or washrooms may be in a protected stairway (provided that it is not used as a cloakroom) and protected shafts.	B(V2) 3.38 B(V2) 8.32

- Escape lighting should be provided in all toilet   B(V2)
  accommodation with a minimum floor area of 8m².   Table 5.1

Note: The term 'a place of assembly, entertainment or recreation' includes public toilets.

4.17 Electrical safety

All electrical work shall be carried out in accordance with Amendment No. 3 (2021) to BS 7671 (*Requirements for Electrical Installations*, commonly referred to as the *IEE Wiring Regulations*, 18th edition), plus (for domestic buildings) in accordance with AD-P.

Reminder: The subsections include details for buildings other than dwellings only where they differ from the requirements for dwellings.

Subsection	Description	Page
Requirements for dwellings		
4.17.1	Access to and use of buildings	539
4.17.2	Ventilation	541
4.17.3	Conservation of fuel and power	541
4.17.4	Combustion appliances and fuel storage systems	543
4.17.5	Drainage and waste disposal	543
4.17.6	Electrical work	543
4.17.7	Charging points for electric vehicles	549
4.17.8	Fire safety	553
4.17.9	Sanitation, hot water safety and water efficiency	557
4.17.10	Site preparation and resistance to contaminants and moisture	558
4.17.11	Resistance to the passage of sound	559
4.17.12	Protection from falling, collision and impact	559

Subsection	Description	Page
Requirements for buildings other than dwellings – Additional requirements		
4.17.13	Access to and use of buildings	559
4.17.14	On-site generation of electricity	560
4.17.15	Electric vehicle charging points	560
4.17.16	Conservation of fuel and power	563
4.17.17	Fire safety	563

Note: Although the main requirements for low and extra-low voltage electrical installations in or attached to a dwelling are contained in AD-P, other associated requirements related to BS 7671 are also scattered throughout the other ADs.

Reminder: Free downloads of all the diagrams and tables referenced in the subsections are available on-line at the Governments' Planning Portal: www.gov.uk/government/collections/approved-documents.

Requirements for dwellings

4.17.1 Access to and use of buildings

4.17.1.1 Services and controls

• To assist people who have reduced reach:
  • switches (and door bells, entry phones, light switches, power sockets, TV aerials and telephone jacks) should be 450–1,200mm above floor level, as shown in AD-M Diagram 1.5;
  • consumer units should be mounted with their switches 1,350–1,450mm above floor level.
	M 1.18

• Power-assisted opening shall be provided where the opening force of the door is more than 30N from 0° to 30° or more than 22.5N from 30° to 60° of the opening cycle.	M 3.14
• A fused spur, suitable for the fitting of a powered door opener, should be provided on the hinge side of the door.	M 3.22l

4.17.1.1.2 SERVICES AND CONTROLS FOR WHEELCHAIR USERS

To assist wheelchair users who have reduced reach, services and controls should comply with all of the following:

- The switches of consumer units should be mounted   M 3.44
  between 1,350mm and 1,450mm above floor level.
- Other than controls to radiators, switches and
  sockets, stopcocks and controls should be located with
  their centre line 700–1,000mm above floor level and
  not positioned *behind* appliances.
- Kitchen appliances in wheelchair-accessible dwellings
  should have isolators located within the same height
  range.
- Light switches should be on individual plates, unless
  wide rocker or full plate fittings are provided.
- Switches to double socket outlets should be located at
  the outer ends of the plate.
- A door entry phone with remote door release facility
  should be provided in the main living space and
  principal bedroom.
- An electrical power socket and telephone point
  should be provided together in the main living space.
- Boiler timer controls and thermostats are either
  mounted 900–1,200mm above the finished floor
  level on the boiler, or as separate, wired or wireless
  controllers in an accessible location within the same
  height range.

Provision should be made (by providing blank sockets, conduit and draw wires) in the principal bedroom to install (at perhaps a later date) bed-head controls, such as a two-way light switch, telephone and broadband socket, TV aerial and power socket outlets and a door entry phone.

4.17.1.2 *Wheelchair storage and transfer areas*

A dwelling should have a storage, battery charging and transfer space that will enable a person to charge and store up to two wheelchairs (one for outside mobility and the other for inside the dwelling).

- A power socket should be provided within a transfer   M 3.25c
  space as shown in AD-M Diagram 3.6.

4.17.1.3 Through-floor lifting device provision

Where the dwelling is defined as wheelchair accessible, a suitable through-floor lift or lifting platform should be installed and commissioned. The following electrical inputs are required:

• The power socket for a stairlift should be provided close to the liftway.	M 3.29d M 3.30d
• Doors to the lifting device should be power operated.	M 3.29g

4.17.2 Ventilation

4.17.2.1 Minor works

• In minor works such as providing self-contained mechanical ventilation or air-conditioning appliances, where the electrical work may be exempt from a requirement, it is best to notify the BCB in advance of the work being carried out.	F 0.10

4.17.3 Conservation of fuel and power

4.17.3.1 On-site generation of electricity

• When replacing an existing system, the installed generation capacity of the new system should be no less than that of the existing system.	L(V1) 5.6 L(V1) 6.66
• Heat pumps should meet the full space heating requirement.	L(V1) 5.12
• On-site electricity generation and storage systems should be appropriate for the size of the site, infrastructure and on-site energy demand.	L(V1) L2 L(V1) 6.64
• Once a system for on-site electricity generation has been commissioned, the person carrying out the work must notify the Local Authority that commissioning has taken place.	L(V1) 6.67
• On-site electricity generation within the conservatory or porch should have independent temperature and on/off controls.	L(V1) 10.8b

4.17.3.1.1   COMMISSIONING AN ON-SITE ELECTRICITY
            GENERATION SYSTEM

• On-site electricity generation systems must be commissioned to ensure that they produce as much electricity as is reasonable in the circumstances.	L(V1) 8.1

• A Commissioning Notice must be given to the relevant Building Control Body confirming that:
  • the commissioning plan has been followed;
  • all systems have been inspected and the test results confirm compliance with the design requirements.
	L(V1) 8.4 & 8.5

4.17.3.2 Electric controls

• Domestic hot water circuits that are supplied from a hot water store should have a time control and an electronic temperature control.	L(V1) 5.16
• For an electrically operated oil-fired vaporising appliance, an integral remote or thermostatic control should be provided.	L(V1) 6.10b

4.17.3.3 Electric space heating systems

• New or replacement electric storage heaters should be capable of controlling input charge and the rate of heat release.	L(V1) 6.12
• New or replacement electric panel heaters should be provided with time and temperature controls for either each room or each appliance.	L(V1) 6.13
• New or replacement electric warm air systems should have a programmable room thermostat or time switch for separately controllable heating zones.	L(V1) 6.10b

4.17.3.4 Electric heat pump heating systems

• Electrically driven air-to-air heat pumps with an output of 12kW or less should follow the Ecodesign Commission Regulation 2016/2281.	L(V1) 6.36

- The electrical input power of the primary pump in   L(V1) 6.45
  the solar water heating system measured in watts
  should be less than the higher of 50W or 2% of the
  peak thermal power of the collector.

4.17.3.5 Underfloor heating systems

Note: Room thermostats for electric underfloor heating systems should
have a manual override.

- Electric cables used for underfloor heating should be   L(V1) 6.33
  installed as individual layers.
- Electric cables used for underfloor heating night-   L(V1) 6.34
  energy storage should have:
  - fast-response systems, such as panel heaters;
  - controls which modify the input charge in response
    to either the room thermostat or floor-temperature
    sensing.
- Programmable room thermostats with an override   L(V1) 6.35
  feature should be provided for all direct electric zones
  of the electric underfloor heating system.

4.17.4 Combustion appliances and fuel storage systems

- LPG storage vessels and LPG-fired appliances that   J 3.5i
  are fitted with automatic ignition devices or pilot
  lights must **not** be installed in cellars or basements.

4.17.5 Drainage and waste disposal

- Care should be taken to ensure that the continuity of   H 1.35
  any electrical earth bonding to pipe fittings and joints
  is maintained.

4.17.6 Electrical work

AD-P **only** applies to electrical installations within dwellings that are
intended to operate at low or extra-low voltage and are:

- in or attached to a dwelling;
- in common parts of a building serving one or more dwellings, but excluding power supplies to lifts;
- in a building that receives its electricity from a source located within (or shared with) a dwelling;
- in a garden or land associated with a building where the electricity is coming from a source located within (or shared with) a dwelling.

AD-P covers electrical installations which are designed and installed so that:

- They afford appropriate protection against mechanical and thermal damage.
- They do not present electric shock and fire hazards to people.

AD-P, however, does **not** apply to electrical installations:

- In business premises located in the same building as a dwelling, but with separate metering.
- That supply the power for lifts in blocks of flats.

 Note: The scope of AD-P is illustrated in Diagram 1.

 All electrical work shall be carried out in accordance with BS 7671 *Requirements for Electrical Installations* (commonly referred to as the *IEE Wiring Regulations*).

4.17.6.1 Provision of information

- Sufficient information should be provided to ensure   P 1.2
  that people can operate, maintain and/or alter an
  electrical installation with reasonable safety.
- This information should comprise items listed in BS
  7671, including:
  - electrical installation certificates and reports;
  - permanent labels on items of electrical equipment, their connections, RCDs and earthing devices;
  - operating instructions and logbooks;
  - detailed plans (for unusually large or complex installations).

4.17.6.2 New dwellings

• Wall-mounted socket outlets, switches and consumer units in new dwellings should be easy to reach (especially important for an ambulant disabled person living in a dwelling).	P 1.4

4.17.6.3 New dwellings formed by a change of use

• Where a material change of use creates a new dwelling in a building, all work should be carried in compliance with the requirements of AD-P.	P 1.5

Although in some cases the existing electrical installation will need to be upgraded to meet current standards, **if** the existing cables are adequate, it is not necessary to replace them, even if they do use old (pre-2004) colour codes!

4.17.6.4 Additions and alterations to existing electrical installations

Note: When extending or altering an electrical installation, **only the new work** must meet current standards. There is **no** obligation to upgrade the existing installation unless:

• The new work adversely affects the safety of the existing installation.
• The state of the existing installation is such that the new work cannot be operated safely.

• **All** new work should be carried out in accordance with BS 7671.
• The existing electrical installation should be checked to ensure that:
  • the rating and condition of the existing equipment is sufficient to carry additional loads arising from the new work;
  • adequate protective measures are used;
  • the earthing and potential bonding arrangements are satisfactory.
	P 1.7

4.17.6.5 Cables to outside buildings

For cables to an outside building (e.g. a garage or shed):

• If run underground, cables should be routed and positioned so as to give protection against electric shock and fire as a result of mechanical damage to a cable.	BS 7671
• If concealed in floors and walls they may need to have an earthed metal covering.	

Note: Although BS 7671 is well structured, it is not the easiest of standards to get to grips with for a particular situation. Occasionally, it can be very confusing and requires the reader to constantly flick backwards and forwards through the book to find what it is all about.

For your assistance, one of Ray Tricker's other Pocket Books – *Wiring Regulations Pocket Book*, 1st edition – is available from: www.routledge.com/ Wiring-Regulations-Pocket-Book/Tricker/p/book/9780367760090.

4.17.6.6 Notifiable work

Note: Notifiable work involves activities (usually electrical) which should be notified by the installer to the Local Authority's Building Control.

• A person intending to carry out building work is required to give a Building Notice or a Full Plans Application if the work consists of:
  • the installation of a new circuit;
  • the replacement of a consumer unit;
  • any addition or alteration to existing circuits in a special location (such as within a room containing a bath or shower – see AD-P Diagram 2);
  • a room containing a swimming pool or sauna heater.
	P 2.5
	P 2.6

4.17.6.7 Non-notifiable work

Note: Non-notifiable work consists of additions and alterations to existing installations outside special locations, and replacements, repairs and maintenance anywhere.

- The following situations are non-notifiable: P 2.8
  - installing fixed electrical equipment;
  - installing a built-in cooker (unless a new cooker circuit is needed);
  - connecting an electric gate or garage door to an existing isolator switch;
  - installing prefabricated, modular wiring (e.g. for   P 2.9 kitchen lighting systems) linked by plug and socket connectors.

4.17.6.7.1 INSPECTION AND TESTING OF NON-NOTIFIABLE WORK

- Non-notifiable electrical installation work should be   P 3.13 designed, installed, inspected, tested and certificated in accordance with BS 7671.

Be careful: if Local Authorities find that non-notifiable work is unsafe and non-compliant, they can take enforcement action!

4.17.6.8 Certifications, inspection and testing

Note: To verify that the design and installation of electrical work is safe to use, maintain and alter, **all** electrical work should be inspected and tested in accordance with the procedures in BS 7671.

- One of the following procedures must be used to   P 3.1 certify that the work complies with the requirements set out in the Building Regulations:
  - self-certification by a registered competent person;
  - certification by a registered third-party certifier;
  - certification by a BCB.

4.17.6.8.1 SELF-CERTIFICATION BY A REGISTERED COMPETENT PERSON

Note: Electrical installers who are registered as competent persons should complete a BS 7671 electrical installation certificate for every job they undertake.

- The installer or the installer's registration body must,   P 3.4
  within 30 days of the work being completed:
  - provide a copy of the Building Regulations
    Compliance Certificate to the occupier;
  - give the certificate to the BCB.

4.17.6.8.2 CERTIFICATION BY A REGISTERED THIRD PARTY

- Before work begins, an installer who is not a   P 3.5
  registered competent person may appoint a registered
  third-party certifier to inspect and test the work as
  necessary.
- Within five days of completing the work, the installer   P 3.6
  must notify the registered third-party certifier who
  will then complete an electrical installation condition
  report and give it to the person ordering the work.

4.17.6.8.3 CERTIFICATION BY A BUILDING CONTROL BODY

If an installer is not a registered competent person and has not appointed
a registered third-party certifier, then before work begins the installer
must notify a BCB who will:

- Determine the extent of inspection and testing   P 3.9
  needed in order to establish that the work is safe,
  based on the nature of the electrical work and the
  competence of the installer.
- Perhaps choose to carry out any necessary inspection
  and testing itself, or it may contract a specialist to
  carry out some or all of the work and furnish it with
  an electrical installation condition report.

4.17.6.9 *Reports and certificates*

An Electrical Installation Certificate has to be provided for any new
installations or changes to existing installations.

• An Electrical Installation Certificate containing details of the installation, together with a record of the inspections made, test results obtained and recommendation when the next periodic inspection should occur, has to be provided for any new installations or changes to an existing installation.	BS 7671: 2018

4.17.7 Charging points for electric vehicles

4.17.7.1 All new homes to have electric car chargers **by law**

With effect from 2022, the Government have announced that:

> In England, **all new** residential buildings with associated parking spaces **must** have access to an electric vehicle charging point.
>
> The government said the move will see up to 145,000 charging points installed across the country each year.
>
> Newbuild supermarkets, workplaces and buildings undergoing major renovations will also come under the new law.

The consequence of this new law means that:

• Electric vehicle charging points must meet the following minimum standards. They shall be:
  • capable of providing a reasonable power output for each parking space for which it is intended to be used;
  • run on a dedicated circuit;
  • compatible with all vehicles which may require access to it.
	S 44J

Note: For your convenience, AD-S Diagram 6.4 shows the minimum space requirements for future floor-mounted electric vehicle charging points, whilst AD-S Diagram 6.5 shows the minimum space requirements for wall-mounted electric vehicle charging points.

4.17.7.2 Underground cable ducts

- Underground cable ducts should meet BS EN 61386-24   S 6.10
  and:
  - cable routes should be laid as straight as possible and
    with suitable access points, so as to enable cables to
    be pulled through at a later date;
  - the termination points of cable ducts should be sited
    where maintenance access is unrestricted.

 Connection points identified for future use should leave space to install electric vehicle charging points and sufficient vehicle access for installing, maintaining and using these charging points (as shown in AD-S Figure NEW S2).

4.17.7.3 Covered car parks

- The requirement to install electric vehicle charging points   S 2.10
  does not apply to parking spaces in a covered car park.   S 2.16

  - Where one or more associated parking spaces are   S 2.17
    within a covered car park, the requirement should be   S 5.4
    met by installing charge points in associated parking
    spaces that are outside the covered car park.

  - If there are not enough associated parking spaces   S 2.18
    outside of the covered car park, then cable routes   S 5.5
    should be installed for associated parking spaces
    within the covered car park.

- When constructing a new mixed-use building or   S 5.1
  completing a major renovation of a mixed-use building,
  electric vehicle charging infrastructure is required to be
  installed.

- Each electric vehicle charging point should be as   S 6.2
  described in BS EN 61851 and should comply with
  the requirements of BS 7671. The untethered electric
  vehicle charging point to be fitted with:
  - an indicator to show the equipment's charging status
    using lights, or a visual display.

 In exceptional circumstances, such as for a self-build property, if the vehicle requirements are already known, a tethered electric vehicle charging point may be acceptable.

4.17.7.4 Accessible parking spaces

• Where accessible parking spaces are associated with a new building (and future electric vehicle charging points are being provided), then at least one accessible parking space should have access to either a future connection location or an electric vehicle charging point. (See example at AD-S Diagram 6.2.)	S 6.4

Note: Future connection locations should be clearly marked on a site plan.

4.17.7.5 Cable routes

• Where a parking space requires cable routes, the following points should be taken into consideration:
  • sufficient space should be allocated to enable the electric vehicle charging point to be safely installed;
  • cable routes should be provided from a metered electricity supply point;
  • the allocated location should be clearly signed (see AD-S Diagram 6.3);
  • the requirements for concrete plinths (or footings), vehicle barriers, how the charge point will be earthed and possible upgrades to the electrical infrastructure should be determined.
	S 6.7

For more information and advice, see AD-S 6.9–6.12.

4.17.7.6 New dwellings

• All new residential buildings with associated parking spaces must have access to electric vehicle charging points.	S S1
• Where associated parking spaces are provided for a new residential building, the number of associated parking spaces will depend on the number of dwellings that the car park serves.	S 1.1 & 1.2

- If the number of associated parking spaces for the   S 1.3
new residential building is both more than 10 and
more than the number of dwellings, then cable
routes must be provided for all associated parking
spaces which do not have access to an electric vehicle
charging point.

4.17.7.7 Dwellings resulting from a material change of use

- Where one or more dwellings with associated parking   S S2
result from a building, or a part of a building,
undergoing a material change of use, at least one
associated parking space should have access to an
electric vehicle charging point for the use of each of
the dwellings.

- If an associated parking space is exempt from the   S 2.3
requirement to install an electric vehicle charging
point, the number of associated parking spaces with
access to either cable routes or an electric vehicle
charging point should be greater than either the
number of associated parking spaces or the number
of dwellings that the car park serves.

The following building types may receive special   S 2.4 & 2.7
consideration regarding installing electric vehicle   S 2.5 & 2.6
charging points:
- buildings of architectural and historical interest
and those identified as non-designated heritage
assets;
- buildings of architectural and historical interest
within National Parks, Areas of Outstanding
Natural Beauty, registered historic parks and
gardens, registered battlefields and World Heritage
Sites;
- Listed Buildings;
- buildings in Conservation Areas;
- scheduled monuments.

- If there are not enough associated parking spaces   S 2.18
outside of the covered car park to meet the
requirement to install electric vehicle charging points,
then cable routes should be installed for associated
parking spaces within the covered car park.

- If the electrical power supply to the building or car   S 2.9
park is insufficient for electric vehicle charging points
to be installed for all associated parking spaces:
  - as many electric vehicle charging points as can be
  accommodated within the existing power supply
  should be installed;
  - cable routes should be provided for the additional
  parking spaces which would have required an
  electric vehicle charging point.

4.17.7.8 Residential buildings undergoing major renovation

- If a residential building undergoing major renovation   S S3 &
will have more than 10 associated parking spaces   Regulation
after the renovation is completed, then at least one   44F
associated parking space must have access to an
electric vehicle charging point and cable routes for
electric vehicle charging points must be installed in all
additional associated parking spaces.

4.17.8 Fire safety

Although the Fire Regulations apply to **new** construction work, they do
not require existing buildings to be brought up to standard. However,
where new work is being carried out to existing buildings, such as alterations, extensions, loft conversions, window replacement, insulation and
so on, then the regulations **do** apply.

4.17.8.1 Smoke alarms

In most cases, the installation of smoke detectors in dwellings can significantly increase the occupant's safety by giving early warning of a fire out-
break, which is the reason now why all new dwellings should be provided
with a fire detection and fire alarm system in accordance with the relevant recommendations of BS 5839-6.

4.17.8.2 General points

It is essential that fire detection and fire alarm systems are properly designed, installed and maintained. Where a fire alarm system is installed, an installation and commissioning certificate should be provided.

Smoke alarms should **not** be located:

- over a stair or any other opening between floors;
- next to or directly above heaters or air-conditioning outlets;
- in bathrooms, showers, cooking areas or garages;
- in any place where steam, condensation or fumes could give false alarms;
- in places that get very hot (such as a boiler room);
- in places that get very cold (such as an unheated porch);
- on surfaces which are normally much warmer or colder than the rest of the space.

• There should be at least one smoke alarm on every storey of a dwelling.	B(V1) 1.9

• Smoke and heat alarms should:
  • be mains-operated;
  • conform to BS EN 14604 (for smoke) or BS 5446-2 (for heat);
  • have a standby power supply, such as a battery (rechargeable or non-rechargeable) or capacitor.
	B(V1) 1.1–1.4

• If a dwelling is extended, smoke alarms should:
  • be provided in all circulation spaces;
  • be positioned in the circulation spaces between sleeping spaces and places where fires are most likely to start (e.g. kitchens and living rooms).
	B(V1) 1.9

Where the kitchen area is not separated from a circulation space (or a stairway) by a door, then an interlinked heat detector or heat alarm should be installed (in the kitchen) in addition to whatever smoke alarms are required in the circulation space(s).

If more than one alarm is installed, the alarms should be linked so that the detection of smoke or heat by one unit operates the alarm signal in all of them.

4.17.8.3 Smoke alarms in thatched roofs

With roofing made of thatch or wood shingles, it is a good idea to extend the smoke alarm installation into the roof spaces.

• In thatched roofs, the smoke alarm installations should be extended into the roof space.	B(V1) 12.9

4.17.8.4 Electrically powered locks and door

• Electrically powered locks should return to the unlocked position if:
  • the fire detection and alarm system operates;
  • there is loss of power or a system error;
  • the security mechanism override is activated.
	B(V1) 3.92

Security mechanism overrides for electrically powered locks should be a Type A call point, as described in BS 7273-4.

• If revolving doors, automatic doors or turnstiles are placed across escape routes they should have:
  • a monitored failsafe system to open the doors if the mains electricity supply fails;
  • a failsafe system that provides outward opening from any open position.
• Non-automatic swing doors should be positioned immediately adjacent to the revolving or automatic door.
	B(V1) 3.98

Note: More guidance on door closing and 'hold open' devices for fire doorsets is contained in Appendix C to AD-B(V1).

4.17.8.5 Openings in cavity barriers and compartment walls or floors

• Openings in cavity barriers are permitted for the passage of cables or conduits containing cables.	B(V1) 5.24
• Openings in compartment walls or floors are permitted for the passage of service cables.	B(V1) 7.20

4.17.8.6 Lighting common escape routes

Note: All escape lighting should conform to BS 5266-1.

• If the mains electricity power supply fails, escape route artificial lighting should illuminate the route (including external escape routes).	B(V1) 3.41

• Escape lighting should be provided to:
  • emergency control rooms;
  • electricity and generator rooms;
  • switch and battery rooms providing emergency lighting systems;
  • toilet accommodation (with a minimum floor area of 8m²).
	B(V1) 3.42

• Escape stair lighting should be on a separate circuit from the electricity supply to any other part of the escape route.	B(V1) 3.43

4.17.8.7 Lighting diffusers

• Lighting diffusers (in the form of a light panel or cover) may be used as part of a luminaire or used below sources of light.	B(V1) 4.15

4.17.8.8 Protected stairways

• The space within a protected stairway should not be used for anything else other than a lift well or for electricity meters.	B(V1) 3.78

Note: All electricity meters should be in securely locked cupboards that are separated from an escape route by a fire-resisting construction.

4.17.8.9 Protected power circuits

• Cables in protected circuits should either be sufficiently robust or, where they may be exposed to damage, be included in a physically protected area.	B(V1) 3.46

- A protected circuit to operate equipment during a fire   B(V1) 3.47
  should:
  - only pass through parts of the building in which
    the fire risk is reasonably negligible;
  - be separate from any other circuit provided for
    another purpose.

Note: Guidance on cables for large and complex buildings is given in BS 5839-1, BS 5266-1 and BS 8519.

4.17.9 Sanitation, hot water safety and water efficiency

Electrical work associated with hot water systems should be carried out in accordance with BS 7671 and AD-P.

4.17.9.1 Water safety devices

- Non-self-resetting energy cut-outs (in compliance   G 3.28
  with BS EN 60335-2-73) may only be used where
  they would have the effect of instantly disconnecting
  the supply of energy to the storage vessel.

Where an energy cut-out is fitted, each heat source should have a separate non-self-resetting energy cut-out. Where relevant, temperature and pressure-activated safety devices should be fitted in addition to a safety device such as an energy cut-out.

- Temperature relief valves should conform to relevant   G 3.36
  national standards such as BS 6283–2.
- Temperature relief valves and combined temperature   C 3.35
  and pressure relief valves should **not** be used in systems
  which do not automatically replenish the stored water.
- Temperature relief valves should: G 3.37
  - give a discharge rating at least equal to the total   G 3.38
    power input to the hot water storage system;
  - be located directly on the storage vessel, to ensure
    the stored water does not exceed 100° C.

The temperature relief valve of a hot water storage system should:

- be factory fitted;
- **not** be disconnected except for replacement;
- **not** be relocated in any other device or fitting installed.

4.17.9.2 *Electric water heating*

• Electric fixed immersion heaters should comply with BS EN 60335-2-73.	G 3.43
• Electric instantaneous water heaters should comply with BS EN 60335-2-35.	G 3.44
• Electric storage water heaters should comply with BS EN 60335-2-21.	G 3.45

4.17.9.3 *Solar water heating*

• Where solar water heating systems are used, an additional heat source should be available in order to maintain the water temperature and thus to restrict microbial growth.	G 3.48
• Factory-made solar water heating systems should comply with BS EN 12976-1.	G 3.46
• Other solar water heating systems should comply with BS EN 12977-1.	G 3.47

4.17.10 Site preparation and resistance to contaminants and moisture

4.17.10.1 *Roofs – resistance to damage from interstitial condensation*

• To avoid roofs being damaged by an inflow of warm air and moisture, penetrations for electrical wiring should be filled and sealed. (This is particularly important in areas of high humidity, such as bathrooms and kitchens).	C 6.12

4.17.11 Resistance to the passage of sound

• Electric cables give off heat when in use and special precautions may be required when they are covered by thermally insulating materials.	E 3.21

4.17.12 Protection from falling, collision and impact

• If the controls for power-operated windows cannot be positioned within safe reach of a permanent stable surface, a safe manual or electrical means of remote operation should be provided.	K 8.1
• Power-operated doors and gates should include: 　• safety features (such as pressure-sensitive door edges which operate the power switch); 　• an identifiable and accessible stop switch; 　• manual or automatic opening if subject to a power failure.	K 10.1d

Requirements for buildings other than dwellings – Additional requirements

Other than for fire safety, the basic requirements for electrical safety in electrical installations are the same as for dwellings, with the addition of the following requirements.

4.17.13 Access to and use of buildings

• The use of self-closing devices should be minimised (particularly in parts of buildings used by the general public). • Where closing devices are required for fire control, electrically powered hold-open devices or swing-free closing devices should be used as appropriate. (See BS 8300 for further guidance on electrically operated door systems.)	M2 3.7

4.17.14 On-site generation of electricity

Note: Operating and maintenance instructions should be provided for on-site electricity generation.

> • On-site electricity generation should have automatic   L(V2) 6.77
>   controls that are suitable (or specifically made for)
>   generation and storage systems.

Note: If the controls are only 'on' and 'off' switches, then this particular service does **not** need to be commissioned.

> • On-site electrical generation within a conservatory   L(V2)
>   or porch should have independent temperature and   10.13
>   on/off controls.

4.17.15 Electric vehicle charging points

While AD-S provides guidance on the installation and location of electric vehicle charging points, the following ADs should also be taken into account:

•   AD-B: ensuring that fire safety requirements are met.
•   AD-K: for guidance on vehicle barriers and loading bays.
•   AD-M: in relation to the provision of manual controls within reasonable reach of the occupants, access requirements and accessible parking spaces.

> • Electric vehicle charging points must meet the   L(V2) Reg
>   following minimum standards: 44J
>   • be capable of providing a reasonable power output
>   for each parking space;
>   • be run on a dedicated circuit;
>   • be compatible with all vehicles which may require
>   access to it.
> • Each electric vehicle charging point should: S 6.2
>   • be designed and installed as described in BS EN
>   61851, and meet the requirements of BS 7671
>   (i.e. the Wiring Regulations);
>   • be fitted with an untethered charging point;
>   • have a minimum nominal rated output of 7kW;

- be fitted with an indicator showing the equipment's
  charging status;
- be suitable for use by electric vehicles with   S 6.3
  charging inlets in different places.

- Cable ducts should: S 6.10
  - meet the requirements of BS EN 61386-24;
  - be laid as straight as possible;
  - be sealed at the point of entry to the building;
  - have a draw rope if they are underground.

4.17.15.1 Accessible parking spaces

- Electric vehicle charging points shall be provided for   S 6.4
  at least one accessible parking space at the building
  (see AD-S Diagram 6.1).
- In some circumstances, the charge point will be   Diagram
  required for more than one parking space, in which   6.2
  case the charge point could be fitted with multiple   S 6.6
  outlets (as shown in AD-S Diagram 6.2).

4.17.15.2 Buildings other than residential buildings undergoing major renovation work

- Where a building – which is not a residential building   S5
  or a mixed-use building – is undergoing major   B 44H
  renovation and will eventually have more than 10
  parking spaces, then:
  - one of those parking spaces must have access to an
    electric vehicle charging point; and
  - cable routes for other charge points must be
    installed for up to a minimum of one-fifth of the
    total number of remaining parking spaces.

A 'minimum of one in every five' means that, for example, if there are
11 parking spaces, two parking spaces must have access to cable routes in
addition to the one parking space with an electric vehicle charging point.

- If all of the parking spaces are inside a covered car   S 4.6
  park, there is no need to install an electric vehicle
  charging point; however, cable routes must still be
  provided for a minimum of one in every five of those
  parking spaces.

4.17.15.3 Mixed-use buildings

- In a new mixed-use building – or a mixed-use   S 5.1–5.5
  building undergoing major renovation that contains
  one or more dwellings – which has parking spaces
  within a covered car park area as well as a number of
  parking spaces outside the building, any requirements
  to install electric vehicle charging points or cable
  routes must first be applied to those parking spaces
  outside the covered car park.

Note: For further information (although rather repetitive), see AD-S 5.1 to 5.5.

4.17.15.4 New buildings other than residential or mixed-use buildings

- If a new building (other than a residential or mixed-   S 3.2–3.4
  use building) has some of the parking spaces within
  the building whilst others are situated within the site
  boundary, electric vehicle charging points and cable
  routes should be made available as follows:
  - one of those parking spaces should have access to
    an electric vehicle charging point; and
  - cable routes for other charge points must be
    installed for up to a minimum of one-fifth of the
    total number of remaining parking spaces.

Note: If all of the parking spaces are situated inside the covered car park, then there is no need to install an electric vehicle charging point; however, cable routes must still be provided for a minimum of one in every five parking spaces.

4.17.16 Conservation of fuel and power

AD-L(V1) provides guidance on controls for fixed building services and on-site electricity generation. Manual controls, where provided, should be within reasonable reach of the occupants and the guidance in AD-K and AD-M should be followed.

4.17.17 Fire safety

4.17.17.1 Cavities

• Electrical wiring may be laid in a cavity on metal trays or metal conduit.	B(V2) 9.12

4.17.17.2 Escape lighting

Note: Escape lighting should conform to BS 5266-1.

• Electricity and generator rooms should be provided with escape lighting.	B(V2) Table 5.1
• All escape routes should have adequate artificial lighting. • If the mains electricity power supply fails, escape lighting should illuminate the routes (see AD-B(V2) Table 5.1).	B(V2) 5.25
• Escape stair lighting should be on a separate circuit from the supply to any other part of the escape route.	B(V2) 5.26

4.17.17.3 Electrically powered locks

• Electrically powered locks should return to the unlocked position if:
  • the fire detection and alarm system operates;
  • there is loss of power or a system error;
  • the security mechanism override is activated (as described in BS 7273-4).
	B(V2) 5.8

4.17.17.4 Fire detection and alarm systems

A fire detection and alarm system warns people when smoke, fire, carbon monoxide or other fire-related emergencies are detected. These alarms may be activated automatically from smoke detectors and heat detectors, or may also be activated via manual fire alarm activation devices such as manual call points or pull stations.

 Note: Depending on the fire strategy of the building, an electrically operated fire alarm system (that is operated by a fire detection system) should be provided in all buildings other than for some small buildings and/or premises (as discussed below).

• 'Residential (institutional)' and 'residential (other)' occupancies should be provided with automatic fire detection and alarm systems in accordance with BS 5839-1.	B(V2) 1.4
• Automatic fire detection and alarm systems should be provided in non-residential occupancies where a fire could break out in an unoccupied part of the premises.	B(V2) 1.5
• Automatic fire detection will also be necessary where fire protection systems (e.g. door releases) need to operate automatically.	B(V2) 1.6

However, every building design should be assessed individually (see BS 5839-1 Table A1 for further guidance) and, where necessary:

• Electrical alarm system call points should comply with either of the following:
  • BS 5839-2;
  • BS EN 54-11 Type A (direct operation).
• Call points should be installed in accordance with BS 5839-1.
	B(V2) 1.10

 BS EN 54-11 Type B (indirect operation) call points should **only** be used with the approval of the Building Control Body.

- Where people are unable to respond quickly to a   B(V2) 1.11
  fire warning, or are unfamiliar with the fire warning
  arrangements, then a voice alarm system complying
  with BS 5839-8 should be considered for certain parts
  of the building.

Note: Clause 18 of BS 5839-1 provides detailed guidance on the
design and selection of fire alarm warning for people with impaired
hearing.

Voice or visual fire alarm systems are particularly relevant:

- In premises where lots of members of the public are   B(V2) 1.12
  present and a general alarm could be undesirable
  (e.g. it could cause 'panic' evacuation!).
- Where phased evacuation is planned, and a staged   B(V2) 1.13
  alarm system is more appropriate.
- In buildings or part of a building where people may   B(V2) 1.15
  be in relative isolation.
- In buildings where the population is managed, a
  vibrating personal paging system could be the ideal
  answer.

4.17.17.4.1 DESIGN OF FIRE DETECTION AND ALARM SYSTEMS

Obviously, with regard to the safety of persons in a building, all fire
detection and alarm systems must be properly designed, installed and
maintained. To ensure this happens:

- A design, installation and commissioning certificate   B(V2) 1.16
  should be provided for every fire detection and alarm
  system.
- As fire detection and alarm systems sometimes trigger   B(V2) 1.17
  other systems (as described in BS 7273), the interface
  between systems must be reliable.

4.17.17.5 Revolving and automatic doors

- Where revolving doors, automatic doors and   B(V2) 5.15
  turnstiles are placed across escape routes, they should
  comply with one of the following:
  - automatic doors should have a failsafe system that
    provides outward opening from any open position
    which includes a monitored failsafe system to open
    the doors if the mains electricity supply fails;
  - non-automatic swing doors should be provided
    immediately adjacent to the revolving or automatic
    door or turnstile.

- Rolling shutters: B(V2) C9
  - should be capable of manual opening and closing
    for firefighting purposes;
  - across a means of escape should only be released
    by a heat sensor in the immediate vicinity of the
    door;
  - used as a means of escape should not be closed by
    smoke detectors or a fire alarm system.

4.18 Combustion appliances

Reminder: The subsections include details for buildings other than
dwellings only where they differ from the requirements for dwellings.

Subsection	Description	Page
Requirements for dwellings		
4.18.1	Combustion appliances and fuel storage systems	567
4.18.2	Conservation of fuel and power	572
4.18.3	Fire safety	572
4.18.4	Ventilation	572
4.18.5	Overheating	573
Requirements for buildings other than dwellings – Additional requirements		
4.18.6	Conservation of fuel and power	574
4.18.7	Ventilation	574

Reminder: Free downloads of all the diagrams and tables referenced in the subsections are available on-line at the Governments' Planning Portal: www.gov.uk/government/collections/approved-documents.

Requirements for dwellings

4.18.1 Combustion appliances and fuel storage systems

Any work associated with a combustion appliance is classed as notifiable building work and the BCB will need to be notified, before any work starts, unless work is carried out under a self-certification scheme or the work is repair or maintenance.

(Details of Competent Person's Schemes can be found at: www.comm unities.gov.uk/planningandbuildingregulations/competentpersons schemes).

In order that combustion appliances continue to work safely and effectively, it is **essential** that they are regularly serviced and maintained.

• Where the building work includes the installation of a combustion appliance, tests should cover flue pipes and the gas-tightness of joints between flue pipes and combustion appliance outlets.	J 1.55 J App E.E8

4.18.1.1 Air supply for combustion installations

Combustion appliances need to be installed so that there is an adequate supply of air to them for the process of combustion, in order for them to prevent overheating and for the efficient working of any flue.

• Air vent sizes, which are dependent upon the type of fuel burned, are for one combustion appliance only and so the air supply will need to be increased where a room contains more than one appliance.	J 1.3
• A room containing an open-flued appliance may need permanently open air vents (see AD-J Diagram 8).	J 1.4

- Compartments that enclose open-flued combustion   J 1.5
  appliances should be provided with vents large
  enough to admit the amount of air required by the
  appliance for combustion purposes and proper flue
  operation (see AD-J Diagrams 8 b and c).

- Where appliances require cooling air, compartments   J 1.6
  should be provided with high- and low-level vents,
  large enough to enable air to circulate (see AD-J
  Diagrams 8 d, e, f and g).

Where appliances are to be installed within balanced compartments, special provisions will be necessary.

In a flueless situation, air for combustion (and to carry away its products) can be achieved as shown in AD-J Diagram 8(h).

- If flued appliances are supplied with combustion air   J 1.9
  through vents opening into adjoining rooms, then the
  air vent in these adjoining rooms should have at least
  the same size of air vents direct to the outside.

Note: Air vents for flueless appliances, however, should open directly to the outside air.

4.18.1.2 Air vents

Permanently open air vents should be non-adjustable, sized to admit sufficient air for the purpose intended, and need to be positioned where they are unlikely to become blocked.

- Air vents should be sufficient for the appliances to be   J 1.11
  installed.

- Air vents should be sited outside fireplace recesses   J 1.11a
  and beyond the hearths of open fires so that dust or
  ash from the fire will not be disturbed by draughts
  (see AD-J Diagram 9 a and b).

In noisy areas, it may be necessary to install noise-attenuated ventilators to limit the entry of noise into the building.

- Grilles or meshes protecting air vents from the entry of animals (particularly rats and mice) or birds should have aperture dimensions no smaller than 5mm. — J 1.15

- Ventilation ducts or vents installed to supply air to a combustion appliance should **not** penetrate a building's structure, which may have airtight membranes (or radonproof membranes) in their floors. — J 1.17

Note: Rooms or spaces containing open-flued combustion appliances may need both permanent and adjustable ventilation.

- Open-flued combustion appliances and extract fans should be able to operate safely whether or not the fans are running. — J 1.20

- If a kitchen contains an open-flued gas appliance, the extract rate of the kitchen extract fan should not exceed 20l/s (72m³/h). — J 1.20a

- When installing ventilation for solid-fuel appliances, avoid installing extract fans in the same room. — J 1.20c

4.18.1.3 Bathrooms and shower rooms

- Open-flued oil-fired appliances should **not** be installed in bathrooms and bedrooms where there is an increased risk of carbon monoxide poisoning. — J 4.2

4.18.1.4 Flues

The most important pollutant(s) will vary between building types (e.g. dwelling, office or factory) and building use. However, combustion products from unflued appliances (e.g. gas, oil or solid-fuel cookers) are common pollutants.

To counteract this possibility, appliances (other than appliances **designed** to be flueless, of course!) should incorporate or be connected to suitable flues which discharge to the outside air.

- Flues should be built straight and narrow to offer the least resistance to the passage of gasses.   J 1.48

- Facilities should be made to enable flues to be swept and inspected. (See AD-J Diagram 14.)   J 1.49

- Where offset components are used, the size of a flue or duct (area, diameter, etc.) should be greater than the minimum required for the combustion appliance (see AD-J Diagram 7).   J 0.5

- New masonry chimneys should be constructed with flue liners (clay, concrete or pre-manufactured) and masonry (bricks, medium-weight concrete blocks or stone).   J 1.27

- Liners should be selected to form the flue (without cutting) and joints should be kept to a minimum.   J 1.28
- Liners should be placed with the sockets or rebate ends uppermost to contain moisture and other condensates within the flue.
- Joints should be sealed with fire cement or refractory mortar.
- Spaces between the lining and the surrounding masonry should **not** be filled with ordinary mortar.

 Chimneys and flues should provide satisfactory control of water condensation.

4.18.1.4.1 RE-USE OF EXISTING FLUES

- If a flue in an existing chimney needs to be re-used with a different type of flue or a different appliance rating, the flue and the chimney should be checked and, if necessary, altered to ensure compliance.   J 1.36

4.18.1.4.2 HEARTHS

- Hearths should be constructed using robust materials so that, in normal use, they prevent combustion appliances setting fire to the building fabric and furnishings. To achieve this, hearths should be made of solid, or non-combustible, material such as tiles, concrete or masonry. (See AD-J Diagram 24.)   J 2.22

4.18.1.5 *Carbon monoxide detectors and alarms*

• Carbon monoxide alarms should be: • located in the same room as the appliance; • provided in the dwellings room where a new or replacement fixed solid-fuel appliance is installed; • in compliance with BS EN 50291.	J 2.34 J 2.35

4.18.1.6 *Liquid fuel storage systems*

• Liquid fuel storage systems (including the pipework connecting them to the combustion appliances in the buildings) should be located and constructed so that they are reasonably protected from fires which may occur in buildings and surrounding areas.	J 5.1a

4.18.1.7 *Oil storage tanks and the pipes*

• Oil storage tanks and the pipes connecting them to combustion appliances shall: • be reasonably resistant to physical damage and corrosion; • minimise the risk of oil escaping during the filling or maintenance of the tank; • incorporate secondary containment when there is a significant risk of pollution; and • be labelled with information on how to respond to a leak.	J 5.1 b

4.18.1.8 *Notices*

• If a hearth, fireplace (including a flue box), flue or chimney is provided or extended as part of refurbishment work, a notice plate (see AD-J Diagram 16) should be permanently posted containing information essential to the correct application and use of these facilities.	J 1.57

4.18.2 Conservation of fuel and power

4.18.2.1 Replacing existing appliances

> • When replacing an existing appliance, the L(V1) 5.4
>   replacement appliance:
>   • should not be less efficient than the one it is
>   replacing;
>   • should either use the same fuel as the previous
>   service or use a different fuel that produces less
> CO_2 emissions per kWh of heat and have a less
>   primary energy demand per kWh of heat than the
>   appliance being replaced.

Note: Prior to actually replacing a heating appliance, consideration should be given to connecting the dwelling to an existing district heat network or community heating system.

4.18.3 Fire safety

Connecting flue pipes and factory-made chimneys should always be guarded if there is a possibility of them being damaged or if they could present a burn hazard.

> • If a flue passes through a compartment wall or floor B(V1) 9.23
>   (or is built into a compartment wall), each wall of
>   the flue should have a fire resistance of at least half
>   that of the corresponding wall or floor (see AD-B1
>   Diagram 16).

It is recommended that hidden voids in the construction should be sealed and sub-divided to inhibit the unseen spread of fire and products of combustion.

4.18.4 Ventilation

Combustion appliances require ventilation to supply them with air for combustion. Ventilation is also required to ensure the proper operation of flues or, in the case of flueless appliances, to ensure that the products of combustion are safely dispersed to the outside air.

If a self-contained mechanical ventilation or air-conditioning appliance is installed in a room that contains an open-flued combustion appliance, then it will **not** be considered minor work!

• Ventilation fans could cause combustion gases to spill from open-flued appliances and, instead of going up the flue or chimney, these gases could fill the room.	F(V1) 0.17 O 0.14
• Combustion appliances must operate safely whether or not fans are running.	F(V1) 0.18
• Ventilation systems should be designed to minimise the intake of external air pollutants.	F(V1) 2.1b

Occasionally, the guidance provided with AD-F(V1) may not be adequate to address pollutants from flueless combustion space heaters or from occasional, occupant-controlled events (such as painting, smoking, cleaning or other high-polluting activities) and occupants should be made aware of these possibilities.

4.18.5 Overheating

4.18.5.1 Ventilation fans

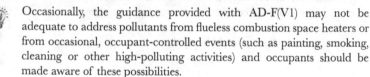

• Ventilation fans might cause combustion gases to spill from open-flued appliances and fill the room, instead of going up the flue or chimney. This can occur even if the combustion appliance and fan are in separate rooms.	O 0.14
• The guidance in AD-J should be followed when installing and testing ventilation appliances, and combustion appliances must operate safely whether or not fans are running.	O 0.15

Requirements for buildings other than dwellings – Additional requirements

This section provides the details for combustion appliances in buildings other than dwellings, only where they differ from the requirements already given above.

4.18.6 Conservation of fuel and power

The Target CO_2 Emission Rate (TER) is the minimum energy performance required for new buildings, and work on combustion appliances remains notifiable building work. So, basically:

- Where a biomass heating appliance is supplemented   L(V2) 2.6a
  by an alternative appliance such as gas, the CO2
  emission factor for the overall heating system should
  be based on a weighted average for the two fuels
  based on the anticipated usage of those fuels.
- If the same appliance is capable of burning both
  biomass fuel and fossil fuel, the CO2 emission factor
  for dual-fuel appliances should be used, except where
  the building is in a smoke-control area, in which case
  the anthracite figure should be used.

4.18.7 Ventilation

- When providing a self-contained mechanical   F(V2) 0.10c
  ventilation or air-conditioning appliance, it must not
  be installed in a room that contains an open-flued
  combustion appliance.

4.19 Hot water storage

Reminder: The subsections include details for buildings other than dwellings only where they differ from the requirements for dwellings.

Subsection	Description	Page
Requirements for dwellings		
4.19.1	Access to facilities	575
4.19.2	Commissioning heating and hot water systems	575
4.19.3	Hot water supply and systems	576
4.19.4	Hot water storage systems	577
4.19.5	Heated water storage for space or domestic hot water	578

Subsection	Description	Page
4.19.6	Sanitation, hot water safety and water efficiency	579
Requirements for buildings other than dwellings – Additional requirements		
4.19.7	Conservation of fuel and power	580

Reminder: Free downloads of all the diagrams and tables referenced in the subsections are available on-line at the Governments' Planning Portal: www.gov.uk/government/collections/approved-documents.

Requirements for dwellings

4.19.1 Access to facilities

- Switches for hot water appliances (especially for the assistance of disabled people) should be located between 400mm and 1,200mm above the floor, unless they are needed at a higher level for particular appliances.   M2 4.30

4.19.2 Commissioning heating and hot water systems

Commissioning must be carried out in such a way as not to prejudice compliance with any applicable health and safety requirements.

The life expectancy of a water heater is about 8 to 12 years according to the location and design of the unit, quality of installation, maintenance schedule and water quality.

- Hot water systems (and fixed building services), including controls, should be commissioned, by testing and adjustment, to ensure that they use no more fuel and power than is considered reasonable in the circumstances.   G 3.71

When commissioning a heating and/or hot water system, the following should be included in the test and inspection schedule:

| • Hot water outlet temperature devices being used to limit the maximum temperature supplied should **not** be capable of being easily altered by building users. | G3 |
| • All pipework has been designed and installed to minimise the transfer time between the hot water storage system and hot water outlets. | G3.7 |

In addition, when commissioning heating and hot water systems, the person carrying out the commission should ensure that:

* Independent temperature and on/off controls to all heating appliances have been provided.
* Both heating and cooling do not operate simultaneously.
* Energy meters have been included.
* Low or Zero Carbon (LZC) systems have been installed.
* Automatic meter reading and data collection have been provided if required.

 The person carrying out the work shall provide the Local Authority with a notice confirming that all fixed building services have been properly commissioned in accordance with a procedure approved by the Secretary of State.

| • The tundish should be vertical and be fitted as close as possible to, and lower than, the safety device, with no more than 600mm of pipe between the valve outlet and the tundish (see AD-G Diagram 1). | G 3.54 |

 Note: The tundish should incorporate a suitable air gap.

4.19.3 Hot water supply and systems

All electrical work associated with hot water systems shall be carried out in accordance with BS 7671 (*The Wiring Regulations*).

4.19.3.1 Limiting heat losses and gains

| • All pipes connected to hot water storage vessels should be insulated for at least 1m from the point at which they connect to the vessel. | L(V1) 4.24c |

• Hot water (heated wholesome water or heated softened water) shall be supplied to the sanitary appliances and locations without waste, misuse or undue consumption of water.	G 0.a
• Where the operating temperature of domestic hot water in a dwelling's storage vessel is capable of exceeding 80°C, it should be fitted with a device to ensure that the temperature supplied to the domestic hot water distribution system does **not** exceed 60°C.	G 3.64

• Hot water system components (including any cistern supplying water to, or from, the hot water system) shall continue to safely contain the hot water:
  • during normal operation of the hot water system;
  • following failure of any thermostat used to control temperature;
  • during the operation of any of the safety devices.
	G 3(2)

4.19.4 Hot water storage systems

Pipework should be designed and installed in such a way that it minimises the transfer time between the hot water storage system and hot water outlets.

• Hot water storage systems should be designed and installed in accordance with BS 6700 and/or BS EN 12897.	G3.10
• Temperature relief valves should be sized to give a discharge rating at least equal to the total power input to the hot water storage system.	G3.37
• Temperature relief valves should be located directly on the storage vessel to ensure that the stored water does not exceed 100°C, and to discharge the water in the event of serious overheating.	G 3.38

• Temperature relief valve(s) in hot water storage system units and packages should not be:
  • disconnected (other than for replacement);
  • relocated, fitted or installed in any other device.
	G 3.39

4.19.5 Heated water storage for space or domestic hot water

• Vessels that store heated water for a central heating or domestic hot water system should have standing losses that do not exceed the maximum heat losses shown in AD-L(V1) Table 4.5 for that particular hot water cylinder.	L(V1) 4.29
• Copper hot water storage units should comply with BS 3198. • Vented cylinders should comply with the heat-loss and heat-exchanger requirements of BS 1566-1or BS EN 12897 as appropriate. • Unvented hot water storage system products should comply with BS EN 12897.	L(V1) 4.30
• Primary storage systems should meet the insulation requirements of the Hot Water Association Performance Specification for Thermal Stores.	L(V1) 4.30
• Solar water heating systems should maximise the useful energy gain from the solar collectors and minimise the accidental loss of stored energy.	L(V1) 6.43b

4.19.5.1 Vented hot water storage systems

• Hot water storage vessels should conform to BS 853-1, BS 1566-1, or other relevant national standards.	G3.11
• Vented hot water storage systems should have a vent pipe connected to the top of the hot water storage system and above the level of the water in the cold water storage cistern.	G 3.12

• The system should incorporate either:
  • a non-self-resetting energy cut-out;
  • an overheat cut-out;
  • a temperature relief valve or a combined temperature and pressure relief valve that will discharge water to a tundish.
	G 3.18

Note: Vent pipes should discharge over a cold water storage cistern conforming to BS 417-2.

4.19.5.2 *Unvented storage system*

The installation of an unvented system is **notifiable** building work which must be reported to the BCB before work commences, **unless** the installer is registered with a Competent Person Scheme, in which case the installer may self-certify that the work complies with all relevant requirements in the Building Regulations.

• A hot water storage system that has an unvented storage vessel shall have at least two independent safety devices that release pressure and prevent the temperature of the stored water exceeding 100°C.	G 3.17
• Any unvented hot water storage system with a power input of more than 45kW (but a capacity of 500 litres or less) should meet the requirements of BS EN 12897.	G 3.22

A warning sign (see the example shown on page 21 of AD-G) should be indelibly marked on the hot water storage system unit or package so that it is visible after installation.

4.19.6 Sanitation, hot water safety and water efficiency

Legionella bacteria are found naturally in freshwater environments, like lakes and streams. The bacteria can become a health concern when they grow and spread in human-made building water systems like hot water tanks and heaters.

• Design measures to avoid legionella contamination should be included.	F 6.4

Requirements for buildings other than dwellings – Additional requirements

The basic requirements for hot water storage in buildings other than dwellings are the same as for dwellings, with the addition of the following requirements.

4.19.7 Conservation of fuel and power

• Domestic hot water storage vessels should meet either of the following:
   • maximum heat losses in AD-L(V2) Table 6.18.1;
   • maintenance consumption values in BS EN 89.
	L(V2) 4.26

• Before a new heating appliance is installed, all central heating and primary hot water circuits should be thoroughly cleaned and flushed out. • In hard water areas, suitable measures should be taken to protect the primary heating circuit against scale and corrosion.	L(V2) 5.13
• The minimum COP for heat pumps in new and existing buildings for domestic hot water heating is 2.0.	L(V2) Table 6.8

4.20 Liquid fuel storage

A service or fitting for fuel storage is classed as a 'controlled service or fitting'.

Reminder: The subsections include details for buildings other than dwellings only where they differ from the requirements for dwellings.

Reminder: Free downloads of all the diagrams and tables referenced in the subsections are available on-line at the Governments' Planning Portal: www.gov.uk/government/collections/approved-documents.

Requirements for dwellings

4.20.1 Fire safety

4.20.1.1 Special fire hazards

The following locations are where the storage and usage of fuel are considered to be places of special fire hazard:

- any room that houses a fixed internal combustion engine;
- an oil-filled transformer room;
- boiler rooms;
- buildings for industrial and commercial activities that sell fuels;
- enclosed car parks;
- storage spaces for fuel or other highly flammable substance(s);
- motorised switch gear rooms.

In these locations, special precautions should be taken and the regulations are quite specific, as shown in the following examples:

• In high fire-risk areas (e.g. kitchens, places of special fire hazard, or in proximity to where hot works occur), only class A1 cored panels should be used.	B(V1) 4.10(b)
• Separate natural smoke outlets should be provided from places of special fire hazard.	B(V1) 16.7
• Where a stair serves an enclosed car park or another place with a special fire hazard (such as a boiler room), the lobby or corridor should either have permanent ventilation or be protected by a mechanical smoke-control system.	B(V1) 3.75

4.20.1.2 Garages

Where a door is provided between a dwelling house and the garage – see AD-B(V1):

• The floor of the garage should be laid so as to allow fuel spills to flow away from the door to the outside.	B(V1) 5.7
• The door opening should be positioned at least 100mm above garage floor level.	

4.20.1.3　Escape routes from flats

• Common stairs forming part of the only escape route from a flat should not serve as fuel storage space.	B(V1) 3.73

4.20.2　Combustion appliances and fuel storage systems

4.20.2.1　Fuel storage areas

• Fuel storage areas shall be equipped with oil retention separators that have: 　• a nominal size (NS) equal to 0.018 times the contributing area; 　• a silt storage volume in litres equal to 100 times NS.	H App H3 A.7

4.20.2.2　Stairs serving ancillary accommodation

• Where a stair serves a place of special fire hazard (e.g. a boiler room), the lobby or corridor should either have permanent ventilation or be protected by a mechanical smoke-control system.	B(V1) 3.75

4.20.2.3　Liquid petroleum gas storage and supply

For oil and LPG fuel storage installations (including the pipework connecting them to the combustion appliances in the buildings they serve):

• LPG storage systems and appliances that are fitted with automatic ignitron devices must **not** be installed in basements and cellars. They should:	J 3.5i
• be located and constructed so that they are reasonably protected from fires that may occur in buildings or surrounding boundaries;	J 5.1a
• be reasonably resistant to physical damage and corrosion;	J 5.1a

- be designed and installed to minimise the risk of oil escaping during the filling or maintenance of the tank; J 5.1bi
- incorporate secondary containment when there is a significant risk of pollution; J 5.1bii
- contain labelled information on how to respond to a leak. J 5.1biii

Note: An LPG storage system of up to 1.1 tonne capacity, comprising one tank standing in the open air, should comply with the UKLPG Code of Practice 1: Bulk LPG Storage at Fixed Installations Part 1 (2009) and BS 5482-1.

- The separation distances and shielding of LPG storage tanks shown in AD-J Table 11 and Diagram 43 should be applied. J 5.14

Note: Drains, gullies and cellar hatches within the separation distances should be protected from gas entry.

- The LPG tank should be installed outdoors and **not** within an open pit. J 5.15
- The tank should be adequately separated from buildings, the boundary and any fixed sources of ignition, to enable safe dispersal in the event of venting or leaks and in the event of fire, to reduce the risk of fire spreading (see AD-J Diagram 43). J 5.15

Firewalls may be freestanding, built between the tank and the building, boundary and fixed source of ignition (see AD-J Diagram 43b), or as part of the building, or a fire-resisting boundary wall belonging to the property.

- Where a firewall is part of the building or a boundary wall, it should be located in accordance with AD-J Diagram 43c. J 5.16
- If the firewall is part of the building, it should be constructed as shown in AD-J Diagram 43d.

- Firewalls should: J 5.17
  - have no holes or openings;
  - be built of solid masonry, concrete or similar construction material;
  - have a fire resistance (insulation, integrity and stability) of at least 30 minutes.
- If firewalls are part of the building (as shown in AD-J Diagram 43d), they should have a fire resistance of at least 60 minutes.
- To ensure good ventilation, firewalls should **not** be built on more than one side of a tank.

- A firewall should be at least as high as the pressure   J 5.18
  relief valve.

Further guidance on LPG pipework can be found in the Gas Safety (Installation and Use) Regulations 1998 (GSIUR), the UKLPG Code of Practice 22 and the Institution of Gas Engineers and Managers standard IG/TD/4.

Where an LPG storage installation is made up of a set of cylinders, a way of meeting the requirements would be to follow the provisions below and as shown in AD-J Diagram 44 (J 5.19):

- Cylinders should: J 5.20
  - stand upright and be secured by straps (or chains) against a wall outside the building, in a well-ventilated position at ground level;
  - be provided with a firm, level base, such as concrete at least 50mm thick or paving slabs bedded on mortar;
  - be readily accessible.

4.20.2.4 Oil pollution

Fixed oil storage tanks and pipes that connect them should be constructed and protected so as to stop any oil spillage or seepage.

- Oil-fired appliances should not be installed in   J 4.2
  bathrooms, owing to the increased risk of carbon
  monoxide poisoning.

• Oil supply pipe systems should have a means of automatic isolation, such as a proprietary fire valve system, in accordance with the relevant recommendations in BS 5410-1.	J 5.6
• Secondary containment should be provided where there is a significant risk of oil pollution.	J 5.8
• Bunds, whether part of prefabricated tank systems or constructed on-site, should have a capacity of at least 110% of the largest tank they contain.	J 5.11
• An oil storage installation should carry a label in a prominent position giving advice on what to do if an oil spill occurs, and the telephone number of the Environment Agency's Emergency Hotline (also see AD-J Appendix F).	J 5.12

Oil storage below ground should only be considered when no other option is available. Underground tanks are difficult to inspect and leaks may not be immediately obvious.

4.20.2.5 *Protective measures against fire*

• To reduce the risk of fuel storage system fires igniting buildings (and also to provide against the installation becoming overgrown), a hard surface beneath the tank such as concrete, or paving slabs (at least 42mm thick and extending out at least 300mm beyond the perimeter of the tank), should be installed.	J 5.4
• Firewalls should be stable so that they do not pose a danger to people around them.	J 5.5
• The outlet from a flue should be fitted externally to the building (for more advice, see AD-J Diagram 41 and Table 41).	J 4.6

Note: In high fire-risk areas, places of special fire hazard, or places in the proximity to where hot works occur, only class A1 cored panels should be used. (See B(V1) Section 7 for the requirements of a load-bearing wall enclosing a place of special fire hazard.)

Requirements for buildings other than dwellings – Additional requirements

The basic requirements for liquid fuel storage in buildings other than dwellings are the same as for dwellings, with the addition of the following requirements.

4.20.3 Fire safety

Buildings for industrial and commercial activities that present a special fire hazard, e.g. those that sell fuels, may require additional fire precautions to those in ADs B(V1) and B(V2); nevertheless, it should be remembered that every place of special fire hazard should be enclosed with fire-resisting construction.

- A pipe that conveys oil (other than in the mechanism   B(V2) 8.36
  of a hydraulic lift) should not be installed in a
  protected shaft containing a protected stairway and/
  or a lift.

4.21 Kitchens and utility rooms

Reminder: The subsections include details for buildings other than dwellings only where they differ from the requirements for dwellings.

Subsection	Description	Page
Requirements for dwellings		
4.21.1	Access to and use of buildings	587
4.21.2	Electrical safety	588
4.21.3	Fire safety	588
4.21.4	Sanitation, hot water safety and water efficiency	590
4.21.5	Site preparation and resistance to contaminants and moisture	591
4.21.6	Resistance to the passage of sound	591
4.21.7	Ventilation	592
Requirements for buildings other than dwellings – Additional requirements		

Subsection	Description	Page
4.21.8	Drainage and waste disposal	595
4.21.9	Fire safety	595
4.21.10	Sanitation, hot water safety and water efficiency	596
4.21.11	Ventilation	596

Reminder: Free downloads of all the diagrams and tables referenced in the subsections are available on-line at the Governments' Planning Portal: www.gov.uk/government/collections/approved-documents.

Requirements for dwellings

4.21.1 Access to and use of buildings

4.21.1.1 Living, kitchen and eating areas

Note: If the dwelling is to be wheelchair adaptable, key parts of the accommodation, including kitchens, should be easily altered to meet the needs of a wheelchair user.

• Kitchen and eating areas in wheelchair-accessible dwellings should comply with all of the following:
  • the interconnection between the kitchen, dining and living areas should be convenient and step-free;
  • the kitchen and principal eating area should be within the same room, or connected to each other, and located within the entrance storey;
  • there should be a clear access zone in front of, and between, all kitchen units and appliances (see AD-M1 Diagram 3.8);
	M1 3.32

• the minimum length of kitchen worktops should meet the requirements of AD-M1 Table 3.3 and as shown in Diagram 3.8;	M1 3.33
• kitchen appliances should have isolators located at the same height range.	M1 3.44c

4.21.2 Electrical safety

AD-P (*Electrical safety*) applies to all electrical installations in shared amenities of blocks of flats (such as shared kitchens).

• Installing prefabricated, modular wiring (such as kitchen lighting systems) linked by plug and socket connectors is notifiable, but only if it involves: • the installation of a new circuit; • the replacement of a consumer unit.	P 2.9

4.21.3 Fire safety

4.21.3.1 Emergency escape routes

Note: Kitchens and utility rooms that are situated not more than 4.5m above ground level and whose only escape route is through another room, should be provided with an emergency egress window (see AD-B(V1) Diagram 2.5 for full details).

• A kitchen or utility room which is classified as an inner room may **not** be used as an escape route.	B(V1) 2.11 & 2.12

4.21.3.2 Insulated cored panels

• Kitchens are considered particularly high fire-risk areas and, similar to other places where hot works occur (or have become a special fire hazard), **only** class A1 cored panels (sometimes referred to as composite insulated panels) should be used.	B(V1) 4.10(b)

4.21.3.3 Kitchen areas

• The maximum number of people a kitchen or a utility room is designed to hold (usually called the 'floor space factor') is 7m^2 per person.	B(V1) Table D1

4.21.3.4 *Kitchen heat alarms*

• If a multi-storey flat does not have its own external entrance at ground level, a heat alarm is required in the kitchen (also see AD-B(V1) Diagram 3.4).	B(V1) 3.21

4.21.3.5 *Lifts*

• If a lift provides an exit into a protected corridor or protected lobby that is part of sleeping accommodation, and the lift also serves a storey containing a high fire-risk element (such as a kitchen), then the lift should also be separated from that high fire-risk area by a protected lobby or protected corridor.	B(V1) 3.103

4.21.3.6 *Smoke alarms*

It has been recommended that smoke alarms should be positioned between sleeping places and where fires are most likely to start (e.g. kitchens and living rooms).

4.21.3.7 *Ventilation and air-conditioning systems*

• In mixed-use buildings, non-domestic kitchens: • should have a separate, independent, extraction system; and • extracted air should not be recirculated.	B(V1) 9.10
• Thermally activated fire dampers and automatically activated fire and smoke dampers should **not** be used for extract ductwork serving kitchens.	B(V1) 9.15

4.21.4 Sanitation, hot water safety and water efficiency

4.21.4.1 Chemical or composting toilets

• Chemical toilets or composting toilets may be used where the waste can be removed from the premises without carrying it through any food preparation areas or kitchens.	G 4.19

4.21.4.2 Cold water supplies

The cold water supply shall:

• Convey wholesome water or softened wholesome water without waste, misuse, undue consumption or contamination of water.	G1
• Have a pressure and flow rate sufficient for the operation of all appliances and locations planned in the building.	

4.21.4.3 Cold water storage systems

• Cold water storage cisterns into which a vent pipe discharges should be supported on a flat, level and rigid platform.	G 3.15
• The cistern should be accessible for maintenance, cleaning and replacement.	G 3.16

4.21.4.4 Sanitary conveniences and washing facilities

• Any room containing a sanitary convenience, a bidet, or any facility for washing hands **must** be separated from any kitchen or any area where food is prepared.	G4(3)
• If a door is used to separate a kitchen (or room being used for the preparation of food), it should be positioned as shown in AD-G Diagrams 2 and 3.	G 4.10

- A suitable sink must be provided in any area where   G 6.5
  food is prepared and this sink should discharge
  through a grating, a trap and a branch discharge pipe
  to an adequate system of drainage.

Note: Where a dishwasher is provided in a separate room that is not the principal place for the preparation of food, an additional sink need not be provided in that room.

4.21.4.5 *Water efficiency*

- The figures shown in AD-G Table A1 should be used   G A6
  calculate the consumption of each kitchen fitting in
  litres per person per day.

4.21.5 Site preparation and resistance to contaminants and moisture

4.21.5.1 *Floor coverings*

- Floorboards in kitchens and utility rooms should be   C 4.15
  moisture resistant.

4.21.5.2 *Roofs*

To avoid excessive moisture transfer to roof voids, the gaps and penetrations for pipes and electrical wiring within kitchens should be filled and sealed.

- A roof will meet the requirement if it is designed   C 6.10
  and constructed in accordance with Clause 8.4 of BS
  5250. Further guidance is given in the BRE Report
  BR 262.

4.21.6 Resistance to the passage of sound

- A wall separating a kitchen and a refuse chute should   E 2.28
  have mass of at least 1,320 kg/m2.

Note: Tests should be completed in unfurnished rooms and available spaces, but only if all kitchen units and wall cupboards, etc., have their doors open and are unfilled during testing.

4.21.7 Ventilation

An adequate ventilation system should be capable of extracting water vapour from areas where it is produced in significant quantities, particularly in kitchens and utility rooms.

4.21.7.1 Control of ventilation

Humidity-controlled devices should regulate the humidity of the indoor air and minimise the risk of condensation and mould growth.

• Continuously running fans (although normally set to operate without occupant intervention) may have manual or automatic controls for selecting a higher rate of operation, set as follows:
  • manual high-rate controls should be provided locally to the actual spaces (e.g. bathrooms and kitchens) being served;
  • automatic controls which can be adjusted according to the amount of humidity, occupancy, usage and pollutant release.
	F(V1) 1.35
	Table 1.4

• Humidity-controlled devices should regulate the humidity of the indoor air and minimise the risk of condensation and mould growth.	F(V1) 1.36

Note: Background ventilators with automatic controls should also have manual override.

4.21.7.2 Extract ventilation

Note: Kitchens, utility rooms, bathrooms and sanitary accommodation should all have extract ventilation to the outside.

- The minimum extract ventilation rates for   F(V1) 1.19
  intermittent extract systems is:
  - kitchen (cooker hood extracting to the outside) –
    30l/s;
  - kitchen (no cooker hood or cooker hood does not
    extract to the outside) – 60l/s;
  - utility room – 30l/s.
- The minimum extract ventilation rates for
  continuous extract systems is:
  - kitchen – 13l/s;
  - utility room – 8l/s.

4.21.7.3 *Natural ventilation*

Note: All rooms with external walls should have background ventilators.

- For kitchens and utility rooms, the extract rate can be   F(V1) 1.47
  met by using an intermittent extract fan.
- In a room with no openable window, an intermittent   F(V1) 1.51
  extract fan should be capable of operating for at least
  15 minutes after the room is vacated.
- All rooms with external walls should have background   F(V1) 1.52
  ventilators.
- If the dwelling has a kitchen and a living room which   F(V1) 1.58
  are ***not*** separate rooms, at least three ventilators
  should be provided in the open-plan space.

A window with a night latch position is not adequate for background
ventilation because of the risk of draughts and the difficulty in measuring
the equivalent area – but mainly from a security viewpoint.

4.21.7.4 *Refurbishing kitchens or bathrooms in existing dwellings*

- Where building work is carried out in a kitchen   F(V1) 3.30
  or bathroom, any existing fans should be retained
  (provided that they are in good working order) or
  replaced.

If the replacement is one of a similar type, and uses the existing cabling, this need not be notified to a BCB.

• If there is no ventilation system in the original room, it is not necessary to provide one in the refurbished room. However, additional ventilation may be necessary if refurbishment work is likely to make the building less compliant with the ventilation requirements of the Building Regulations than it was before the work was carried out.	F(V1) 3.31
• If an extractor fan or cooker hood is replaced and it uses the existing cabling (and assuming that it still meets the requirements of the Wiring Regulators), then this does not need to be notified to the local BCB.	F(V1) 3.32

4.21.7.5 Work on existing dwellings

• When building work is carried out that will affect the ventilation of the existing dwelling, it should either meet the standards in the relevant AD or be no less satisfactory than before the work was carried out.	F(V1) 3.2
• Other ventilation solutions than those detailed here may be used, as long as it can be demonstrated to a BCB that they comply with the requirements.	F(V1) 3.4
• Replacing a window is likely to increase the airtightness of the dwelling. If ventilation is not provided via a mechanical ventilation with heat recovery system, then increasing the airtightness of the building may reduce beneficial ventilation in the building. In these circumstances, incorporating background ventilators in the replacement windows may become a necessity.	F(V1) 3.15

Requirements for buildings other than dwellings – Additional requirements

The basic requirements for kitchens and utility rooms in buildings other than dwellings are the same as for dwellings, with the addition of the following requirements.

4.21.8 *Drainage and waste disposal*

• Drainage serving kitchens in commercial hot food premises should be fitted with a grease separator, or other effective means of removing grease, to comply with BS EN 1825-1.	H 2.21

4.21.9 *Fire safety*

4.21.9.1 *Floor space factor*

• The 'floor space factor' for a kitchen is $7m^2$ per person.	B(V2) D1

4.21.9.2 *Residential care homes – ancillary accommodation*

• In care homes, ancillary accommodation such as a kitchen should be enclosed by fire-resisting construction (minimum REI 30).	B(V2) 2.44

4.21.9.3 *Small premises*

• Any kitchen or other open cooking arrangements should be at the extremity of any dead end that is remote from the exits.	B(V2) 4.2
• Other than in kitchens (and ancillary offices and stores), floor areas should be undivided so that exits are clearly visible from all parts.	B(V2) 4.4

4.21.9.4 *Storeys divided into different uses*

• If a storey contains areas for consuming food and/ or drink, and this is not the main use of the building, then:
  • a minimum of two escape routes should be provided from each area; and
  • the escape routes should lead directly to a storey exit without entering an area of high fire hazard such as an associated kitchen.
	B(V2) 2.16

4.21.9.5 Ventilation

• Non-domestic kitchens should have separate and independent extraction systems.	B(V2) 10.10

Note: Extracted air should not be recirculated!

4.21.9.6 Wall linings

• In high fire-risk areas, such as kitchens, only class A1 cored panels should be used.	B(V2) 6.11b

4.21.10 Sanitation, hot water safety and water efficiency

• In small buildings (other than dwellings), a sink should be provided in any kitchen or place used for the preparation of food.	G 6.1, G 6.3
• Sanitary conveniences and/or associated hand-washing facilities should be separated by a door from any place used for the preparation of food – ***including*** a kitchen. (For further guidance see AD-G Diagram 3 and BS 6465-2.)	G 4.17
• In addition to any hand-washing facilities associated with WCs, separate hand-washing facilities should also be provided in kitchens.	G 6.4

4.21.11 Ventilation

Pollutants will vary between building types, building use (e.g. shop or commercial kitchen), and from room to room within a building (e.g. kitchen or photocopier room).

• The ventilation for catering and commercial kitchens should comply with: • HSE Catering Information Sheet No. 10: Ventilation in catering kitchens (2017); • BESA DW 172 Specification for Kitchen Ventilation Systems (2018); • CIBSE Guide B2 Ventilation and Ductwork (2016).	F(V2) Table 1.1

- When incorporating background ventilators in the replacement windows for a domestic-type kitchen, you should achieve a minimum of 8,000mm² equivalent area.   F(V2) 6.3a
- Extract ventilation should be provided in food and beverage preparation areas in offices.   F(V2) 1.24, 1.28, B7
- Food and drink preparation areas should have an intermittent air extract rate of:
  - in areas only for using a microwave and preparing drinks – 15 litres per second;
  - in areas for using a domestic-type hob or cooker and the extract ventilator is adjacent to the hob/cooker – 30 litres per second;
  - in areas for using a domestic-type hob or cooker and the extract ventilator is remote from the hob/cooker – 60 litres per second.

Note: For food and beverage preparation areas, the extract rates used for dwellings have been applied.

Note: Extract ventilation rates for food and beverage preparation areas and commercial kitchens are shown in AD-F Table 5.1a.

4.22 Loft conversions

Reminder: The subsections include details for buildings other than dwellings only where they differ from the requirements for dwellings.

Subsection	Description	Page
Requirements for dwellings		
4.22.1	Cavity insulation	598
4.22.2	Fire safety	598
4.22.3	Ventilation	599
4.22.4	Conservation of fuel and power	599
4.22.5	Protection from falling, collision and impact	600
4.22.6	Resistance to contaminants	601
4.22.7	Resistance to the passage of sound	601

Subsection	Description	Page
4.22.8	Sanitation, hot water safety and water efficiency	602
4.22.9	Security	602
Requirements for buildings other than dwellings – Additional requirements		
4.22.10	Access to and use of buildings	603
4.22.11	Fire safety	603

Reminder: Free downloads of all the diagrams and tables referenced in the subsections are available on-line at the Governments' Planning Portal: www.gov.uk/government/collections/approved-documents.

Requirements for dwellings

4.22.1 Cavity insulation

Note: Precautions should be taken to prevent the permeation of any toxic fumes from insulating material into a cavity in a cavity wall.

4.22.2 Fire safety

- If a new storey is created above 4.5m: B(V1) 2.21
  - the full extent of the escape route should be addressed;
  - fire-resisting doors (minimum E 20) and partitions (minimum REI 30) should be provided;
  - upgrade the existing doors where necessary;
  - new partitions should be provided to enclose the escape route if the layout is open plan.

- Alternatively, the conversion would comply by   B(V1) 2.23
  providing:
  - sprinkler protection to the open-plan areas;
  - a fire-resisting partition (minimum REI 30) and door (minimum E 20) to separate the ground storey from the upper storey.

> **Note:** Where it is undesirable to replace existing doors B(V1) 2.22
> because of their historical or architectural merit, the
> possibility of retaining or upgrading them should be
> investigated.

4.22.2.1 Floors in loft conversions

> • If an additional storey to a two-storey single family B(V1) 5.4
>   dwelling is added, then the following should have a
>   fire resistance minimum rating of R 30:
>   • all new floors;
>   • any floor that is part of the enclosure to the
>   circulation space between the loft conversion and
>   the final exit;
>   • the existing first-storey construction.

4.22.3 Ventilation

4.22.3.1 Loft insulation

> • All of the following are classed as minor works: F(V1) 3.8
>   • renewing loft insulation (including effective edge
>   sealing at junctions and penetrations);
>   • changing a cold loft (insulation at ceiling level) to a
>   warm loft (insulation at roof level);
>   • replacing a loft hatch with a sealed/insulated unit.

4.22.4 Conservation of fuel and power

4.22.4.1 Change of energy status

> • A 'change to energy status' is when parts of a heated L(V1) 11.6
>   dwelling (such as a converted loft) have now become
>   part of the dwelling. In these cases, that space (which
>   was previously exempt from the energy-efficiency
>   requirements) has now become part of the heated
>   dwelling and, therefore, a change to energy status
>   now applies to that particular space.

If a previously unheated loft is converted into a flat, and its energy status has changed, this may be classified as a material change of use.

4.22.4.2 Renovated and retained elements

• Elements being retained in existing dwellings (e.g. through a loft conversion) with a U-value that is higher than the threshold value should be upgraded. (See AD-L(V1) Table 4.3.)	L(V1) 4.12b

4.22.4.3 Loft insulation

• Loft insulation at the eaves should extend beyond the wall insulation without any reduction in thickness due to the pitch of the roof.	L(V1) 4.17 i
• If the eaves are still accessible, roof insulation should be installed.	
• At gables and party walls, insulation should extend to the wall.	

 Note: If the space between the wall and joist is less than 100mm, perimeter insulation may be required.

Where the roof is insulated at ceiling level, loft hatches should be installed to ensure optimum airtightness.	L(V1) 4.21 i
• When changing the ceiling to a cold loft space, existing insulation at ceiling level will need to be removed as part of the works, but you should provide loft insulation (such as 250mm of mineral or cellulose fibre) as a 'quilt' between and across ceiling joists.	L(V1) Table

 Note: This fibre insulation may be boarded over if required.

4.22.5 Protection from falling, collision and impact

• All escape routes should be provided with a minimum clear headroom of 2m (except in doorways).	K1 1.13

Note: If there is not enough space for a clear headroom of 2m, the headroom may be reduced as shown in AD-K Diagram 14.

* The construction of an alternating tread stair should   K 1.30
  conform to AD-K Diagram 1.10, and should also
  have:
  * uniform steps with parallel nosings;
  * slip-resistant surfaces on treads;
  * tread sizes over the wider part of the step in line
    with AD-K Table 1.1;
  * a suitable construction so that a 100mm diameter
    sphere cannot pass through the open risers;
  * a minimum clear headroom of 2m.

Note: This requirement applies to stairs that form part of the building.

4.22.6 Resistance to contaminants

* All new buildings, extensions and conversions,   C 2.39
  whether residential or non-domestic, that are built in
  areas where there may be elevated radon emissions
  (e.g. the West Country) may need to incorporate
  precautions against this chemical element.

4.22.7 Resistance to the passage of sound

* Internal walls between a bedroom (or a room   E 2
  containing a water closet) and another room should
  provide reasonable resistance to sound.

New walls and floors within a dwelling should provide a minimum sound insulation value of 40 R_W dB.

4.22.7.1 Properties sold before fitting out

* Loft apartments sold before being fitted out with   E 1.28
  internal walls and other fixtures and fittings should
  ensure that the fitting out does not interfere with
  existing sound insulation.

4.22.8 Sanitation, hot water safety and water efficiency

• There must be a suitable installation for the provision of wholesome water.	G1
• A hot water system, including any cistern or other vessel that supplies water to or receives expansion water from a hot water system, should resist any effects of temperature and pressure that occur either in normal use or in the event of a malfunction.	G 3(2)
• A hot water system that has a hot water storage vessel shall incorporate precautions to prevent the temperature of the water stored in the vessel at any time exceeding 100° C; and ensuring that any discharge from safety devices is safely conveyed to where it is visible but will not cause a danger to persons in or about the building.	G 3(3)

4.22.9 Security

• Accessible doorsets that provide access into a dwelling should be secure doorsets.	Q 1.1
• Ground floor, basement and other easily accessible windows (including rooflights) should have secure windows.	Q 2.1
• Any glazing (which if broken, would permit someone to insert their hand and release the locking device on the inside of the door) should be a minimum of class P1A in accordance with BS EN 356:2000.	Q B.11
• Double- or triple-glazed units need to incorporate only one pane of class P1A glass.	

Requirements for buildings other than dwellings – Additional requirements

This section provides the details for the requirements for loft conversions in buildings other than dwellings, only where they differ from the requirements already given above.

4.22.10 Access to and use of buildings

• Wheelchair platform stairlifts should only be considered for conversions and alterations where it is not practicable to install a conventional passenger lift or a lifting platform.	M2 3.44
• Wheelchair platform stairlifts should not be installed where their operation restricts the safe use of the stair by other people.	

4.22.11 Fire safety

• When an existing building is converted into flats and there is a material change of use, despite potential difficulties in meeting the provisions for fire resistance, they ***must*** be overcome.	B(V2) 7.9

4.23 Extensions and additions to buildings

Reminder: The subsections include details for buildings other than dwellings only where they differ from the requirements for dwellings.

Subsection	Description	Page
Requirements for dwellings		
4.23.1	Access to and use of buildings (Optional requirements)	604
4.23.2	Fire safety	605
4.23.3	Conservation of fuel and power	606
4.23.4	Ventilation	608
4.23.5	Sanitation, hot water safety and water efficiency	609
4.23.6	Security	610
4.23.7	Resistance to contaminants	610
4.23.8	Resistance to the passage of sound	611
Requirements for buildings other than dwellings – Additional requirements		
4.23.9	Access to and use of buildings	612

| 4.23.10 | Conservation of fuel and power | 613 |
| 4.23.11 | Ventilation | 614 |

Reminder: Free downloads of all the diagrams and tables referenced in the subsections are available on-line at the Governments' Planning Portal: www.gov.uk/government/collections/approved-documents.

Requirements for dwellings

4.23.1 Access to and use of buildings (Optional requirements)

- For the assistance of disabled people, the principal   M4(2) 2.20
  private entrance, or the alternative private entrance
  where step-free access cannot be achieved, should (if
  possible) comply with all of the following:
  - there should be a level external landing with a
    minimum width and depth of 1,200mm;
  - the landing should be covered for a minimum
    900mm (width) and 600mm (depth);
  - lighting should use fully diffused luminaires,
    activated automatically by a dusk to dawn timer or
    by detecting motion;
  - the door has a minimum clear opening width of
    850mm;
  - where there are double doors, the main (or leading)
    leaf provides the required minimum clear opening
    width;
  - a minimum 300m door nib is provided to the
    leading edge of the door, and the extra width
    created by this nib is maintained for a minimum
    distance of 1,200mm beyond it;
  - the depth of the reveal on the leading side of the
    door (usually the inside) is a maximum of 200mm;
  - the threshold is an accessible threshold;
  - where there is a lobby or porch, the doors are a
    minimum of 1,500mm apart and there is at least
    1,500mm between door swings.

- All other external doors – including doors to and
  from a private garden, balcony, terrace, garage,
  carport, conservatory or storage area that is integral
  with, or connected to, the dwelling – should, where
  possible, also comply with the provisions above, and
  should have a minimum 300m nib to the leading edge
  of the door, with the extra width created by this nib
  extending for a minimum 1,800mm beyond it.

  M4(2) 2.21
  M4(3) 3.23

4.23.2 *Fire safety*

The chances of fire spreading across an open space between buildings
should be limited as much as possible.

4.23.2.1 *Escape routes*

- Escape routes into an enclosed courtyard or garden
  should be of sufficient size to exceed the height of
  the dwelling and any extensions. (See AD-B(V1)
  Diagram 2.5.)

  B(V1) 2.10

4.23.2.2 *Fire detection and alarm systems*

- Where new habitable rooms are provided, a fire
  detection and alarm system should be installed.

  B(V1) 1.8

- Smoke alarms should be provided in the circulation
  spaces of the dwelling.

  B(V1) 1 1.9

4.23.2.3 *Ventilation of protected shafts conveying gas*

- Any extension of the storey floor into the protected
  shaft should not compromise the free movement of
  air throughout the entire length of the shaft.

  B(V1) 7.28

For further guidance on protected shafts conveying piped flammable gas,
including the size of ventilation openings, see BS 8313.

4.23.3 Conservation of fuel and power

4.23.3.1 Extensions to and work on existing dwellings

For all buildings (particularly extensions to existing buildings), consideration must be given to the conservation of fuel and power in buildings by:

• Limiting heat gains and losses through thermal elements, building fabric, pipes, ducts, space heating, space cooling and hot water services.	L(V1) 0.7
• Providing fixed building services which are energy efficient, have effective controls, and are commissioned to ensure they use no more fuel and power than is necessary.	
• Limiting the U-values of new fabric elements in extensions to existing dwellings. (For details of these values, see AD-L(V1) Table 4.2.)	L(V1) 4.9

• Ensuring that new and replacement windows, roof windows, rooflights and doors:
  • are draught-proofed;
  • have insulated cavity closers, where appropriate;
  • meet the minimum standards shown in AD-L(V1) Table 4.2.
	L(V1) 10.3

• If a door is enlarged or a new one is created, either the area of windows, roof windows, rooflights and doors should not exceed 25% of the total floor area of the dwelling, or compensating measures should be taken to improve the energy efficiency of the dwelling.	L(V1) 10.5

For an existing dwelling with a total useful floor area of over 1000m², additional work may be required to improve the overall energy efficiency of the dwelling, particularly if the proposed work consists of, or includes, an extension.

The term 'controlled fitting' refers to the **entire** unit of a window, roof window, rooflight or door, including the frame. Replacing glazing, a window or a door in its **existing** frame is **not** providing a controlled fitting and such work, therefore, does not need to meet energy-efficiency requirements.

- When extending an existing dwelling with a total useful floor area of over 1,000m², consequential energy-efficiency improvements may be required.   L(V1) 10.11

4.23.3.2 *Conservatories and porches in new dwellings*

- Where a conservatory or porch is installed as part of the construction of a new dwelling, providing that there is adequate thermal separation between the dwelling and the conservatory or porch, and the dwelling's heating system has not been extended into the conservatory or porch, it can be treated as if it were an extension being added onto an existing building.   L(V1) 0.6
- If either or both of the above provisions have not been achieved, then the conservatory or porch should be treated as a room in the new dwelling.

4.23.3.3 *Exemptions – conservatories, porches and covered areas*

- Where a dwelling is extended by adding a conservatory or porch, the work is exempt from the energy-efficiency requirements provided that all of the following apply: L(V1) 0.14
  - the extension is at ground level;
  - the floor area of the extension does not exceed 30m²;
  - the glazing complies with Approved Document K;
  - any wall, door or window that separates the extension from the dwelling has been retained or, if removed, has been replaced with a wall, door or window that meets the required standards;
  - the heating system of the dwelling is not extended into the conservatory or porch, and nor does the extension have its own fixed heating appliance.

- Where a dwelling is extended by adding a carport that is open on at least two sides, a covered yard, a covered walkway or a covered driveway, the work is exempt from the energy-efficiency requirements if the extension is at ground level and the floor area of the extension does not exceed 30m².   L(V1) 0.15

4.23.3.4 Historic and traditional buildings

When undertaking work on any historic or traditional building, the aim should always be to improve energy efficiency as far as is reasonably practicable **without** prejudicing the character of the host building or increasing the risk of long-term deterioration of the building fabric or fittings.

> • The energy efficiency of historic and traditional   L(V1) 0.10
>   dwellings should be improved *only* if this will not
>   cause long-term deterioration of the building's
>   existing fabric or fittings.

Note: This particularly applies to historic and traditional buildings with a vapour permeable construction (such as wattle and daub, cob or stone, and constructions using lime render or mortar) that both absorbs moisture and also readily allows moisture to evaporate.

> • New extensions to historic and traditional dwellings   L(V1) 0.11
>   should comply fully with the energy-efficiency
>   standards, unless there is a need to match the
>   external appearance or character of the extension to
>   that of the host building.

Before making a decision as to whether full energy-efficiency improvements should be implemented, you are recommended to have a chat with your local authority's conservation officer.

4.23.4 Ventilation

Note: New extensions to historic and traditional dwellings should comply with all ventilation standards written in the ADs, unless there is a need to match the external appearance or character of the extension to that of the host building.

4.23.4.1 Addition of a habitable room to an existing dwelling

The addition of a habitable room (not including a conservatory) requires adequate ventilation, either through its own openings to the outside (which can provide background or purge ventilation) or ventilation through another room.

Section F(V1) provides a full description of how this can be achieved, as listed below.

• Adequate ventilation can be met if the additional room is connected to an existing habitable room that is already equipped with sufficient background ventilation, and there is a permanent opening which is a minimum area of 1/20th of the combined floor area.	F(V1) 3.17
• If the existing habitable room has no windows opening to the outside, then purge ventilation or background ventilation will have to be used.	
• If the dwelling already has mechanical ventilation, the centralised system could be extended into the additional room.	F(V1) 3.18

4.23.4.2 *Addition of a wet room to an existing dwelling*

• If a wet room is added to an existing dwelling, whole dwelling ventilation should be extended, and either intermittent extract or background extract ventilation should be provided.	F(V1) 3.25
• If a continuously running single room heat recovery ventilator is used in a wet room, it should use the minimum high rate given in AD-F(V1) Table 1.2, and 50% of this value as the continuous rate. A background ventilator is not required in the same room as the single room heat recovery ventilator.	F(V1) 3.26

4.23.5 Sanitation, hot water safety and water efficiency

The regulations relating to sanitation – G1, G3(2) and G3(3) – apply equally to extensions where they share facilities with another building which is covered by the Regulations.

• There must be a suitable installation for the provision of wholesome water.	G1

- A hot water system, including any cistern or other   G 3(2)
  vessel that supplies water to or receives expansion
  water from a hot water system, shall be designed,
  constructed and installed so as to resist the effects of
  temperature and pressure that may occur.

- A hot water system that has a hot water storage vessel   G 3(3)
  shall incorporate precautions to:
  - prevent the temperature of the water stored in the
    vessel at any time exceeding 100°C;
  - ensure that any discharge from water system or
    storage devices is safely conveyed to where it is
    visible but will *not* cause a danger to persons in or
    about the building.

4.23.6 Security

- Easily accessible doorsets that provide access into a   Q 1.1
  dwelling (or into a building containing a flat) should
  be 'secure' doorsets.

- Ground floor, basement and other easily accessible   Q 2.1
  windows and rooflights should be 'secure' windows.

- Any glazing which, if broken, could enable someone   Q App
  to insert their hand and release the locking device on   B.11
  the inside of the door, should be a minimum of class
  P1A in accordance with BS EN 356.
- Double- or triple-glazed units should incorporate
  only one pane of class P1A glass.

4.23.7 Resistance to contaminants

When assessing gas risks (such as radon) in the context of traditional
housing, two possibilities need to be considered, namely:

- Gas entering the dwelling through a substructure.
- Householder exposure in garden areas, including where extensions
  and outbuildings (e.g. garden sheds and greenhouses) are
  constructed.

To counteract this possibility:

- All new buildings, extensions and conversions (residential or non-domestic) built in areas where there may be elevated radon emissions (e.g. the West Country) may need to incorporate precautions against radon.   C 2.39

4.23.7.1 Cavity insulation

- Reasonable precautions shall be taken to prevent the permeation of toxic fumes from insulating material into the cavity of a cavity wall in or facing an extension.   D1

4.23.8 Resistance to the passage of sound

- Internal walls and floors in extensions between a bedroom (or a room containing a water closet) and another room should provide reasonable resistance to sound.   E2

Requirements for buildings other than dwellings – Additional requirements

This section provides the details for extensions and additions to buildings other than dwellings, only where they differ from the requirements already given above.

- An extension should be regarded as a **new** building if it has a total useful floor area that is both greater than 100m² and more than 25% of the total useful floor area of the actual existing building.   L(V2) 10.7

Large extensions that are impractical to seal off will be treated as a large, complex building.

- Buildings (including extensions) that are not dwellings and are being treated as **new** buildings must be pressure tested, except for buildings with less than 500m² total useful floor area. (See AD-L(V2) 7.5 for compete details of this requirement.)   L(V2) 7.4

- When a building is extended, elements should satisfy   L(V2) 10.9
  all of the following:
  - new (or replacement) thermal elements should
    meet standards in AD-L(V2) Table 4.1;
  - new windows, roof windows, rooflights and doors
    (controlled fittings) should meet the standards listed
    in AD-L(V2) Table 4.1;
  - thermal elements (e.g. existing fabric elements)
    should meet the limiting standards in AD-L(V2)
    Table 4.2;
  - the area of openings in the extension should not
    exceed that given in AD-L(V2) Table 10.1.

4.23.9 Access to and use of buildings

Independent access must be provided to the extension where practicable.

- An extension to a non-domestic building should be   M2 0.5
  treated as a *new* building.
- If sanitary conveniences are already provided in the   M2 0.8
  building, then provision should also be made in an
  extension for sanitary conveniences.
- The principal entrance and any lobby should be   M2 0.5
  accessible.

Note: Where the principle entrance cannot be made accessible, an alternative accessible entrance should be provided.

- Corridors and passageways should be wide enough   M2 3.11
  to allow people with buggies, people carrying cases or
  people on crutches to pass others on the access route.
- Wheelchair users should be able to:
  - have access to adjacent rooms and spaces;
  - pass other people;
  - where necessary, turn through 180°.

4.23.10 *Conservation of fuel and power*

Note: New buildings that include a freestanding building on an existing site (e.g. a new outpatients building at an existing hospital site, or a new classroom block at a school) are not classified as an extension but must be treated (in all respects) as a new building.

4.23.10.1 *Exemptions from the energy-efficiency requirements*

• Other than dwellings, the following classes of buildings are exempt from energy-efficiency requirements:
  • places of worship;
  • temporary buildings with a total planned time of use of two years or less;
  • buildings with low energy demand (such as industrial sites, workshops or non-residential agricultural buildings);
  • new and existing standalone buildings other than dwellings, with a total useful floor area of less than 50m²;
  • carports, covered yards and covered ways.
	L(V2)
	0.11 &
	0.19

Note: Listed Buildings, historic and traditional dwellings, buildings in Conservation Areas and monuments will only need to comply with the energy-efficiency requirements if this would not alter the character or appearance of the building in an unacceptable way.

4.23.10.2 *Consequential improvement*

• If an existing building (with a floor area in excess of 1,000m²) is being extended (or the area of a habitable room within the building is being increased), consequential improvements (such as increasing the size of the boiler) should be considered. For more information, see AD-L(V2) Appendix D, Table D1.	L(V2) 12.3

Note: The minimum value of any 'consequential improvement' works should **not** be less than 10% of the total value of the principal works.

4.23.10.3 Limiting standards for new or replacement elements

• New or replacement fabric elements in extensions must meet the standards shown in AD-L(V2) Table 4.1.	L(V2) 4.5
• If fully glazed pedestrian doors cannot meet the requirements of an existing building, such as owing to the need to maintain the character of the building, then the following standards should be met:	L(V2) 4.6
• fittings should not exceed a centre pane U-value of 1.2W/(m²·K);	
• single glazing should be supplemented with low-emissivity secondary glazing.	

4.23.10.4 Historic and traditional buildings

New extensions to historic and traditional dwellings should comply with the updated requirements with respect to energy efficiency and ventilation unless there is a need to match the external appearance or character of the extension to the host building.

4.23.11 Ventilation

• New extensions to historic and traditional buildings should comply with all current ventilation standards shown in AD-F and other relevant Approved Documents, unless there is a need to match the external appearance or character of the extension to that of the host building.	F(V2) 0.7
• Building work in an existing building includes work on ventilation and hence should meet the current regulations.	F(V2) 3.1

4.23.11.1 Conservatories and porches in new buildings

• If a conservatory or porch is installed as part of a new building:	L(V2) 0.7
• there must be adequate thermal separation between the building and the conservatory or porch; and	
• the building's heating system must not have been extended into the conservatory or porch.	

Note: If both of the above conditions already exist, then the conservatory or porch should be treated as if it were an **extension** being added onto an existing building. On the other hand, if one or both of the above conditions have not been achieved, then the conservatory or porch should be treated as a **room** in the new building.

4.23.11.2 *New freestanding buildings*

New buildings that include a freestanding building on an existing site (e.g. a new outpatients building at an existing hospital site, or a new classroom block at a school) are not classified as an extension, but must be treated (in all respects) as a **new** building.

4.24 Conservatories

Conservatories, porches and domestic greenhouses that share their electricity with a dwelling are **not** exempt from AD-P (*Electrical safety*) and must comply with its requirements. On the other hand: AD-R (*Physical infrastructure for high-speed electronic communications networks*) does **not** apply to conservatories and other small detached buildings which have no sleeping accommodation.

Reminder: The subsections include details for buildings other than dwellings only where they differ from the requirements for dwellings.

Subsection	Description	Page
Requirements for dwellings		
4.24.1	Access to and use of buildings (Optional requirements)	616
4.24.2	Addition of a conservatory to an existing building	617
4.24.3	Fire safety	618
4.24.4	Sanitation	619
4.24.5	Security in new dwellings	619
4.24.6	Ventilation	620

Subsection	Description	Page
Requirements for buildings other than dwellings – Additional requirements		
4.24.7	Access to and use of buildings	621
4.24.8	Fire safety	621
4.24.9	Addition of a conservatory to a new building	622
4.24.10	Addition of a conservatory to an existing building	622
4.24.11	Material change of use	622
4.24.12	Exemptions from the energy-efficiency requirements	623
4.24.13	On-site generation of electricity	624
4.24.14	Continuity of insulation	624

Reminder: Free downloads of all the diagrams and tables referenced in the subsections are available on-line at the Governments' Planning Portal: www.gov.uk/government/collections/approved-documents.

Requirements for dwellings

4.24.1　Access to and use of buildings (Optional requirements)

If accessible doorways for ambulant disabled persons are required to conservatories from dwellings, then the following guidance should be followed:

• The principal private entrance, or the alternative private entrance where step-free access cannot be achieved to the principal private entrance, should: 　• have a level external landing with a minimum width and depth of 1,200mm; 　• have the landing covered for a minimum width of 900mm and a minimum depth of 600mm; 　• be illuminated by lighting which uses fully diffused luminaires activated automatically by a dusk to dawn timer or by motion detector;	M 2.20

- have a door with a minimum clear opening width of 850mm;
- have a minimum 300m nib provided to the leading edge of the door and for a minimum distance of 1,200mm beyond the door;
- ensure the depth of the reveal on the leading side of the door is a maximum of 200mm;
- ensure the threshold is accessible;
- have the doors leading to a conservatory a minimum of 1,500mm apart with at least 1,500mm between door swings.

- All other external doors that are integral with, or connected to, the dwelling should comply with the provisions above and should have a minimum 300m nib to the leading edge of the door which extends for a minimum 1,800mm beyond the door.

M 2.21
M 3.23

4.24.2 Addition of a conservatory to an existing building

- A conservatory or porch must have thermal separation from the existing dwelling.

L(V1)
10.12

- If the thermal separation is removed or the dwelling's heating system is extended into the conservatory or porch, the conservatory or porch will then be treated as an extension of the existing building.

- If the conservatory or porch is **not** exempt from the energy-efficiency requirements, it should meet all of the limiting U-values for new fabric elements in existing dwellings (as shown in AD-L(V1) Table 4.2).

L(V1)
10.13

Note: These requirements are discussed in greater detail in Subsections 4.6 (Floors), 4.7 (Walls), 4.12 (Windows) and 4.13 (Doors).

4.24.2.1 Exemptions for conservatories and porches

- Where a dwelling is extended by adding a   L(V1) 0.14
  conservatory or porch, the work is exempt from the
  energy-efficiency requirements if **all** of the following
  apply:
  - the extension is at ground level;
  - the floor area of the extension does not exceed 30m^2;
  - any wall, door or window that separates the
    extension from the dwelling has been retained or
    (if removed) has been replaced with a wall, door or
    window that meets the required standards;
  - the heating system of the dwelling is not extended
    into the conservatory or porch;
  - the extension does not have its own fixed heating
    appliances.

4.24.2.2 New dwellings

- If a conservatory or porch has been installed as part   L(V1) 0.6
  of the construction of a new dwelling, the treatment
  of the conservatory or porch depends on whether
  both of the following have been achieved:
  - there is adequate thermal separation between the
    dwelling and the conservatory or porch;
  - the dwelling's heating system has not been
    extended into the conservatory or porch.
- If both the above been achieved, the conservatory
  or porch should be treated as if it was an extension
  being added onto an existing dwelling.
- If either of the above have not been achieved, then
  the conservatory or porch should be treated as **a
  room** which is part of the new dwelling.

4.24.3 Fire safety

Some roof coverings (such as plastic rooflights) do not provide protection
against the spread of fire and are, therefore, not permitted adjacent to
a boundary.

- Plastic rooflights in a conservatory with a maximum   B(V1)
  floor area of 40m² shall (depending on their structure)   Table 12.2
  conform to the following limitations regarding the
  minimum distance from any point on a relevant
  boundary to the rooflight:
    - type EROOF(t4) or DROOF(t4) – 6m;
    - type FROOF(t4) – 20m.
- Thermoplastic rooflights in a conservatory with a   B(V1)
  maximum floor area of 40m² shall be a minimum   Table 12.3
  distance of 6m from any point on a relevant boundary.

See AD-B Tables 12.1 and 12.2 for further limitations on the use of plastic rooflights.

4.24.4 Sanitation

The Regulations do not require hot or cold water systems to be provided to conservatories, but if systems are provided, they must meet the minimum hygiene and safety requirements of the ADs, namely:

- The water that a conservatory (including those   G 0.ii
  under 30m²) receives from a shared building shall be
  wholesome water.
- If hot water is supplied to a conservatory, it shall be
  designed and installed so as to resist the effects of
  temperature and pressure.
- Any hot water stored shall not exceed 100°.

4.24.5 Security in new dwellings

The new AD-Q requires secure doors and windows to be fitted in new dwellings and parts of a building which enable access to be gained from outside to within the building.

- All easily accessible doorsets that provide access into   Q 1.1
  a dwelling or into a building containing a dwelling
  should be secure doorsets.

- Ground floor, basement and other easily accessible   Q 2.1
  windows (including easily accessible rooflights) should
  have secure windows.

- Any glazing which, if broken would permit someone   Q B.11
  to insert their hand and release the locking device on
  the inside of the door, should be a minimum of class
  P1A in accordance with BS EN 356.
- Double- or triple-glazed units need to incorporate
  only one pane of class P1A glass.

4.24.6 Ventilation

Adding a conservatory with a floor area of over 30m² to a new dwelling
requires a fixed system for mechanical ventilation.

Note: The guidance here only applies to conservatories with a floor area
that exceeds 30m². Conservatories with a floor area less than 30m² are
exempt from these requirements.

- The general ventilation rate for a new conservatory   F(V1) 3.22
  (and if necessary, adjoining rooms) could be achieved
  using background ventilators.
- A system for purge ventilation (delivered through   F(V1) 3.23
  windows, doors or a mechanical extract ventilation
  system) should be provided in each habitable room
  and be capable of extracting at least four air changes
  per hour (4 ach) per room directly to the outside.
- For a new conservatory, performance testing should   F(V1) 3.24
  comply with the standards laid out in AD-F(V1)
  Table 1.5.

4.24.6.1 Ventilation of a habitable room through
              another room

Note: If the new conservatory is, in effect, a habitable room, then the
following requirements need to be considered:

- A habitable room without windows that can be   F(V1)
  opened may be ventilated through either another   1.40–1.44
  habitable room or a conservatory which has openings
  to the outside, providing both purge and background
  ventilation.

- Between the two rooms there should be a permanent opening with a minimum area of 1/20th of the combined floor area of the two rooms. (See AD-F(V1) Diagram 1.3.)

Requirements for buildings other than dwellings – Additional requirements

The basic requirements for conservatories in buildings other than dwellings are the same as for dwellings, with the addition of the following requirements.

4.24.7 Access to and use of buildings

For buildings other than dwellings, the ADs require you to provide suitable accessible entrances for **all** users.

- The principal entrance, any main staff entrance, and any lobby should be accessible.   M2 2.2
- Where it is not possible for the principal or main staff entrance to be accessible, an alternative accessible entrance should be provided.

4.24.8 Fire safety

4.24.8.1 Resisting fire spread over roof coverings

- If plastic rooflights are used in a conservatory with a maximum floor area of 40m^2, they must be a minimum distance (from any point on a relevant boundary to the rooflight):   B(V2) Table 14.2
  - type EROOF(t4) or DROOF(t4) – 6m;
  - type FROOF(t4) – 20m.

- If thermoplastic rooflights are used in a conservatory with a maximum floor area of 40m^2, they must be a minimum distance of 6m from any point on a relevant boundary.   B(V2) Table 14.3

If either or both of these requirements are **not** followed, then the conservatory or porch will be treated as a room in the new building!

4.24.9 Addition of a conservatory to a new building

If a conservatory or porch is being installed as part of a new building, it will be treated as an extension to the building if:

• There is sufficient thermal separation between the building and the extension.	L(V2) 0.7
• The building's heating system has **not** been extended into the extension.	

If either or both of these requirements are **not** followed, then the conservatory or porch will be treated as a room in the new building!

4.24.10 Addition of a conservatory to an existing building

If a conservatory or porch is being added to an existing building:

• The additional conservatory or porch must be thermally separated from the existing dwelling.	L(V2) 10.12
• If the thermal separation is removed or the building's heating system is extended into the conservatory or porch, the conservatory or porch will then be treated as an extension of the existing building.	

4.24.11 Material change of use

• Any window that separates a conservatory or porch from the building which has been retained or, if removed, been replaced with a window of similar characteristics, is exempt from the requirements.	L(V2) 0.18d
• New windows or roof windows in a conservatory or porch that is thermally separated from the existing building (and its heating system does not extend into it), should meet the minimum standards shown in AD-L(V2) Table 4.1.	L(V2) Table 4.1 & 10.13
• Windows in a conservatory or porch should be insulated and draught-proofed to at least the same extent as in the existing building.	

- If all of the windows, roof windows, rooflights or L(V2)10.3
  doors are replaced, then all units should:
  - be draught-proofed;
  - meet the minimum standards in AD-L(V2) Table 4.1;
  - have insulated cavity closers installed where
    appropriate.

4.24.12 Exemptions from the energy-efficiency requirements

- Where a building is extended by adding a L(V2) 0.18
  conservatory or porch, the work is exempt from the
  energy-efficiency requirements – but only if **all** of
  the following apply:
  - the extension is at ground level;
  - the floor area of the extension does not exceed
    $30m^2$;
  - the building's heating system has not been
    extended into the conservatory or porch;
  - the extension does not have its own fixed heating
    appliances;
  - if a wall, door or window separating the extension
    from the building has been retained, or if removed
    has been replaced with a similar wall, door or
    window.

Note: If the conservatory or porch is not exempt from the energy-efficiency requirements, then the following components should meet the minimum standards listed in AD-L(V2) Table 4.1:

- All new or replacement thermal elements. L(V2)
- All new windows, roof windows, rooflights and doors. 10.13
- In addition:
  - walls, doors and windows should be insulated
    and draught-proofed to the same extent as in the
    existing dwelling;
  - fixed building services and on-site electricity
    generation within the conservatory or porch should
    meet the standards set in L(V2) Sections 5 and 6.

4.24.13 On-site generation of electricity

• On-site electricity generation within a conservatory or porch should have independent temperature and on/off controls. (See L(V2) Section 5 for further information.)	L(V2) 10.13

Note: If these controls are only 'on' and 'off' switches, then this particular service does not need to be commissioned.

4.24.14 Continuity of insulation

• Any wall in a conservatory or porch should be insulated and draught-proofed to the same extent as in the existing building.	L(V2) 10.13

Acronyms

ach	air changes per hour
AD	Approved Document
ATTMA	Air Tightness Testing and Measurement Association
BCB	Building Control Body
BCO	building control officer
BRE	Building Research Establishment
BRIB10	*Building Regulations in Brief* – 10th edition
BSI	British Standards Institution
CHP	combined heat and power
CO	carbon monoxide
CO_2	carbon dioxide
COP	coefficient of performance
DCER	dwellings carbon emission rate
DCLG	Department for Communities and Local Government
DER	dwelling (CO_2) emission rate
DFE	decorative fuel effect
DPM	damp-proof membrane
DQRA	Detailed Quantitative Risk Assessment
DSER	doorset energy rating
DSMA	Door and Shutter Manufacturers' Association
EVC	emergency voice communication
FENSA	Fenestration Self-Assessment Scheme
FPA	Fire Protection Association
GQRA	Generic Quantitative Risk Assessment
GSIUR	Gas Safety (Installation and Use) Regulations
H&S	health and safety
HBF	Home Builders Federation
HDM	house dust mite
HETAS	Heating Equipment Testing and Approval Scheme
HIP	Home Information Pack

HSE	Health and Safety Executive
IEE	Institute of Electrical Engineers
ILFE	inset live fuel effect
ILU	inter-language unification
LPG	liquid petroleum gas
LZC	low or zero carbon
MCS	Microgeneration Certification Scheme
MDD	Medical Devices Directive
MEV	mechanical extract ventilation
micro-CHP	micro combined heat and power
MVHR	mechanical ventilation with heat recovery
NCM	National Calculation Methodology
NS	nominal size
OFTEC	Oil Firing Technical Association
PAS	product assessment specification
PB	polybutylene
PE-X	cross-linked polyethylene
PHE	Progressive Horizontal Evacuation
PIR	passive infra-red detectors
PSV	passive stack ventilation
PTFE	polytetrafluoroethylene
RCD	residual current device
RE	load-bearing capacity and integrity (of fire extinguishers)
REI	Recreational Equipment Incorporated
RIBA	Royal Institute of British Architects
SAP	Standard Assessment Procedure
SO_2	sulphur dioxide
SRHRV	single room heat recovery ventilator
TER	target emissions rate
TFEE	target fabric energy-efficiency
TP	thermoplastic
TPO	Tree Preservation Order
TRADA	Timber Research and Development Association
UF	urea-formaldehyde foam
UKAEA	United Kingdom Atomic Energy Authority
UKLPG	the trade association for the liquefied petroleum gas (LPG) industry in the UK
uPVC	unplasticised polyvinyl chloride
VOC	volatile organic compounds
WAUILF	Workplace Applied Uniform Indicated Low Frequency (application)
WER	window energy rating
YFR	yearly forecast rational

Bibliography

British, European and International Standards

Title	Standard
Acoustics. Measurement of sound insulation in buildings and of building elements Part 3 Laboratory measurement of airborne sound insulation of building elements	BS EN ISO 140
Acoustics. Rating of sound insulation in buildings and of building elements Part 1 Airborne sound insulation	BS EN ISO 717
Actions on structures. General actions. Densities, self-weight, imposed loads for buildings Actions on structures. General actions. Wind actions	BS EN 1991-1-1 BS EN 1991-1-4
Aggregates for concrete	BS EN 12620
Air admittance valves for drainage systems. Requirements, test methods and evaluation of conformity	BS EN 12380
Anti-flooding devices	BS EN 13564
Automatic electrical controls for household and similar use. Particular requirements for temperature-sensing controls	BS EN 60730-2-9
Barriers in and about buildings. Code of practice	BS 6180
British Standards publication PAS 24	PAS 24

Title	Standard
Building hardware. Panic exit devices operated by a horizontal bar. Requirements and test methods	BS EN 1125
Cement Part 1 Composition, specifications and conformity criteria for common elements Part 2 Conformity evaluation	BS EN 197
Chimneys. Requirements for metal chimneys. Part 1 Chimneys serving one appliance Part 2 Metal liners and connecting flue pipes	BS EN 1856
Code of practice for accommodation of building services in ducts	BS 8313
Code of practice for fire door assemblies with non-metallic leaves	BS 8214
Code of practice for mechanical ventilation and air-conditioning in buildings	BS 5720
Code of practice for oil firing. Installations up to 44kW output capacity for space heating and hot water supply purposes. AMD 3637	BS 5410-1
Code of practice for use of masonry Part 3 Materials and components, design and workmanship	BS 5628
Code of practice for ventilation principles and designing for natural ventilation. AMD 8930	BS 5925
Concrete Part 1 Guide to specifying concrete Part 2 Method for specifying concrete mixes Part 3 Specification for the procedures to be used in producing and transporting concrete Part 4 Specification for the procedures to be used in sampling, testing and assessing compliance of concrete	BS 5328
Copper indirect cylinders for domestic purposes. Open-vented copper cylinders. Requirements and test methods	BS 1566-1
Design of masonry structures	BS EN 1996-2

Title	Standard
Design of masonry structures Part 1.1 General rules for reinforced and unreinforced masonry structures	BS EN 1996-1-1
Drain and sewer systems outside buildings Part 4 Hydraulic design and environmental aspects	BS EN 752
Electrical apparatus for the detection of carbon monoxide in domestic premises	BS EN 50291-2
Emergency lighting Part 1 Code of practice for the emergency lighting of premises	BS 5266

Fire classification of construction products and
   building elements
Part 2 Classification using data from fire resistance
   tests, excluding ventilation services
Part 5 Classification using data from external fire
   exposure to roof tests
	BS EN 13501

Fire detection and alarm systems for buildings
Part 1 Code of practice for system design, installation
   and servicing
Part 6 Code of practice for the design, installation
   and maintenance of fire detection and fire alarm
   systems in dwellings
	BS 5839

Fire detection and fire alarm devices for dwellings. Specification for heat alarms	BS 5446-2
Fire detection and fire alarm systems. Manual call points	BS EN 54-11

Fire precautions in the design, construction and use
   of buildings
Part 5 Code of practice for firefighting in stairs
   and lifts
Part 6 Code of practice for places of assembly
Part 7 Code of practice for the incorporation of atria
   in buildings
Part 8 Code of practice for means of escape for
   disabled people
Part 9 Code of practice for ventilation and
   air-conditioning ductwork
	BS 5588

Fire resistance tests for door and shutter assemblies Part 1 Fire doors and shutters	BS EN 1634

Title	Standard
Fire tests on building materials and structures Part 3 Classification and method of test for external fire exposure to roofs Part 11 Method for assessing the heat emission from building materials Part 20 Method for determination of the fire resistance of elements of construction (general principles)	BS 476
Flue blocks and masonry terminals for gas appliances Part 1 Specification for precast concrete flue blocks and terminals	BS 1289
General actions Part 1 Densities, self-weight, imposed loads for buildings	BS EN 1991-1-1
Glass in building. Pendulum test. Impact test method and classification for flat glass	BS EN 12600
Glass in building. Security glazing. Testing and classification of resistance against manual attack	BS EN 356
Graphical symbols and signs. Safety signs, including fire safety signs. Specification for geometric shapes, colours and layout	BS 5499-1
Gravity drainage systems inside buildings Part 1 Scope, definitions, general and performance requirements Part 2 Wastewater systems, layout and calculation Part 3 Roof drainage layout and calculation Part 4 Effluent lifting plants, layout and calculation Part 5 Installation, maintenance and user instructions	BS EN 12056
Grease separators. Principles of design, performance and testing, marking and quality control	BS EN 1825-1
Guide to the conservation of historic buildings	BS 7913:2013
Heating boilers. Heating boilers with forced draught burners. Terminology, general requirements, testing and marking	BS EN 303-1
Household and similar electrical appliances. Safety	BS EN 60335-2-21

Title	Standard
Hygrothermal performance of building components and building elements	BS EN ISO 13788
Installation and maintenance of flues and ventilation for gas appliances of rated input not exceeding 70kW net	BS 5440-1
Installations for separation of light liquids (e.g. petrol or oil)	BS EN 858
Mechanical thermostats for gas-burning appliances	BS EN 257
Method of test for ignitability of fabrics used in the construction of large tented structures	BS 7157
Method of test for resistance to fire of unprotected small cables for use in emergency circuits	BS EN 50200
Requirements for electrical installations (IET Wiring Regulations, 18th edition)	BS 7671:2018
Safety and control devices for use in hot water systems Part 2:1991 Specification for temperature relief valves for pressures from 1 bar to 10 bar	BS 6283
Sanitary installations Part 2 Code of practice for space requirements for sanitary appliances	BS 6465
Smoke alarm devices	BS EN 14604
Smoke and heat control systems Part 6 Specification for pressure differential systems	BS EN 12101
Specification for ancillary components for masonry Part 1 Ties, tension straps, hangers and brackets	BS EN 845
Specification for galvanized low carbon steel cisterns, cistern lids, tanks and cylinders. Metric units	BS 417-2
Specification for impact performance requirements for flat safety glass and safety plastics for use in buildings	BS 6206
Specification for installation of gas-fired boilers of rated input not exceeding 70kW	BS 6798

Title	Standard
Specification for masonry units Part 1 Clay masonry units Part 2 Calcium silicate masonry units Part 3 Aggregate concrete masonry units Part 4 Autoclaved aerated concrete masonry units Part 5 Manufactured stone masonry units Part 6 Natural stone masonry units	BS EN 771
Specification for metal ties for cavity wall construction	BS 1243
Specification for mortar for masonry Part 2 Masonry mortar	BS EN 998
Specification for safety of household and similar electrical appliances	BS EN 60335-2-35
Specification for safety of household and similar electrical appliances. Particular requirements for fixed immersion heaters	BS EN 60335-2-73
Specification for vessels for use in heating systems. Calorifiers and storage vessels for central heating and hot water supply	BS 853-1
Specification for vitreous-enameled low-carbon-steel flue pipes, other components and accessories for solid-fuel-burning appliances with a maximum rated output of 45kW	BS 6999
Sprinkler systems for residential and domestic occupancies. Code of practice	BS 9251
Stainless steels. List of stainless steels	BS EN 10088-1
Stairs, ladders and walkways Part 2 Code of practice for the design of helical and spiral stairs Part 3 Code of practice for the design of industrial type stairs, permanent ladders and walkways	BS 5395
Steel plate, sheet and strip. Carbon and carbon manganese plate, sheet and strip. General specifications	BS 1449-1
Structural design of low-rise buildings Part 3 Code of practice for timber floors and roofs for housing	BS 8103-3

Title	Standard
Thermal solar systems and components	BS EN 12977
Thermal solar systems and components. Factory made systems. General requirements	BS EN 12976-1
Ventilation for buildings. Performance testing of components/products for residential ventilation Part 1 Externally and internally mounted air-transfer devices	BS EN 13141
Wastewater lifting plants for buildings and sites – principles of construction and testing Part 1 Lifting plants for wastewater containing faecal matter Part 2 Lifting plants for faecal-free wastewater Part 3 Lifting plants for wastewater containing faecal matter for limited application	BS EN 1205
WC pans and WC suites with integral trap	BS EN 997
Windows and doors. Product standard, performance characteristics. Windows and external pedestrian doorsets without resistance to fire and/or smoke leakage characteristics	BS EN 14351
Workmanship on building sites Part 6 Code of practice for slating and tiling of roofs and claddings	BS 8000

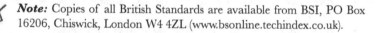

Note: Copies of all British Standards are available from BSI, PO Box 16206, Chiswick, London W4 4ZL (www.bsonline.techindex.co.uk).

Other publications

Air Tightness Testing and Measurement Association (ATTMA)
www.attma.org

- *Measuring Air Permeability of Building Envelopes*

Association for Petroleum and Explosives Administration (APEA)
www.apea.org.uk

- *Code of Practice for Ground Floor, Multi-Storey and Underground Car Parks*

Association for Specialist Fire Protection (ASFP)
www.asfp.org.uk

- *ASFP Red Book – Fire Stopping and Penetration Seals for the Construction Industry*
- *ASFP Yellow Book – Fire Protection for Structural Steel in Buildings*
- *ASFP Grey Book – Fire and Smoke Resisting Dampers*
- *ASFP Blue Book – Fire Resisting Ductwork*

British Automatic Fire Sprinkler Association (BAFSA)
www.bafsa.org.uk

- *Sprinklers for Safety: Use and Benefits of Incorporating Sprinklers in Buildings and Structures*

Building and Engineering Services Association (B&ES)
www.b-es.org

- *A Practical Guide to Ductwork Leakage Testing*
- *DW/144, Specification for Sheet Metal Ductwork*

Building Research Establishment Ltd (BRE)
www.bre.co.uk

- *BRE 128, Guidelines for the Construction of Fire Resisting Structural Elements*
- *BRE 135, Fire Performance of External Thermal Insulation for Walls of Multi-Storey Buildings*
- *BRE 187, External Fire Spread: Building Separation and Boundary Distances*
- *BRE 208, Increasing the Fire Resistance of Existing Timber Floors*
- *BRE 262, Thermal Insulation: Avoiding Risks*
- *BRE 274, Fire Safety of PFTE Based Materials Used in Buildings*
- *BRE 364, Solar Shading of Buildings*
- *BRE 369, Design Methodologies for Smoke and Exhaust Ventilation*
- *BRE 436, Wind Loading on Buildings: Brief Guidance for Using BS 6399-2*
- *BRE 437, Industrial Platform Floors: Mezzanine and Raised Storage*
- *BRE 443, Conventions for U-value Calculations*
- *BRE 498, Selecting Lighting Controls*
- *BRE 454, Multi-Storey Timber Frame Buildings: A Design Guide*
- *Information Paper 1P1/06, 'Assessing the effects of thermal bridging at junctions and around openings in the external elements of buildings'*
- *Information paper 1P14103, 'Preventing hot water scalding in bathrooms: Using TMVs'*

Building Services Research and Information Association (BSRIA)
www.bsria.co.uk

* *Application Guide AG19*
* *Application Guide AG20*
* *Application Guide AG21*

British Standards Institute (BSI)
www.bsigroup.com

* *PD 6693-1*
* *PD 6697*
* *PD 6693-1*

Centre for Window and Cladding Technology (CWCT)
www.cwct.co.uk

* *Thermal Assessment of Window Assemblies, Curtain Walling and Non-Traditional Building Envelopes*

Chartered Institution of Building Services Engineers (CIBSE)
www.cibse.org

* *CIBSE Commissioning Code M, Commissioning Management*
* *CIBSE Guide A, Environmental Design*
* *CIBSE Guide B:2005*
* *CIBSE TM13, Minimising the Risk of Legionnaires' Disease*
* *TR/19 – Guide to Good Practice – Internal Cleanliness of Ventilation Systems*
* *AM 10, Natural Ventilation in Non-Domestic Buildings*
* *Solar Heating Design and Installation Guide*
* *TM 31, Building Log Book Toolkit*
* *TM 36, Climate Change and the Indoor Environment: Impacts and Adaptation*
* *TM 37, Design for Improved Solar Shading Control*
* *TM 39, Building Energy Metering*
* *Ventilation Hygiene Toolkit*

Department for Communities and Local Government (DCLG)
www.gov.uk/government/organisations/department-for-communities-and-local-government

* *Fire Safety in Adult Placements: A Code of Practice*
* *A Householder's Planning Guide for the Installation of Satellite Television Dishes*
* *Domestic Heating Compliance Guide*
* *Outdoor Advertisements and Signs: A Guide for Advertisers*

- *Protected Trees: A Guide to Tree Preservation Procedures*
- *Tree Preservation Orders: A Guide to the Law and Good Practice*

Department for Education (DFE)

www.education.gov.uk

- *Building Bulletin 101, Ventilation of School Buildings, School Building and Design Unit*

Department for Environment, Food and Rural Affairs (Defra)

www.archive.defra.gov.uk

- *The Government's Standard Assessment Procedure for Energy Rating of Dwellings: SAP 2005 (available at: www.bre.co.uk/sap2005)*
- *Rainwater and Greywater: Technical and Economic Feasibility*
- *Rainwater and Greywater: A Guide for Specifiers*
- *Rainwater and Greywater: Review of Water Quality Standards and Recommendations for the UK*
- *Protocol on Design, Construction and Adoption of Sewers in England and Wales*

Department of Health (DH)

www.gov.uk/government/organisations/department-of-health

- *Health Technical Memorandum 05-02, Guidance to Support of Functional Provisions in Healthcare Premises*

Door and Hardware Federation (DHF)

www.dhfonline.org.uk

- *Code of Practice for Fire-Resisting Metal Doorsets*
- *Hardware for Fire and Escape Doors*

Electrical Contractors' Association (ECA) and National Inspection Council for Electrical Installation Contracting (NICEIC)

www.eca.co.uk and www.niceic.org.uk

- *ECA Comprehensive Guide to Harmonised Cable Colours*
- *Electrical Installers' Guide to the Building Regulations*
- *New Fixed Wiring Colours: A Practical Guide*

Energy Saving Trust (EST)

www.energysavingtrust.org.uk

- *CE66, Windows for New and Existing Housing*
- *CE129, Reducing Overheating: A Designer's Guide*

- GIL20, *Low Energy Domestic Lighting*
- GPG268, *Energy Efficient Ventilation in Dwellings: A Guide for Specifiers*

English Heritage
www.english-heritage.org.uk

- *Building Regulations and Historic Buildings*

Environment Agency
www.gov.uk/government/organisations/environment-agency

- *Pollution Prevention Guidelines (PPG18), Managing Fire Water and Major Spillages*

Fire Protection Association (FPA)
www.thefpa.co.uk

- *Design Guide*

Food Standards Agency (FSA)
www.food.gov.uk

- *Code of Practice: Food Hygiene – A Guide for Businesses*

Glass and Glazing Federation (GGF)
www.ggf.org.uk

- *A Guide to Best Practice in the Specification and Use of Fire-Resistant Glazed Systems*

Health and Safety Executive (HSE)
www.hse.gov.uk

- *L24 – Workplace (Health, Safety and Welfare) Regulations*
- *Legionnaires' Disease: Control of Legionella Bacteria in Water Systems*
- *Radon in the Workplace (www.rcr.ac.uk/sites/default/files/bfcr152_irmer.pdf)*

Institution of Engineering and Technology (IET)
www.theiet.org

- *Electrician's Guide to the Building Regulations*
- *IEE Guidance Note 1, Selection and Erection of Equipment, 4th edition*
- *IEE Guidance Note 2, Isolation and Switching, 4th edition*
- *IEE Guidance Note 3, Inspection and Testing, 4th edition*
- *IEE Guidance Note 4, Protection Against Fire, 4th edition*
- *IEE Guidance Note 5, Protection Against Electric Shock, 4th edition*
- *IEE Guidance Note 6, Protection Against Overcurrent, 4th edition*

- *IEE Guidance Note 7, Special Locations, 2nd edition*
- *IEE Wiring Regulations, 18th edition*
- *New Wiring Colour*

Institution of Structural Engineers (IstructE)
www.istructe.org

- *Dynamic Performance Requirements for Permanent Grandstands Subject to Crowd Action: Recommendations for Management, Design and Assessment*
- *Report on Aspects of Cladding*
- *Guide to the Structural Use of Adhesives*

Metal Cladding and Roofing Manufacturers Association (MCRMA)
www.mcrma.co.uk

- *Guidance for Design of Metal Cladding and Roofing to Comply with Approved Document L2*

Modular and Portable Building Association (MPBA)
www.mpba.biz

- *Energy Performance Standards for Modular and Portable Buildings*

National Association of Rooflight Manufacturers (NARM)
www.narm.org.uk

- *Use of Rooflights to Satisfy the 2002 Building Regulations for the Conservation of Fuel and Power*

Passive Fire Protection Federation (PFPF)
www.pfpf.org

- *Ensuring Best Practice for Passive Fire Protection in Buildings*

Planning Portal
www.planningportal.co.uk

- *Copies of Approved Documents*

Steel Construction Institute (SCI)
www.steel-sci.org

- *SCI P197, Designing for Structural Safety: A Handbook for Architects and Engineers*
- *SCI P288, Fire Safe Design: A New Approach to Multi-Storey Steel-Framed Buildings, 2nd edition*
- *SCI P313, Single-Storey Steel-Framed Buildings in Fire Boundary Conditions*

Thermal Insulation Manufacturers and Suppliers Association (TIMSA)
www.timsa.org.uk

- *HVAC Guidance for Achieving Compliance with Part L of the Building Regulations*

Timber Research and Development Association (TRADA)
www.trada.co.uk

- *Timber Fire-Resisting Doorsets: Maintaining Performance Under the New European Test Standard*
- *Span Tables for Solid Timber Members in Floors, Ceilings and Roofs (Excluding Trussed Rafter Roofs) for Dwellings*

Water Regulations Advisory Scheme (WRAS)
www.wras.co.uk

- *Water Regulations Advisory Scheme: Water Regulations Guide*
- *WRAS Information and Guidance Note No. 9-02-05, Marking and Identification of Pipework for Reclaimed (Greywater) Systems*

Wood Protection Association (WPA)
www.wood-protection.org

- *Industrial Wood Preservation: Specification and Practice*

Other useful websites

British Geographical Survey
http://shop.bgs.ac.uk/georeports

Building Control Alliance
www.buildingcontrolalliance.org

Competent Person Scheme
www.communities.gov.uk/planningandbuildingregulations/competentpersonsschemes

Legislation
www.legislation.gov.uk

Microgeneration Certification Scheme
www.microgenerationcertification.org

Planning Inspectorate
www.gov.uk/government/organisations/planning-inspectorate

Radon Association
www.radonassociation.co.uk

UK Radon
www.UKradon.org

Water UK
www.water.org.uk/consumers/find-your-supplier

Appendix – Other books associated with Building Regulations

Title	*Extracts from book reviews*	*ISBN*
Wiring Regulations in Brief (Fourth Edition)  WIRING REGULATIONS IN BRIEF FOURTH EDITION RAY TRICKER	This newly updated edition of *Wiring Regulations in Brief* provides a user-friendly guide to the newest amendments to BS 7671 and the IET Wiring Regulations. Topic-based chapters link areas of working practice – such as earthing, cables, installations, testing and inspection, and special locations – with the specifics of the regulations themselves. This allows quick and easy identification of the official requirements relating to the situation in front of you. The requirements of the regulations, and of related standards, are presented in an informal, easy-to-read style to remove confusion.

Packed with useful hints and tips, and highlighting the most important or mandatory requirements, this book is a concise reference on all aspects of the eighteenth edition of the IET Wiring Regulations. This handy guide provides an on-the-job reference source for electricians, designers, service engineers, inspectors, builders and students. | Routledge Taylor & Francis Group Paperback – ISBN 9780367431983 Hardback – ISBN 9780367432010 e-Book – ISBN 9781003001829 |

Title	Extracts from book reviews	ISBN
Building Regulations in Brief (Tenth Edition) 	This tenth edition of the most popular and trusted guide reflects all the latest amendments to the Building Regulations, Planning Permission and the Approved Documents in England and Wales. This includes coverage of the new Approved Document P on security, and a second part to Approved Document M which divides the regulations for 'dwellings' and 'buildings other than dwellings'. A new chapter has been added to incorporate these changes and to make the book more user friendly. Giving practical information throughout on how to work with (and within) the regulations, this book enables compliance in the simplest and most cost-effective manner possible. The no-nonsense approach of *Building Regulations in Brief* cuts through any confusion and explains the meaning of the regulations. Consequently, it has become a favourite for anyone in the building industry or studying, as well as those planning to have work carried out on their home.	Paperback – ISBN 9780367774233 Hardback – ISBN 9781032007618 e-Book – ISBN 9781003175483

(continued)

Title	*Extracts from book reviews*	*ISBN*
Wiring Regulations Pocket Book (First Edition) 	This new Routledge Pocket Book provides a user-friendly guide to the latest amendments to the eighteenth edition of IET Wiring Regulations (BS 7671:2018). This Pocket Book contains topic-based chapters that link areas of working practice with the specifics of the regulations themselves. The requirements of the regulations are presented in an informal, easy-to-read style that strips away confusion. Packed with useful hints and tips that highlight the most important or mandatory requirements, the book is a concise reference on all aspects of the eighteenth edition of the IET Wiring Regulations. This handy guide provides an on-the-job reference source for electricians, designers, service engineers, inspectors, builders and students.	Paperback – ISBN 9780367431983 Hardback – ISBN 9780367760304 e-Book – ISBN 9781003165170

Title	Extracts from book reviews	ISBN
Scottish Building Standards in Brief (First Edition) 	*Scottish Building Standards in Brief* takes the highly successful formula of Ray Tricker's *Building Regulations in Brief* and applies it to the requirements of the Building (Scotland) Regulations 2004. With the same no-nonsense and simple-to-follow guidance, but written specifically for the Scottish Building Standards, it's the ideal book for builders, architects, designers and DIY enthusiasts working in Scotland.	Paperback – ISBN 9780750685580 Hardback – ISBN 9781138162365 e-Book – ISBN 9780080942513

The book explains the meaning of the regulations, their history, current status, requirements, associated documentation and how Local Authorities view their importance, and emphasises the benefits and requirements of each one.

There is no easier or clearer guide to help you to comply with the Scottish Building Standards in the simplest and most cost-effective manner possible.

(*continued*)

Title	Extracts from book reviews	ISBN
Water Regulations in Brief (First Edition) 	*Water Regulations in Brief* is a unique reference book, providing all the information needed to comply with the regulations, in an easy-to-use, full colour format. Crucially, unlike other titles on this subject, this book doesn't just cover the Water Regulations, but also clearly shows how they link in with the Building Regulations, Water By-laws **AND** the Wiring Regulations, providing the only available complete reference to the requirements for water fittings and water systems. Structured in the same logical, time-saving way as the author's other bestselling '...in Brief' books, *Water Regulations in Brief* will be a welcome change to anyone tired of wading through complex, jargon-heavy publications in search of the information they need to get the job done.	Paperback – ISBN 9781856176286 Hardback – ISBN 9781138408661 e-Book – ISBN 9780080950945

Title	Extracts from book reviews	ISBN
Quality Management Systems *A Practical Guide to Standards Implementation* (First Edition) 	This book provides a clear, easy-to-digest overview of Quality Management Systems (QMS). Critically, it offers the reader an explanation of the International Standards Organization's (ISO) requirement that in future all new and existing Management Systems Standards will need to have the same high-level structure, commonly referred to as Annex SL, with identical core text, as well as common terms and definitions. In addition to explaining what Annex SL entails, this book provides the reader with a guide to the principles, requirements and interoperability of Quality Management System Standards, how to complete internal and external management reviews, third-party audits and evaluations, as well as how to become an ISO Certified Organisation once your QMS is fully established. As a simple and straightforward explanation of QMS Standards and their current requirements, this is a perfect guide for practitioners who need a comprehensive overview to put theory into practice, as well as for undergraduate and postgraduate students studying quality management as part of broader Operations and Management courses.	 Paperback – ISBN 9780367223533 Hardback – ISBN 9780367223519 e-Book – ISBN 9780429274473

(*continued*)

Title	Extracts from book reviews	ISBN
ISO 9001:2015 in Brief (Fourth Edition) 	*ISO 9001:2015 in Brief* provides an introduction to quality management systems for students, newcomers and busy executives, with a user-friendly, simplified explanation of the history, the requirements and benefits of the new standard.　　This short, easy-to-understand reference tool also helps organisations to quickly set up an ISO 9001:2015 compliant Quality Management System for themselves at minimal expense and without high consultancy fees.	Paperback – ISBN 9781138025868　Hardback – ISBN 9781138025851　e-Book – ISBN 9781315774831

Title	Extracts from book reviews	ISBN
ISO 9001:2015 for Small Businesses (Sixth Edition) 	Small businesses face many challenges today, including the increasing demand by larger companies for ISO 9001 compliance, a challenging task for any organisation, and in particular for a small business without quality assurance experts on its payroll. Ray Tricker has already guided hundreds of businesses through to ISO accreditation, and this sixth edition of his life-saving ISO guide provides all you need to meet the new 2015 standards. This edition includes an example of a complete, generic Quality Management System consisting of a Quality Manual plus a whole host of Quality Processes, Quality Procedures and Word Instructions; **AND** access to a **FREE** software copy of these generic QMS files to give you a starting point from which to develop your own documentation.	 Paperback – ISBN 9781138025868 Hardback – ISBN 9781138025820 e-Book – ISBN 9781315774855

(*continued*)

Title	Extracts from book reviews	ISBN

ISO 9001:2015 Audit Procedures (Fourth Edition) 	Revised and fully updated, *ISO 9001:2015 Audit Procedures* describes the methods for completing management reviews and quality audits and describes the changes made to the standards for 2015 and how they are likely to impact on your own audit procedures.
	
	    ISO 9001:2015 Audit Procedures is for auditors of small businesses looking to complete a quality audit review for the 2015 standards. This book will also prove invaluable to all professional auditors completing internal, external and third-party audits.
	
	    The book also includes access to a **FREE** software copy of these generic ISO 9001:2015 audit files to give you a starting point from which to develop your own audit procedures.
		
		Paperback – ISBN 9781138025899
		Hardback – ISBN 9781138025882
		e-Book – ISBN 9781315774817

Title	*Extracts from book reviews*	*ISBN*
Environmental Requirements for Electromechanical and Electronic Equipment (First Edition) 	This book contains background guidance, typical ranges, details of recommended test specifications, case studies and regulations covering the environmental requirements required by designers and manufacturers of electrical and electromechanical equipment worldwide. The implementation of the EMC directive is just one aspect of the requirements placed upon manufacturers and designers of electrical equipment. Factors that must be taken into account include temperature, solar radiation, humidity, pressure, weather and the effects of water and salt, pollutants and contaminants, mechanical stresses and vibration, ergonomic considerations, electrical safety including EMC, reliability and performance.	 Paperback – ISBN 9780080973586 Hardback – ISBN 9780750639026 e-Book – ISBN 9780080505817

Index

Printed in the United States
by Baker & Taylor Publisher Services